I0882338

Governance for Justice and Environmental Sustainability

Understanding the governance of complex social-ecological systems is vital in a world faced with rapid environmental change, conflicts over dwindling natural resources, stark disparities between rich and poor and the crises of sustainability. Improved understanding is also essential to promote governance approaches that are underpinned by justice and equity principles and that aim to reduce inequality and benefit the most marginalised sectors of society.

This book is concerned with enhancing the understanding of governance in relation to social justice and environmental sustainability across a range of natural resource sectors in sub-Saharan Africa. By examining governance across various sectors, it reveals the main drivers that influence the nature of governance, the principles and norms that shape it, as well as the factors that constrain or enable achievement of justice and sustainability outcomes. The book also illuminates the complex relationships that exist between various governance actors at different scales, and the reality and challenge of plural legal systems in much of sub-Saharan Africa.

The book comprises 16 chapters, 12 of them case studies recounting experiences in the forest, wildlife, fisheries, conservation, mining and water sectors of diverse countries: Madagascar, Zimbabwe, Botswana, Namibia, South Africa, Zambia, Mozambique, Sierra Leone and Cameroon. Through insights from these studies, the book seeks to draw lessons from the praxis of natural resource governance in sub-Saharan Africa and to contribute to debates on how governance can be strengthened and best configured to meet the needs of the poor, in a way that is both socially just and ecologically sustainable.

Merle Sowman is the Director of the Environmental Evaluation Unit and Associate Professor in the Department of Environmental and Geographical Science at the University of Cape Town (UCT), South Africa.

Rachel Wynberg is Deputy Director of the Environmental Evaluation Unit and Associate Professor in the Department of Environmental and Geographical Science at the University of Cape Town (UCT), South Africa, where she holds a Research Chair on Social and Environmental Dimensions of the Bio-economy.

Governance for Justice and Environmental Sustainability

Lessons across natural resource sectors in sub-Saharan Africa

Edited by
Merle Sowman and Rachel Wynberg

Routledge
Taylor & Francis Group
LONDON AND NEW YORK

earthscan
from Routledge

First published 2014
by Routledge

2 Park Square, Milton Park, Abingdon, Oxon OX14 4RN
and by Routledge
711 Third Avenue, New York, NY 10017

First issued in paperback 2016

Routledge is an imprint of the Taylor & Francis Group, an informa business

British Library Cataloguing-in-Publication Data
A catalogue record for this book is available from the British Library

Library of Congress Cataloging-in-Publication Data
Governance for justice and environmental sustainability : lessons across natural resource sectors in Sub-Saharan Africa / edited by Merle Sowman and Rachel Wynberg.
pages cm
Includes bibliographical references and index.
1. Natural resources—Africa, Sub-Saharan—Management. 2. Sustainability—Government policy—Africa, Sub-Saharan. 3. Sustainable development—Government policy—Africa, Sub-Saharan. 4. Social justice—Government policy—Africa, Sub-Saharan. 5. Environmental justice—Government policy—Africa, Sub-Saharan.
I. Sowman, Merle, editor of compilation. II. Wynberg, Rachel, editor of compilation.
HC800.Z65G68 2014
333.70967—dc23
2013030531

ISBN 13: 978-0-415-52359-2 (hbk)
ISBN 13: 978-1-138-68006-7 (pbk)
ISBN 13: 978-0-203-12088-0 (ebk)

Typeset in Baskerville
by Swales & Willis Ltd, Exeter, Devon

Table of contents

List of illustrations

Figures

Tables

List of contributors

Mijasoa M. Andriamarovolonona – Department of Social and Cultural Anthropology, VU University, Amsterdam, Netherlands

Kent Buchanan – Lund University Centre for Sustainability Studies (LUCSUS), Lund, Sweden

Marlène Buchy – Independent Researcher, Langport, United Kingdom

Brian Child – Department of Geography, University of Florida, Gainesville, Florida, USA

Tessa Cousins – Association for Water and Rural Development (AWARD), Acornhoek, South Africa

Barry Ferguson – Centre Ecologique de Libanona, Fort Dauphin, Madagascar

Charlie J. Gardner – Durrell Institute of Conservation and Ecology, University of Kent, Canterbury, United Kingdom

Mathilde Gingembre – Institute of Development Studies, University of Sussex, Brighton, United Kingdom

Claire Gupta – National Science Foundation SEES Postdoctoral Fellow, Yale School of Forestry and Environmental Studies

Ralph Hamann – Graduate School of Business, University of Cape Town, South Africa

Melissa Hansen – Lund University Centre for Sustainability Studies (LUCSUS) and Lund University Centre of Excellence for the Integration of Social and Natural Dimensions of Sustainability (LUCID), Lund, Sweden

Tim Healy – Aquaterre, Antananarivo, Madagascar

Neal Hockley – School of Environment, Natural Resources and Geography, Bangor University, Wales, United Kingdom

Arthur Hoole – A+LTG Environmental Planning and Resource Management, Ottawa, Ontario, Canada

Witness Kozanayi – Department of Environmental and Geographical Science, University of Cape Town, Cape Town, South Africa

Rodgers Lubilo – School of Anthropology and Conservation, University of Kent, United Kingdom

Roy Maconachie – Department of Social and Policy Sciences, University of Bath, United Kingdom

Frank Matose – Department of Sociology, University of Cape Town, South Africa

Shadrack Mogotsi – BIOKAVANGO Project, Okavango Research Institute, University of Botswana, Botswana

Belda Mosepele – Okavango Research Institute, University of Botswana, Botswana

Ketlhatlogile Mosepele – Okavango Research Institute, University of Botswana, Botswana

Patricia Mupeta – School of Natural Resources and Environment, University of Florida, Gainesville, Florida, USA

Frank Muttenzer – Department of Ethnology, Luzern University, Switzerland

Shylock Muyengwa – School of Natural Resources and Environment, University of Florida, Gainesville, Florida, USA

Kabiri Ngeta – United Nations University – Institute of Advanced Studies (UNU-IAS), Yokohama, Japan

Sharon Pollard – Association for Water and Rural Development (AWARD), Acornhoek, South Africa

Serge Raemaekers – Environmental Evaluation Unit, University of Cape Town, South Africa

Vasnar Ramasar – Lund University Centre for Sustainability Studies (LUCSUS) and Lund University Centre of Excellence for the Integration of Social and Natural Dimensions of Sustainability (LUCID), Lund, Sweden

Shirley M. Smith – School of Agriculture, Policy and Development, University of Reading, United Kingdom

Merle Sowman – Environmental Evaluation Unit and Department of Environmental and Geographical Science, University of Cape Town, South Africa

Jackie Sunde – Department of Environmental and Geographical Science, University of Cape Town, South Africa

Douglas Thamage – Cheetah Conservation Botswana, Mokolodi Nature Reserve, Botswana

Penny Urquhart – Independent climate change adaptation and sustainable development specialist, South Africa

Jaci van Niekerk – Environmental Evaluation Unit, University of Cape Town, South Africa

Rachel Wynberg – Environmental Evaluation Unit and Department of Environmental and Geographical Science, University of Cape Town, South Africa

Preface

This book stemmed from a three-year project to enhance understanding, knowledge and capacity regarding environmental governance in sub-Saharan Africa through collaborative, transdisciplinary research across natural resource sectors and institutions in the region. The project was prompted by an increasing realization of how important it is to understand the governance of complex human-ecological systems in a world faced with rapid environmental change, conflicts over dwindling natural resources, and crises of economic, social and ecological sustainability. Developing intellectual capacity in this arena is fundamental to improving governance, which is a critical requirement for achieving sustainability, reducing inequality and setting in place social justice approaches that benefit the most marginalized sectors of society.

To achieve these aims, the project set about analysing and documenting insights, findings and lessons from the governance arena across a range of natural resource sectors in sub-Saharan Africa, including the freshwater, forest, fisheries, wildlife and agriculture sectors. We cast the net wide, and circulated an invitation for abstracts from researchers working in the region on various aspects of natural resource governance, justice and sustainability. Of the 40 abstracts received, 14 were selected for further development and inclusion in this edited volume. A workshop of project partners and chapter contributors was held in March 2011 to give authors the opportunity to present and discuss their work, to discuss the theoretical ideas underpinning the notion of 'governance for justice and sustainability', and to explore some of the thematic issues to be included in the book.

Through the process of compiling this book, new research linkages and collaborations have been fostered among researchers working in sub-Saharan Africa. We have also used the platform established by the project to help develop the research capacity and profile of postgraduate students, many of whom have co-authored the chapters in this book.

We are very grateful to the University of Cape Town Vice-Chancellor's Strategic Fund for supporting this project. We also thank the South African Research Chairs Initiative of the national government's Department of Science and Technology and its National Research Foundation. Many other organizations and individuals have played an important role in bringing the project to completion. Jaci van Niekerk is warmly acknowledged for her sterling coordination and editing efforts, Paul Wise set his meticulous copy-editing skills to work, Fahdelah Hartley provided super-efficient secretarial support, and Paula Wood assisted with illustrations. We extend our gratitude to contributing authors for sharing their experiences and insights and to Tim Hardwick and Ashley Wright of Earthscan/Routledge for their patience and support in the process of finalizing this book.

DISCLAIMER

1 Governance, equity and sustainability in sub-Saharan Africa

An introduction to the discourse

Merle Sowman and Rachel Wynberg

Worldwide, natural resources are faced with increasing pressures from globalization, growing consumption levels, stark disparities between the rich and the poor and contestation for use by different stakeholders (Agardy et al 2005; Turner and Fisher 2008; German et al 2010). The consequences of this approach are well-known: runaway climate change, accelerated biodiversity loss and deforestation, a decline in well-being for billions of people and a planet that many describe as being at a tipping point of irreversible change (World Health Organization [WHO] 2005; Rockström et al 2009; Biermann et al 2012; Organization for Economic Co-operation and Development [OECD] 2012). These threats have had a severe impact on local communities and indigenous people, many of whom are exposed to abject poverty and rely on natural resources for food security, income and livelihoods (Fabricius et al 2004; Béné 2008; Nelson 2010; Barrett et al 2011). This is especially the case in Africa, where biologically rich forests, drylands, savannahs, rivers, lakes and seas underpin the livelihoods of millions of people.

Histories of colonialism, independence and 'modernization' have fundamentally changed the way in which natural resources are accessed and used in Africa, and sub-Saharan Africa in particular. A growing consensus among scholars and development practitioners points to governance as the most significant obstacle to enabling sustainable use of these resources and, more widely, securing a development path for developing countries that is sustainable and equitable (Ribot 2004; United Nations Development Programme [UNDP] 2005; Cronkleton et al 2008; Adger and Jordan 2009; Ratner and Allison 2012). Indeed, governance has taken central stage in unfolding debates about the nature and role of the state in managing natural resources, the way in which the state interacts with other social actors, the evolving systems of rules that shape these actions and the plurality of social and legal systems – more especially in rural areas where customary rules often still hold strong (Kooiman et al 2005; Treib et al 2007). Governance, at its most general level, is about steering and guiding society and the economy through collective action to address problems and achieve common goals (Kooiman et al 2005; Torfing et al 2012). It is a slippery concept with a diversity of definitions and understandings. In the natural resource arena, it is thus increasingly interpreted in its widest sense as comprising 'political, institutional, and cultural frameworks through which diverse interests in natural and cultural resources are coordinated and controlled' (Cronkleton et al 2008: 1) with interactions between and across a wide diversity of governing actors (Kooiman et al 2005).

What, then, does governance mean in the context of natural resource management in sub-Saharan Africa, and how can it be configured to best meet the needs of the poor, in a way that is both socially just and ecologically sustainable? This question underpins the rationale for this book, which seeks to understand the nature of governance across a range of natural resource sectors and governance approaches. The book builds on a spate of recent books on governance and natural resources in Africa (e.g. Keeley and Scoones 2003; Fabricius et al 2004; Nelson 2010; German et al 2010; Ojha et al 2013), but seeks to provide a more holistic, interdisciplinary and multisectoral perspective on governance by including case studies from a host of natural resource sectors across sub-Saharan Africa, which focus on different dimensions and modes of governance.

The intention is to enhance the understanding of governance across diverse natural resource sectors, in order to learn lessons about the complex interactions and institutional arrangements that constitute governance. It is also to provide guidance on the principles, approaches, processes and mechanisms that lead to or mitigate against equitable, socially just and environmentally sustainable outcomes.

A key assumption underlying this book is that both social justice and environmental sustainability are desirable societal objectives. Yet, they may not always be compatible. Our central thesis is that it is the nature and mode of governance that will ultimately determine the extent to which these objectives can be fulfilled and where trade-offs need to be negotiated.

Of particular interest are governance modes involving interactions and processes which work towards outcomes that do not exacerbate the situation of the most vulnerable members of society. The chapters of this book thus seek to explore the extent to which considerations of justice and sustainability inform governance interactions and interventions, as well as the institutional arrangements and decision processes that determine who gets what, when and how.

Natural resource use in sub-Saharan Africa: Ecology, economy and context

The geographies of sub-Saharan Africa – the region's physical attributes, ecological composition, social and economic profile, political economy and history – are integrally tied to the governance and use of natural resources in the region. Key physical characteristics include an erratic and variable climate, highly erodible and often nutrient-poor soil and severely constrained water resources. In southern Africa, which is the focus of many of the chapters in this book, most of the region receives less than 600 millimetres of rainfall per year, and aridity and drought are central features of the landscape (Southern African Development Community [SADC] 2008). At the same time, the region is extremely rich in biodiversity, with floral species forming the foundation of more than 70 of the major vegetation units that are nested within the subcontinent's six floristically distinct biomes (Cowling and Hilton-Taylor 1997). The diversity of terrestrial, freshwater and marine ecosystems includes an array of wildlife, ranging from little-known insects, reptiles, birds and amphibians through to the spectacular and somewhat iconic group of large mammals that typify the region and which have often been at the centre of community-based conservation initiatives.

The importance of these natural resources, especially for the world's poor, is unequivocal. Despite Africa's rapid urbanization (United Nations Human Settlements Programme 2008; Turok and Parnell 2009), with almost 50 per cent of the population now living in cities (Dietz 2011), the use of water, land, crops, forests (including non-timber forest products), fisheries and wildlife has been and continues to be an integral part of the day-to-day lives of millions of African people, generating substantial food and livelihood benefits (Fabricius et al 2004; German et al 2010; Nelson 2010). Furthermore, natural resources provide a vital supplement to other livelihood activities and are an important safety net in times of environmental stress or economic crisis, thereby reducing vulnerability and risk (Béné 2003; Jul-Larsen 2003; Shackleton and Shackleton 2004).

Natural resources also underpin many African economies, providing the economic basis for the region's expanding agriculture, mining, tourism, forestry and fisheries industries. Although less than 25 per cent of southern Africa is considered arable, at least 70 per cent of the population relies on agriculture for their food, income and employment, with crop production contributing 34 per cent of southern Africa's gross domestic product (GDP) (SADC 2008). Similarly, inland and marine fisheries contribute to the food security of an estimated 200 million Africans and the income of over ten million people who engage in production, processing and trade (New Partnership for Africa's Development [NEPAD] 2004; Béné and Heck 2005). Forests, although comprising only 3 per cent of GDP in southern Africa, which is largely from exotic plantations and commercial indigenous timber, also contribute significantly to rural incomes. Over two-thirds of the continent's inhabitants rely on forest products, either for direct subsistence use or as a form of cash income (Timko et al 2010). Some studies suggest that contributions from forest products is as high as 40 per cent of the total income for the poorest households (Shackleton et al 2011), with the use of natural products for medicinal purposes common amongst some 80 per cent of the African population (Roe 2008; Payyappallimana 2010).

Natural resources, moreover, frequently go way beyond their utilitarian value, remaining deeply embedded in the ways of life and cultural identity of many indigenous and local communities (Jentoft 2003; Bernard and Kumalo 2004; Turner 2004; Sunde et al 2013). Indeed, in some cases, peoples' need to access resources and live off the land are intrinsic not only to the welfare of such communities, but also to their cultural survival (Stevens 1997).

This wealth in natural resources is not, however, mirrored by the economic profile of the region's inhabitants. Seventy per cent of the population in southern Africa, for example, live below the internationally recognized poverty line of US$2 a day (SADC 2008), while in sub-Saharan Africa some 51 per cent of the population live on less than US$1 a day (Chen and Ravallion 2008). While the past decade has seen a reduction in poverty levels on the subcontinent and significant economic growth in many African countries, this has been accompanied by growing levels of economic inequality. In its 2012 annual report, the prestigious African Progress Panel (2012) notes: 'The deep, persistent and enduring inequalities in evidence across Africa have consequences. They weaken the bonds of trust and solidarity that hold societies together. Over the long run, they will undermine economic growth, productivity and the development of markets.'

Understanding inequality in Africa, including its historical roots, is an integral part of comprehending governance. Colonial interventions in particular had a profound influence on economic justice and environmental sustainability, shaping rights and access to natural resources and in many cases altering or eradicating local systems of governance, including the appropriation of resources from indigenous populations (Beinart and McGregor 2003; Hara et al 2009). Colonialism also imposed new forms of centralized political authority over access to land and natural resources, often undermining local institutions or, in some cases, supporting and relying on customary institutions to promote the interests of the colonial state and the European settlers (Mamdani 1996; Lemos and Agrawal 2006). As Hara et al (2009) remark, this led to a fragmentation of integrated resource systems, in contrast to customary resource governance approaches and norms that coordinated resource use and overlapping claims. Increased centralization also led to situations of dual governance, or legal pluralism, with customary and statutory systems existing side by side, often in conflict, which is a situation that prevails today (Okoth-Ogendo 2008).

The political independence of many African states in the second half of the twentieth century reinforced the strategy of state centralization, as states sought to consolidate their political authority and drive modernization plans (Nelson 2010). Declining economic growth in the 1970s and 1980s, however, combined with the shock of rising petroleum prices and an increase in inflation and public debt throughout the world, led to new approaches of global integration and trade liberalization and the increased dominance of market forces (Independent Commission on Population and Quality of Life 1996). Increased debt in African countries formed the setting for the introduction of structural adjustment programmes, often requiring deep cuts in social spending and a diminished role for the state. This was matched by a realization that the state had limited capacity. Increasingly, therefore, states began to adopt decentralization strategies to devolve natural resource management to the local level.

These decentralization reforms evolved alongside increasing concerns that central state agencies had mismanaged natural resources. Despite state claims of ownership, a lack of capacity to enforce restrictions on use led to open access situations, or the misappropriation of public assets for private benefit (Nelson 2010). Increasing calls were made by environmental and civil society organizations for more participatory resource governance systems, which were buoyed by the democratization of several African states. The 1990s thus witnessed the emergence of a range of community-orientated initiatives. Community-based natural resource management (CBNRM), for example, arose out of a desire 'to rectify the human costs associated with coercive conservation' and sought 'to return the stewardship of biodiversity and natural resources to local communities through participation, empowerment and decentralization' (Dressler et al 2010, p. 5). Many of the underlying principles of CBNRM were also applied in cooperative or co-management arrangements which emerged as a result of concerns about resource management and the need for greater involvement of resource users in decision-making. The emphasis was on sharing rights and responsibilities between government, resource users and other stakeholders in order to achieve sustainable and equitable resource governance (Peluso 1992; Borrini-Feyerabend et al 2000; Roe et al 2000).

Despite initial optimism, however, many of these initiatives have seen limited benefits for local resource users, with decisions regarding who gains access to land and resources, especially high-value resources such as land, wildlife and forests, remaining vested in the state (Murombedzi 2001; Murphree 2005; Jones 2010; Nelson 2010). A key obstacle to implementing these decentralization and devolution programmes has been the reluctance of central governments to relinquish power over valuable natural resources (Keeley and Scoones 2003; German et al 2010; Nelson 2010). Furthermore, the role of customary laws and institutions remains unclear and is largely subservient to formal state policies and legislation, despite being the de facto governance regime for the majority of Africa's rural poor in local resource allocation and decision-making (Claassens 2011; Wicomb and Smith 2011).

New forces are now emerging with completely different sets of dynamics. Since the 1990s, increased pressures have been brought to bear on natural resources in sub-Saharan Africa from continued trade liberalization, broader global economic forces such as globalization and increasing demands for commodities. This in turn is exacerbating declines in natural resources and promoting contestation over access to and use of resources. In particular, increasing levels of foreign direct investment in the region to gain access to agricultural land, forest products and other raw materials, especially from China, India and other emerging economies (Cotula et al 2009; Karsenty 2010; Nelson 2010), are undermining the environmental resource base upon which many of these economies depend. The tendency of countries such as China to offer financial assistance to resource-rich African countries, in exchange for guaranteed access to specific strategic resources, is threatening governance reforms and commitments to bilateral agreements and other voluntary instruments aimed at sustainable use and development of resources (Capistrano 2010; Karsenty 2010). Land-grabbing for the purposes of growing food crops and agrofuels, among other activities, has emerged as a significant threat, with foreign investors targeting smallholders' land in particular. This is especially prevalent in Africa. Of the 46,000,000 hectares that have changed hands in 756 verified land deals to date, about half took place in Africa, especially in Ethiopia and Mozambique (Land Matrix 2013).

These geo-political and economic processes, as well as the ongoing drive to address poverty through short-term economic growth initiatives, have resulted in natural resources being threatened by overexploitation, degradation and pollution. This in turn undermines natural resource requirements for current and future generations, and has a severe direct impact on the livelihoods of local communities and indigenous peoples, many of whom suffer abject poverty (Fabricius et al 2004; Béné 2008; Brown et al 2008). Communities that are dependent on natural resources are thus bearing the brunt of neoliberal economic forces and processes that drive decisions, investments, development approvals and trade agreements.

The governance of natural resource use in sub-Saharan Africa

Many of these trends have been influenced by the way in which governance has been configured. The central question remains: just how can we ensure that new, unfolding systems uphold the rights of the poor and entrench ecological sustainability? Chapter 2 of this book presents some theoretical perspectives on governance, but we now turn to a brief review of the concept of governance, how it has evolved over time and how it relates to justice and sustainability.

The notion of governance, despite having wide interpretation, is now a central theme in the fields of international relations, development policy, public administration and finance, political studies and, more recently, natural resource management (Hyden and Court 2002; Kooiman et al 2005; Treib et al 2007; Cronkelton et al 2008; Adger and Jordan 2009; Nelson 2010; Torfing et al 2012). While approaches differ, there is a growing consensus among scholars and development practitioners that inadequate governance is the most significant obstacle to achieving sustainable development and the sustainable use of natural resources (Ribot 2004; UNDP 2005; Cronkleton et al 2008; Adger and Jordan 2009; Ratner and Allison 2012).

Young (2005) argues that there is no grand theory of governance, but rather various discourses, normative prescriptions and interpretations that enable us to better understand governance in all its complexity. Notwithstanding the variety of narratives on governance theory, the overriding common element among them, which distinguishes governance from government and management, is that governance is concerned with interactions and processes which occur between a diverse group of actors, including non-state actors often with diverging interests, in the process of addressing societal issues and creating solutions (Bavinck et al 2005; Kooiman et al 2005; Lemos and Agrawal 2006; Jentoft 2007; Torfing et al 2012). Governance is basically concerned with embracing a wider range of governance actors, beyond the state, working in 'networked configurations' (Walters 2004) with and through a host of private sector, parastatal, customary, voluntary and other civil society groupings. Such 'new governance spaces' (Taylor 2007) offer opportunities for dialogue, deliberation and empowerment among different actors (Swyngedouw 2005).

These interactions may or may not be characterized by shared values and visions, deliberative processes of policy formulation and implementation, collaborative rule-making and enforcement, joint planning and decision-making, mechanisms for conflict resolution and the design and operation of institutional arrangements to support these governing activities. The intended focus of these interactions is to collectively address societal problems and create solutions (Kooiman et al 2005). This interpretation of governance recognizes the enormous value and potential of involving a range of different actors in addressing societal issues by bringing new knowledge, fresh insights, expertise and a range of perspectives to the deliberations and decision-making process.

Governance scholars have also been debating and seeking conceptual clarity on the various 'modes of governance' that occur in different contexts and across a range of scales (Hyden and Court 2002; Kooiman et al 2005; Swyngedouw 2005; Treib et al 2007). There are many different conceptions and categorizations of modes of governance in the literature. Most recognize a range of governance modes, including hierarchies (state-centric governance), networks or co-governance (a constellation of actors in varying partnership arrangements), markets (market-based instruments and incentives), voluntarism (non-binding agreements and instruments) and self-governance (including customary governance). Treib et al (2007), although providing a much more nuanced categorization than given above, emphasize that these categories should be seen as ideal and not actual, as elements of one form of governance may be found in another.

Kooiman et al (2005) propose that these modes of governance operate at different levels, or what they refer to as 'orders of governance'. First-order governance is concerned with the day-to-day management that governing actors are engaged in to address problems and find solutions. Second-order governance concerns the design of the institutions, including rules, required to address problems and serve public needs. Third-order governance or meta-governance (Kooiman et al 2005) is concerned with the values and principles that guide first- and second-order governance. Regardless of the mode of governance, all orders of governance are required for effective and legitimate governance (Bavinck et al 2005).

'Governance' has also been used by various agencies in a prescriptive way to characterize a particular mode of governance that, if adhered to, will lead to a set of socially desirable outcomes (UNDP 2002; World Bank 2002; International Union for Conservation of Nature [IUCN] 2011). Consequently, several scholars and international development agencies have sought to identify a set of universal principles which should underpin governance, while others have characterized what they consider constitutes 'good governance' (Hyden and Court 2002; Weiss 2000; World Bank 2002; Graham et al 2003; Greiber et al 2009; Lockwood et al 2010).

However, there are concerns about the appropriateness and value of defining such a set of universal principles, given that these are not value-free and are largely fashioned according to neoliberal political and economic systems of Western societies that do not reflect an appreciation of the plurality of socio-legal and cultural systems (Parthasarathy 2005; Eberhard 2008; Davis and Ruddle 2012). For example, many international agencies promote 'market efficiencies' and 'adherence to the rule of law' as desirable outcomes, yet these may have a negative impact on local livelihoods and customary practices. Davis and Ruddle (2012) in particular present a harsh critique of the rhetoric surrounding new resource governance alternatives, or in Swyngedouw's (2005, p. 251) terms 'governance-beyond-the-state', arguing that in reality, it simply advances neoliberal prescriptions about private ownership and rarely reflects an understanding of conditions on the ground.

In a similar vein, various scholars (Bourdieu 2005; Brand 2005; Swyngedouw 2005; Ratner and Allison 2012; Torfing et al 2012) note that the governance discourse fails to address the complex histories of domination and power relations which exist among actors, as well as the broader international drivers and political frameworks within which deliberations and interventions take place. Opening up new governance spaces may promote greater engagement from non-state actors, but it also provides opportunities for powerful political and economic actors to impose their agendas and influence. Swyngedouw (2005) argues that these new governing processes are questionable in terms of 'democratic content', as there are no generally agreed upon principles for, and rules of, engagement. While governance theorists recognize the inevitability of conflict in these new governance spaces (Kooiman et al 2005; Swyngedouw 2005; Chuenpagdee and Jentoft 2007), the dominant narratives emerging in the literature support a normative view of governance as a space of interaction, collaboration, deliberation and problem solving.

In reality, however, and as the chapters in this book attest, governance in sub-Saharan Africa does not fall neatly into any of these categories, but instead embraces properties of several of these idealized modes. Our intention, therefore, is not to try to define or categorize governance, but rather to explore governance in all its diversity and complexity, with a view to better understanding which modes of governance, or aspects of a particular mode of governance, offer promise for promoting justice and environmental sustainability.

Inequality and justice

Much like governance, the concept of justice is a vigorous and ongoing debate, and there is no universally accepted definition. However, there seems to be broad agreement that justice focuses on fairness and equity. In essence, justice, fairness and equity are concerned with fair treatment or due reward and the equitable distribution of 'things' (Dobson 1998; Maiese, 2003; Schroeder and Pisupati 2010). In general terms, philosophers have identified five categories of justice. As Table 1.1 elucidates, these include justice in exchange, distributive justice, procedural justice, restorative or corrective justice and retributive justice.

The notion of an environmental justice that encompasses elements of the above 'justices' is growing in importance in the environmental management and sustainable development literature, as scholars examine how environmental impacts – 'goods' and 'bads'– are distributed among citizen groups and the processes informing such decisions. In the context of natural resource governance in sub-Saharan Africa, all these categories of justice are relevant.

Despite broad agreement on the basic meaning of justice, conceptions and definitions of justice are not fixed by some natural law, but are always contested and in flux. Cook and Hegtvedt (1983), for example, argue that justice should be viewed as both strategic and normative, emerging from 'the process of people interacting

Table 1.1 Categories of justice

Categories of justice	*Definition*
Justice in exchange	Establishes the fairness and equity of a transaction between parties and concerns the mutually beneficial exchange of things – giving one thing and receiving what is due in return. Such exchanges must be voluntary and all parties must consent to the exchange, and be competent to do so.
Distributive justice	Regulates the fair allocation of existing, scarce resources among qualifying recipients, as well as the allocation of rights, responsibilities, costs and burdens. This includes intergenerational justice.
Restorative, reparative or corrective justice	Is concerned with righting the wrongs that were brought on another. This could be determined through court decisions on just remedies or by administrative actions and receipt of just compensation.
Retributive justice	Establishes the nature of punishment that would be appropriate for a given crime or whether the punishment allocated is fair in terms of the crime committed.
Procedural justice	Is concerned with fair procedures and fair practices, as well as perceived fairness in reaching an agreement or making a decision, as well as the more informal need for honesty, transparency and respect in decision-making processes.

Sources: Cook and Hegtvedt 1983; Dobson 1998; Floyd and Johnson 2002; Whiteman and Mamen 2002; Greiber et al 2009; Schroeder and Pisupati 2010

. . . and attempting to negotiate some agreement or compromise over what will be accepted as just and fair' (Sampson 1981, p. 11). In reality, different perspectives, frames of reference and socio-cultural contexts will influence whether people see processes and decisions as fair or unfair.

As Sen (2009) argues, given the impossibility of achieving consensus on what justice is, and the difficulties of ascertaining whether justice has been achieved or not, it may well be more prudent to focus effort on reducing injustice – a condition which is much easier to discern and agree upon.

Evolving from these various debates and interpretations, justice is now a key principle in the natural resource governance lexicon, and is incorporated into many multilateral and bilateral agreements dealing with environmental issues broadly, and natural resource management in particular. The Convention on Biodiversity (CBD), for example, commits its 193 signatories to the fair and equitable sharing of benefits arising from the use of genetic resources and traditional knowledge, a commitment now articulated in its Nagoya Protocol on Access to Genetic Resources and the Fair and Equitable Sharing of Benefits Arising from their Utilization. Several other provisions of the CBD include strong recognition of indigenous peoples and traditional knowledge holders. Justice is thus central to the CBD, both between generations and between the providers and users of biological resources. Schroeder and Pisupati (2010) consider the incorporation of justice principles in the CBD to be a 'breakthrough in international politics, which puts common concerns for humankind and their ethical resolution at the forefront of international negotiations'.

Despite the adoption of the CBD and a range of other binding and guideline agreements, such as the 2007 UN Declaration on the Rights of Indigenous Peoples, in the natural resource policy rhetoric of most countries, mechanisms for making them operational in a manner that achieves both justice and sustainability are proving difficult to achieve in practice (Keeley and Scoones 2003; Nelson 2010; Ojha et al 2013). For example, balancing the pressures for economic growth against calls for the expansion of conservation areas presents an enormous challenge for those concerned with governance for justice, especially when this involves a loss of access to customary land and resources. Yet, as the next section explains, it is vital that steps be taken to redress the environmental crisis that faces Africa.

Environmental sustainability

It is now widely acknowledged that humans are transforming the planet in ways which could undermine development goals and poverty alleviation (Griggs et al 2013). As Rockström et al (2009) remark, the earth has entered a new epoch, the Anthropocene, in which humans constitute the dominant driver of change to Earth systems – the atmosphere, oceans, forests, waterways, biodiversity and biogeochemical cycles – with deleterious or even catastrophic impacts on human well-being. Natural resources are fundamental to food security, poverty alleviation and socio-economic development, but the levels of resource exploitation and the scale of infrastructural development, as well as the quality of economic growth, are resulting in overexploitation, habitat destruction and the loss of ecosystem services (Agardy et al 2005; Ojha et al 2013).

The findings of the Millennium Ecosystem Assessment (MEA) (2005) provide a sobering view of the state of the earth's resources and ecosystem health. Despite gains in human well-being in some areas, the findings indicate that these gains have resulted in a significant and rapid transformation of ecosystems, a substantial degradation of ecosystem services and an exacerbation of poverty for some groups (MEA 2005). Predictions are that the degradation of ecosystems could grow significantly worse in the first half of the twenty-first century unless these trends are reversed (MEA 2005). Similarly, widespread evidence demonstrates the decline in biodiversity the world over: a quarter of the world's plants are threatened with extinction; the abundance of vertebrate species has fallen by a third over the past 30 years; forests, rivers and other ecosystems continue to fragment and degrade and there is sustained erosion of crop and livestock genetic diversity in agricultural systems (CBD 2010).

Many of these trends are mirrored in sub-Saharan Africa, with the most significant threats emerging from habitat loss as a result of land transformation for uses such as agriculture, mining and the expansion of human settlements. Indeed, in many instances there is even greater cause for concern in the region. Rates of deforestation, for example, are twice the global average, resulting in the loss of more than 4,000,000 hectares of forest per annum and associated fauna and flora

(United Nations Environment Programme [UNEP] 2009). Wildlife populations are similarly affected by habitat transformation, with cattle grazing, management and enforcement problems, poaching and conflict between humans and animals playing a significant role in species declines.

Notwithstanding the widespread recognition of these environmental problems, and decades of work at different levels to resolve them, only limited successes have been reported (CBD 2010). In a similar vein, notwithstanding extensive endorsement of the Millennium Development Goals (MDGs) – which include eradicating extreme poverty and hunger, promoting gender equality and empowering women, ensuring environmental sustainability and developing a global partnership for development – there is still a considerable way to go to attain the MDG targets (Africa Progress Panel 2012). These multiple failures are due in part to the intricate governance challenges of managing natural resources. According to Jentoft and Chuenpagdee (2009), these are 'wicked problems', in that they cannot be remedied or resolved through positivistic science-based methods or by conventional resource management approaches. A range of dimensions needs to be taken into account in their redress – social, cultural, economic, ecological, physical, institutional and political – across a range of scales, from local to global.

With these complexities in mind, an increasing and rapidly evolving body of literature is pointing towards the importance of developing governance systems which bring together approaches in ways that recognize the diverse, complex, dynamic, adaptive and unpredictable nature of these so-called social-ecological systems (e.g. Holling 2001; Berkes et al 2003; Folke et al 2005; Kooiman et al 2005; Folke 2006; Jentoft 2007). Folke (2006, p. 260) observes incisively that ecosystems are typically treated as a 'black box' in the social science literature, with 'the assumption that if the social system performs adaptatively or is well organized institutionally it will also manage the environmental resource base in a sustainable fashion'. Similarly, an undue focus on ecological principles as a basis for decision-making on sustainability has led to 'narrow and wrong conclusions' (Folke 2006, p. 260). Adopting conventional resource management approaches to deal with complex social-ecological systems presents problems due to their diverse biological or physical characteristics, political histories and socio-economic contexts, as well as the plurality of legal systems used to regulate use and management (Berkes et al 2001, 2003; Garcia et al 2008; McConney and Charles 2009; Ribot 2010). Adaptive goverance is thus increasingly considered to provide a more nuanced and holistic approach to understanding and managing these complex natural resource systems (Berkes et al 2003; Symes 2006; Armitage et al 2008).

An important gap in the literature concerns the integration of justice into this thinking. Quite clearly, environmental sustainability and justice are integrally woven together, but this is not necessarily a happy marriage. Concerns have been raised in particular about the uncritical application of the normative concepts of sustainability and justice at the local level (Claassens and Cousins 2008; Greiber et al 2009; Fraser 2010). In the context of Africa, a continent struggling with food insecurity, poverty, habitat degradation and natural resource decline, a key conundrum in determining what constitutes justice is how to reconcile local food and livelihood priorities with global and national environmental priorities. How, it is asked, can local development needs be balanced with the need to conserve 'public goods' which serve the broader interests of the public at large, the nation as a whole and/or the international community?

It is of increasing concern that global and national normative discourses on either conservation or economic growth may take precedence over values and needs at the local level. While there has long been talk of finding win-win solutions to environment-development conflicts, Brown (2009) argues that tackling trade-offs across competing interests, scales and sectors, as well as long- and short-term timescales, is where our attention should now be focused. 'Lack of attention to losers has undermined human welfare, and ultimately the losers have been the poor, the vulnerable and the powerless' (O'Brien and Leichenko 2003, p. 47). It is at this interface – between balancing the demands of equity and social justice against concerns for environmental sustainability – that choices and decisions need to be made.

Overview of chapters

The intention of this book is to begin to shed light on these issues, by exploring the interplay between governance, justice and sustainability in a range of natural resource sectors. The book comprises 16 chapters, 12 of them case studies recounting experiences in the forest, wildlife, fisheries, conservation, mining and water sectors of diverse countries: Madagascar, Zimbabwe, Botswana, Namibia, South Africa, Zambia, Mozambique, Sierra Leone and Cameroon.

Following this introduction, the book begins with an overview of selected theoretical deliberations on natural resource governance in Chapter 2. Here, Kabiri Ngeta notes the absence of a common understanding of governance and provides an overview of some of the conceptual underpinnings on which such an understanding could be based. He offers a series of building blocks to help analyse governance and its constituent elements. With a focus on Africa, Ngeta identifies possible modes of governance and argues that these should be conceptualized as interactive rather than oppositional. He further advises a focus on the effectiveness of governance, based on the stated objectives and outcomes of governance, such as equity and environmental sustainability, rather than on good governance, which may be evaluated quite differently by different actors, depending on their world views.

As Chapter 3 demonstrates, however, the effectiveness of governance is as much about scale as it is about values, including the uncritical application of international sustainability and conservation discourses and norms at the local level. Using the example of a World Heritage Site – the iSimangaliso Wetland Park in KwaZulu-Natal, South Africa – Melissa Hansen, Vasna Ramasar and Kent Buchanan describe how the internationally driven narratives of biodiversity conservation, growth-led economic development and public participation have influenced policy rhetoric at national level and management actions at local level. The case study illustrates how iSimangaliso's status as a site of ecological world heritage raises it above the level of the global conservation discourse – or, as the authors describe, 'above the level of the local territory and often beyond the decision-making authority of local people' – raising questions about who should bear the costs of the park's designation.

External influences of the global conservation agenda have also had negative impacts in Madagsacar, combined with pressures from foreign investors and international agencies such as the World Bank. In Chapter 4, Barry Ferguson, Charlie Gardner, Mijasoa Andriamarovololona, Tim Healy, Frank Muttenzer, Shirley Smith, Neal Hockley and Mathilde Gingembre explore the impacts of policy reforms across the agriculture, mining, forest and conservation sectors on justice in that country. Despite the diversity of these sectors, all share flaws in the conception, design and implementation of policy, largely due to the asymmetries between powerful economic and political actors, and local communities that often lack the capacity to make meaningful contributions to these processes. Decades of well-intentioned environmental governance reforms have thus failed to protect the rights of Malagasy communities effectively.

Power relations are also a central theme of Chapter 5, which explores two case studies in West Africa: the mining sector in Sierra Leone and the forestry sector in Cameroon. Marlène Buchy and Roy Maconachie draw upon the concept of the three-dimensional 'power cube' – visible, hidden and invisible – to emphasize the importance of taking unequal power relations into account in decentralized natural resource management. Community participation, they argue, must consider the web of powers that exists, as well as prevalent social structures and norms, ethnicity, gender, histories and the relevant laws and management institutions.

The complexities of CBNRM are elaborated in the next three chapters. Clare Gupta, in Chapter 6, provides a critical analysis of CBNRM in the Chobe Enclave, Botswana, and argues that the CBNRM model in that country is flawed, has failed to generate the collective action required for such models to work, is not downwardly accountable and is in need of revision to make it more appropriate to local realities. Assumptions about resource tenure, market access, cultural plasticity and social organization, she contends, are seldom met, and despite 16 years of project efforts, villagers still associate wildlife with the decline of agricultural livelihoods. Residents are not dependent on wildlife-related incomes and instead rely strongly on social security and remittances from the highly centralized welfare state. State-led mechanisms thus play a major role in ameliorating human-wildlife conflicts in this study.

In Chapter 7, Art Hoole explains the dichotomies of state- and community-led conservation for a very different situation in Namibia, long held up as one of the leaders in CBNRM in the region. He considers linkages between CBNRM in the Ehi-rovipuka Conservancy and the neighbouring state-centric Etosha National Park. In the colonial period the Herero people were displaced from the area now designated as Etosha, and this led to a 'decoupling' of the social-ecological systems. Although the conservancy has yielded some concrete benefits for local people, in the form of meat and through income from wildlife and projects, people still feel excluded from the benefits of the park. Hoole argues for 'recoupling' to restore social and ecological systems, allowing improved integration in wider regional landscapes along with greater linkages between community-based and state-led conservation.

In Chapter 8, Brian Child, Patricia Mupeta, Shylock Muyengwa and Rodgers Lubilo take us to the heart of CBNRM: the distribution of benefits at a local level. Drawing on case studies from Zimbabwe, Zambia, Namibia and Botswana, they examine what they refer to as 'micro-governance' by comparing participatory and representational governance and ways in which revenues are allocated in each. The authors find profound differences between the two approaches, with only participatory governance resulting in better financial flows, equitable benefit sharing and high levels of participation. They conclude that for CBNRM to succeed, it must be rooted in well-designed, face-to-face processes of decision-making and accountability that are 'constructed from the bottom up and to serve the bottom'.

Moving to freshwater systems in Chapter 9, Belda Mosepele, Ketlhatlogile Mosepele, Shadrack Mogotsi, and Douglas Thamage examine the formation of a fisheries co-management committee in the Okavango floodplains in Botswana. Developing a co-management strategy brought together actors who had previously been in disagreement and helped resolve conflicts and enhance local participation in fisheries governance. While this suggests a more promising approach to floodplain management than a centralized regime, the authors highlight a number of challenges at implementation and operational levels, in particular maintaining its momentum.

The bringing together of different actors to effect policy changes is also the topic of Chapter 10, which reviews a new governance framework for small-scale fisheries in South Africa. Merle Sowman, Serge Raemaekers and Jackie Sunde provide an overview of the five-year policy formulation process largely driven by civil society, non-governmental organizations (NGOs) and researchers, but including ongoing engagement with the fisheries management authority. Underpinned by human rights, sustainability and good governance principles, the policy is a radical departure from the technocratic, resource-centred and highly regulatory approach of the past. The authors argue that despite structural and implementation challenges, the policy offers much greater potential for redress, and for the promotion of social justice and ecological sustainability.

A theme that recurs in many of the chapters is the multiplicity of governance systems that are found within and across different sectors. The next two chapters look explicitly at situations in which customary governance systems exist alongside statutory regulatory systems – a condition often referred to as legal pluralism. In Chapter 11, Sharon Pollard and Tessa Cousins identify tensions and complementarities between statutory and customary systems of water governance in Mozambique, Zimbabwe, Zambia and South Africa. They conclude that local level customary governance systems are an apposite choice, given their adaptable and resilient nature, their holistic and systems-based perspective, and their grounding in the notion of community membership and responsibility. The lack of capacity within southern African states to manage and regulate water resources gives this conclusion further credence.

Witness Kozanayi, Rachel Wynberg and Frank Matose describe in Chapter 12 how the co-existence of customary and statutory systems governing the use and protection of the baobab tree offer a glimpse into the conundrum of multilayered governance systems that exist in communal areas of Africa. In the Chimanimani district of eastern Zimbabwe, customary systems govern the use and management of the baobab through such measures as taboos and the belief in autochthonous spirits. Concerns about the overharvesting of the tree for fibre and fruits, however, have increased the state's role in regulation. This has been accompanied by political and economic upheavals, alongside an erosion of traditional authorities, which, as the authors explain, have now become 'functionaries of the state'. Yet degradation of the baobab resource continues, exacerbated, no doubt, by multiple governance systems and the lack of clarity regarding which norms and rules apply and which institutions hold decision-making authority.

The next three chapters begin to explore the role of the market in determining the different actors involved and the temporal and geographical scale of natural resource governance. In Chapter 13, Ralph Hamann explores the unravelling of a partnership between an international conservation NGO and a mining company. The case, located in Namaqualand in the far north-west of South Africa, was once hailed as a constructive example of how such collaboration could fulfil both conservation objectives and local community interests. The sale of the mine, however, led to the abandoning of community projects and financial responsibilities for rehabilitation, as well as unresolved land claims. The case demonstrates why collaborative governance arrangements require a clear regulatory framework within which to operate.

In Chapter 14, Rachel Wynberg and Jaci van Niekerk examine two plant species endemic to southern Africa and used commercially: *Hoodia gordonii*, succulent plants developed as appetite suppressants based on traditional knowledge of the indigenous San peoples, and *Pelargonium sidoides*, incorporated into a widely used global phytomedicine and harvested by rural communities in the region. Interventions initiated by the state, transformations in the private sector and mediations offered by NGOs offer great promises of benefit distribution, sustainable resource use and poverty alleviation based on these species. However, as the authors relate, local realities are quite different from global promises, and these initiatives may well get caught up in a policy snare that is self-serving in nature and ultimately detrimental to rural producers and traditional knowledge holders.

In the final thematic chapter, Chapter 15, Penny Urquhart takes a step back to examine linkages between climate, justice and adaptation, and in particular the distributional issues of sharing the costs and benefits of climate change. As the author notes, the adaptive capacity of communities relies on their ability to act collectively, through flexible and robust institutions that govern social relations and build in requirements for equity. This is especially relevant given the vulnerabilities of marginalized communities in Africa and the fragile ecosystems in which people reside. Adaptive governance is absolutely critical as the continent heads towards periods of climatic unpredictability and significant change.

Chapter 16, the final chapter, brings the threads together and highlights key findings and governance trends emanating from the various cases. It concludes with a summary of the main enabling and constraining factors that limit governance for justice and sustainability and points towards future areas of action.

Bibliography

Adger, W. N. and Jordan, A. (2009) 'Sustainability: Exploring the processes and outcomes of governance', in W. N. Adger and A. Jordan (eds) *Governing Sustainability*, Cambridge University Press, Cambridge, pp. 3–31

African Progress Panel (2012) *Africa Progress Report 2012: Jobs, Justice and Equity – Seizing Opportunities in Times of Global Change*, African Progress Panel, Geneva, http://africaprogresspanel.org/en/publications/annual-reports/annual-report-2012/africa-progress-report-2012-documents/

Agardy, T., Alder, J., Dayton. P., Curran, S., Kitchingman, A., Wilson, M., Catenazzi, A., Restrepo, J., Birkeland, C., Blaber, S., Saifullah, S., Branch, B., Boersma, D., Nixon, S., Dugan, P., Davidson, N. and Vörösmarty, C. (2005) 'Coastal systems', in R. Hassan, R. Scholes and N. Ash (eds) *Ecosystems and Human Well-Being: Volume 1: Current State and Trends: Analytical Approaches for Assessing Ecosystems and Human Well-Being*, Island Press, Washington, DC, pp. 513–549

Armitage, D., Marschke, M. and Plummer, R. (2008) 'Adaptive co-management and the paradox of learning', *Global Environmental Change*, vol 18, pp. 86–98

Barrett, C. B., Travis, A. J. and Dasgupta, P. (2011) 'On biodiversity conservation and poverty traps', *Proceedings of the National Academy of Sciences of the United States of America*, vol 108, no 34, pp. 1,3907–1,3912

Bavinck, M., Chuenpagdee, R., Diallo, M., van der Heijden, P., Kooiman, J., Mahon, R. and Williams, S. (2005) *Interactive Governance for Fisheries: A Guide to Better Practice*, Eburon, Delft, Netherlands

Beinart, W. and McGregor, J. (2003) *Social History and African Environments*, Ecology and History Series, Ohio University Press, Athens, OH, and James Currey, Oxford

Béné, C. (2003) 'When fishery rhymes with poverty: A first step beyond the old paradigm on poverty in small-scale fisheries', *World Development*, vol 31, no 6, pp. 949–75

Béné, C. (2008) 'Assessing vulnerability to poverty in fishing communities', in *Proceedings of the Fourteenth Biennial Conference of the International Institute of Fisheries Economics and Trade*, 22–25 July 2008, Nha Trang, Vietnam

Béné, C. and Heck, S. (2005) 'Fish and food security in Africa', *Naga: The World Fish Center Quarterly*, vol 28, pp. 8–12

Berkes, F., Mahon, R., McConney, P., Pollnac, R. C. and Pomeroy, R. S. (2001) *Managing Small-Scale Fisheries: Alternative Directions and Methods*, International Development Research Centre, Ottawa

Berkes, F., Colding, J. and Folke, C. (eds) (2003) *Navigating Social-Ecological Systems: Building Resilience for Complexity and Change*, Cambridge University Press, Cambridge

Bernard, P. and Kumalo, S. (2004) 'Community-based natural resource management, traditional governance and spiritual ecology in southern Africa: The case of chiefs, diviners and spirit mediums', in C. Fabricius, E. Koch, H. Magome and S. Turner (eds) *Rights, Resources and Rural Development: Community-Based Natural Resource Management in South Africa*, Earthscan, London, pp. 115–26

Biermann, F., Abbott, K., Andresen, S., Bäckstrand, K., Bernstein, S., Betsill, M. M., Bulkeley, H., Cashore, B., Clapp, J., Folke, C., Gupta, A., Gupta, J., Haas, P. M.,

Jordan, A., Kanie, N., Kluvánková-Oravská, T., Lebel, L., Liverman, D., Meadowcroft, J., Mitchell, R. B., Newell, P., Oberthür, S., Olsson, L., Pattberg, P., Sánchez-Rodríguez, R., Schroeder, H., Underdal, A., Camargo Vieira, S., Vogel, C., Young, O. R., Brock, A. and Zondervan, R. (2012) 'Navigating the Anthropocene: Improving earth system governance', *Science*, vol 335, no 6074, pp. 1,306–7

Borrini-Feyerabend, G., Farvar, M. T., Nguinguiri, J. C. and Ndangang, V. A. (2000) *Co-management of Natural Resources: Organizing, Negotiating and Learning-by-Doing*, GTZ and IUCN, Kasparek Verlag, Heidelberg

Bourdieu, P. (2005) 'Political, social and journalistic fields' in R. Benson and E. Neveu (eds) *Bourdieu and the Journalistic Field*, Polity Press, Cambridge, MA, pp. 29–46,

Brand, U. (2005) 'Order and regulation: Global governance as a hegemonic discourse of international politics?', *Review of International Political Economy*, vol 12, no 1, pp. 155–76

Brown, K. (2009) 'Human development and environmental governance: A reality check', in N. Adger and A. Jordan (eds) *Governing Sustainability*, Cambridge University Press, Cambridge, pp. 32–51

Brown, K., Daw, T., Rosendo, S., Bunce, M. and Cherrett, N. (2008) *Ecosystem Services for Poverty Alleviation: Marine and Coastal Situational Analysis*, Synthesis Report, UK Ecosystem Services for Poverty Alleviation (ESPA) Programme, R4D, Department for International Development

Capistrano, D. (2010) 'Forest governance and decentralization in Africa: Issues and emerging trends', in A. A. German, A. Karsenty and A. M. Tiani (eds) *Governing Africa's Forests in a Globalized World*, Earthscan, London, pp. 379–96

CBD (2010) *Global Biodiversity Outlook 3*, Secretariat of the Convention on Biological Diversity, Montreal, http://cbd.int/doc/publications/gbo/gbo3-final-en.pdf

Chen, S. and Ravallion, M. (2008) *The Developing World is Poorer than We Thought, But No Less Successful in the Fight Against Poverty*, Policy Research Working Paper Series, World Bank, Washington, DC

Chuenpagdee, R. and Jentoft, S. (2007) 'Step zero for fisheries co-management: What precedes implementation', *Marine Policy*, vol 31, no 6, pp. 657–68

Claassens, A. (2011) 'Contested power and apartheid tribal boundaries: The implications of "living customary law" for indigenous accountability mechanisms', *Acta Juridica*, pp. 141–74

Claassens, A. and Cousins, B. (eds) (2008) *Land, Power and Custom: Controversies Generated by South Africa's Communal Land Rights Act*, University of Cape Town Press, Cape Town

Cook, K. S. and Hegtvedt, K. A. (1983) 'Distributive justice, equity, and equality', *Annual Review of Sociology*, pp. 217–41

Cotula, L., Vermeulen, S., Leonard, R. and Keeley, J. (2009) *Land Grab or Development Opportunity? Agricultural Investment and International Land Deals in Africa*, IIED, FAO and IFAD, London and Rome

Cowling, R. M. and Hilton-Taylor, C. (1997) 'Phytogeography, flora and endemism', in R. M. Cowling, D. M. Richardson and S. M. Pierce (eds) *Vegetation of Southern Africa*, Cambridge University Press, Cambridge, pp. 43–61

Cronkleton, P., Taylor, P. L., Barry, D., Stone-Jovicich, S. and Schmink, M. (2008) 'Environmental governance and the emergence of forest-based social movements', Occasional Paper No. 49, Centre for International Forestry Research, Bogor, Indonesia

Davis, A. and Ruddle, K. (2012) 'Massaging the misery: Recent approaches to fisheries governance and the betrayal of small-scale fisheries', *Human Organization*, vol 71, no 3, pp. 244–54

Dietz, T. (2011) 'Silverlining Africa: From images of doom and gloom to glimmers of hope, from places to avoid to places to enjoy', Inaugural address, Leiden University, African Studies Centre, 14 January

Dobson, A. (1998) *Justice and the Environment: Conceptions of Environmental Sustainability and Dimensions of Social Justice*, Oxford University Press, Oxford

Dressler, W., Büscher, B., Schoon, M., Brockington, D., Hayes, T., Kull, C., McCarthy, J. and Streshta, K. (2010) 'From hope to crisis and back? A critical history of the global CBNRM narrative', *Environmental Conservation*, vol 37, pp. 5–15

Eberhard, C. (2008) 'Law in a globalizing world: Towards "good governance" through intercultural dialogue?', English version of 'O direito em um mundo globalizante: em busca da "boa governança" por meio do diálogo intercultural', *Revista Direito, Estado e Sociedade*, vol 33, Editora PUC-Rio, Rio de Janeiro, Brazil

Fabricius, C., Koch, E., Magome, H. and Turner, S. (eds) (2004) *Rights, Resources and Rural Development: Community-Based Natural Resource Management in Southern Africa*, Earthscan, London

Floyd, M. F. and Johnson, C. Y. (2002) 'Coming to terms with environmental justice in outdoor recreation: A conceptual discussion with research implications', *Leisure Sciences*, vol 24, no 1, pp. 59–77

Folke, C. (2006) 'Resilience: The emergence of a perspective for social-ecological systems analyses', *Global Environnemental Change*, vol 16, no 3, pp. 253–67

Folke, C., Hahn, T., Olsson, P. and Norberg, J. (2005) 'Adaptive governance of social-ecological systems', *Annual Review of Environment and Resources*, vol 30, pp. 441–73

Fraser, N. (2010) 'Who counts? Dilemmas of justice in a postwestphalian world', *Antipode*, vol 41, pp. 281–97

Garcia, S. M., Allison, E. H., Andrew, N. J., Béné, C., Bianchi, G., de Graaf, G. J., Kalikoski, D., Mahon, R. and Orensanz, J. M. (2008) 'Towards integrated assessment and advice in small-scale fisheries: principles and processes', Fisheries Technical Paper No. 515, Food and Agriculture Organization of the United Nations, Rome

German, L. A., Karsenty, A. and Tiani, A. M. (2010) *Governing Africa's Forests in a Globalized World*, Earthscan, London

German, L. A., Colfer, C. J. P., Barrow, E., Küchli, C., Blaser, J. and Wardojo, W. (2010) 'Forest governance and decentralization in Africa: Linking local, regional and global dialogues', in A. A. German, A. Karsenty and A. M. Tiani (eds) *Governing Africa's Forests in a Globalized World*, Earthscan, London, pp. 1–28

Graham, J., Amos, B. and Plumptre, T. (2003) *Principles for Good Governance in the 21st Century*, Policy Brief No 15, Institute of Governance, Ottawa

Greiber, T., Janki, M., Orellana, M., Savaresi, A. and Shelton, D. (2009) *Conservation with Justice: A Rights-based Approach*, IUCN Environmental Law and Policy Paper No 71, International Union for Conservation of Nature, Gland, Switzerland

Griggs, D., Stafford-Smith, M., Gaffney, O., Rockström, J., Öhman, M. C., Shyamsundar, P., Steffen, W., Glaser, G., Kanie, N. and Noble, I. (2013) 'Sustainable development goals for people and planet', *Nature*, vol 495, pp. 305–7

Hara, M., Turner, S., Haller, T. and Matose, F. (2009) 'Governance of the commons in southern Africa: Knowledge, political economy and power', *Development Southern Africa*, vol 26, no 4, pp. 521–37

Holling, C. S. (2001) 'Understanding the complexity of economic, ecological, and social systems', *Ecosystems*, vol 4, pp. 390–405

Hyden, G. and Court, J. (2002) *Governance and Development*, World Governance Survey Discussion Paper 1, United Nations University

Independent Commission on Population and Quality of Life (1996) *Caring for the Future*, Oxford University Press, Oxford

IUCN (2011) 'Position paper on the institutional framework for sustainable development', Rio+20 United Nations Conference on Sustainable Development, Rio de Janeiro, 20–22 June 2012 https://cmsdata.iucn.org/downloads/position_paper_on_ifsd_22march2012.pdf

Jentoft, S. (2003) 'Co-management: The way forward', in D. C. Wilson, J. R. Nielsen and P. Dengbol (eds) *The Fisheries Co-management Experience: Accomplishments, Challenges, and Prospects*, Kluwer Academic Publishers, Dordrecht, pp. 1–14

Jentoft, S. (2007) 'The limits of governability: Institutional implications for fisheries and coastal governance', *Marine Policy*, vol 31, pp. 360–70

Jentoft, S. and Chuenpagdee, R. (2009) 'Fisheries and coastal governance as a wicked problem', *Marine Policy*, vol 33, pp. 553–60

Jones, B. (2010) 'The evolution of Namibia's communal conservancies', in F. Nelson (ed) *Community Rights, Conservation and Contested Land: The Politics of Natural Resource Governance in Africa*, Earthscan, London, pp. 106–20

Jul-Larsen, E. (ed) (2003) *Management, Co-management or No Management? Major Dilemmas in Southern African Freshwater Fisheries, 2: Case Studies*, Food and Agriculture Organization of the United Nations, Rome

Karsenty, A. (2010) 'The new economic "great game" in Africa and the future of governance reforms in the forest sector', in A. A. German, A. Karsenty and A. M. Tiani (eds) *Governing Africa's Forests in a Globalized World*, Earthscan, London, pp. 79–100

Keeley, J. and Scoones, I. (2003) *Understanding Environmental Policy Processes: Cases from Africa*, Earthscan, London

Kooiman, J., Bavinck, M., Jentoft, S. and Pullin, R. (eds) (2005) *Fish for Life: Interactive Governance for Fisheries*, Amsterdam University Press, Amsterdam

Land Matrix, (2013) *Land Matrix Newsletter, June*, http://landmatrix.org/media/filer_public/2013/06/10/lm_newsletter_june_2013.pdf

Lemos, M. C. and Agrawal, A. (2006) 'Environmental governance', *Annual Review of Environment and Resources*, vol 31, pp. 297–325

Lockwood, M., Davidson, J., Curtis, A., Stratford, E. and Griffith, R. (2010) 'Governance principles for natural resource management', *Society and Natural Resources*, vol 23, no 10, pp. 986–1,001

Maiese, M. (2003) 'Distributive justice', in G. Burgess and H. Burgess (eds) *Beyond Intractability*, Conflict Information Consortium, University of Colorado, Boulder, http://www.beyondintractability.org/bi-essay/distributive-justice

Mamdani, M. (1996) *Citizen and Subject: Contemporary Africa and the Legacy of Late Colonialism*, Princeton University Press, Princeton, NJ

McConney, P. and Charles, A. (2009) 'Managing small-scale fisheries: Moving towards people-centred perspectives', in R. Q. Grafton, R. Hilborn, D. Squires, M. Tait and M. Williams (eds) *Handbook of Marine Fisheries Conservation and Management*, Oxford University Press, Oxford

MEA (2005) 'Millennium Ecosystem Assessment (Program)' and 'Millennium Ecosystem Assessment', *Ecosystems and Human Well-Being: Our Human Planet: Summary for Decision Makers*, vol 5, Island Press, Washington, DC

Murombedzi, J. (2001) 'Committees, rights, costs and benefits: Natural resource stewardship and community benefits in Zimbabwe's CAMPFIRE programme', in D. Hulme and M. W. Murphree (eds) *African Wildlife and Livelihoods: The Promise and Performance of Community Conservation*, James Currey, Oxford, pp. 244–55

Murphree, M. W. (2005) 'Congruent objectives, competing interests, and strategic compromise: Concept and process in the evolution of Zimbabwe's CAMPFIRE, 1984–1996', in J. P. Brosius, A. L. Tsing and C. Zerner (eds) *Communities and Conservation: Histories and Politics of Community-based Natural Resource Management*, AltaMira Press, Walnut Creek, CA, pp. 105–47

Nelson, F. (ed) (2010) *Community Rights, Conservation and Contested Land: The Politics of Natural Resource Governance in Africa*, Earthscan, London

NEPAD (2004) Background document to 'Fish for All' high-level conference, Abuja, Nigeria, August 2005, New Partnership for Africa's Development, 30 September

O'Brien, K. L. and Leichenko, R. M. (2003) 'Winners and losers in the context of global change', *Annals of the Association of American Geographers*, vol 93, no1, pp. 89–103

OECD (2012) *OECD Environmental Outlook to 2050: The Consequences of Inaction*, Organisation for Economic Co-operation and Development, Paris

Ojha, H. R., Hall, A. and Sulaiman, R. (eds) (2013) *Adaptive Collaborative Approaches in Natural Resource Governance: Rethinking Participation, Learning and Innovation*, Routledge, London and New York

Okoth-Ogendo, H. W. O. (2008) 'The nature of land rights under indigenous law in Africa', in A. Claassens and B. Cousins (eds) *Land, Power and Custom: Controversies Generated by South Africa's Communal Land Rights Act*, University of Cape Town Press, Cape Town

Parthasarathy, D. (2005) 'Taking responsibilities and participation seriously: A critique of "good governance"', in C. Eberhard (ed) *Droit,* gouvernance *et développement durable, numéro spécial des Cahiers d'Anthropologie du Droit*, Éditions Karthala, Paris, pp. 307–21

Payyappallimana, U. (2010) 'Role of traditional medicine in primary health care: An overview of perspectives and challenges' *Yokohama Journal of Social Sciences*, vol 14, no 6, pp. 57–77

Peluso, N. L. (1992) *Rich Forests, Poor People: Resource Control and Resistance in Java*, University of California Press, Berkeley and Los Angeles

Ratner, B. D. and Allison, E. H. (2012) 'Wealth, rights, and resilience: An agenda for governance reform in small scale fisheries', *Development Policy Review*, vol 30, no 4, pp. 371–98

Ribot, J. C. (2004) *Waiting for Democracy: The Politics of Choice in Natural Resource Decentralization*, World Resources Institute, Washington, DC

Ribot, J. C. (2010) 'Forestry and democratic decentralization in sub-Saharan Africa: A rough review', *Governing Africa's Forests in a Globalized World*, pp. 29–55

Rockström, J., Steffen, W., Noone, K., Persson, A., Chapin, F. S., Lambin, E. F., Lenton, T. M., Scheffer, M., Folke, C., Schellnhuber, H. J., Nykvist, B., de Wit, C. A., Hughes, T., van der Leeuw, S., Rodhe, H., Sörlin, S., Snyder, P.K., Costanza, R., Svedin, U., Falkenmark, M., Karlberg, L., Corell, R. W., Fabry, V. J., Hansen, J., Walker, B., Liverman, D., Richardson, K., Crutzen, P. and Foley, J. A. (2009) 'A safe operating space for humanity', *Nature*, vol 461, pp. 472–5

Roe, D. (2008) *Trading Nature: A Report, with Case Studies, on the Contribution of Wildlife Trade Management to Sustainable Livelihoods and the Millennium Development Goals*, TRAFFIC International, Cambridge, and WWF International, Gland, Switzerland

Roe, D., Mayers, J., Greig-Gran, M., Kothari, A., Fabricius, C. and Hughes, R. (2000) *Evaluating Eden: Reporting the Myths and Realities of Community-based Wildlife Management*, International Institute for Environment and Development, London

SADC (2008) 'Chapter 1: Regional overview', in *Southern Africa Environment Outlook*, SADC, SARDC, IUCN and UNEP, Gaborone, Harare and Nairobi

Sampson, E. E. (1981) 'Social change and the contexts of justice motivation', in M. J. Lerner and S. C. Lerner (eds) *The Justice Motive in Social Behavior*, Plenum, New York, pp. 97–124

Schroeder, D. and Pisupati, B. (2010) *Ethics, Justice and the Convention on Biological Diversity*, United Nations Environment Programme and University of Central Lancashire, Preston

Sen, A. (2009) *The Idea of Justice*, Penguin, London

Shackleton, S. and Shackleton, C. (2004) 'Everyday resources are valuable enough for community-based natural resource management programme support: Evidence from South Africa', in C. Fabricius, E. Koch, H. Magome and S. Turner (eds) *Rights, Resources and Rural Development: Community-Based Natural Resource Management in South Africa*, Earthscan, London

Shackleton, S., Delang, C. O., Angelsen, A. (2011) 'From subsistence to safety nets and cash income: Exploring the diverse values of non-timber forest products for livelihoods and poverty', in S. Shackleton, C. Shackleton and P. Shanley (eds) *Non-timber Forest Products in the Global Context*, Springer, Heidelberg, pp. 55–81

Stevens, S. (ed) (1997) *Conservation through Cultural Survival*, Island Press, Washington, DC

Sunde, J., Sowman, M., Smith, H. and Wicomb, W. (2013) 'Emerging proposals for governance of tenure in small-scale fisheries in South Africa', *Land Tenure Journal*, vol 1, no 13, July, Food and Agriculture Organization of the United Nations, Rome, pp. 117–46

Swyngedouw, E. (2005) 'Governance innovation and the citizen: The Janus face of governance-beyond-the-state', *Urban Studies*, vol 42, no 11, pp. 1,991–2,006

Symes, D. (2006) 'Fisheries governance: A coming of age for fisheries social science?', *Fisheries Research*, vol 81, pp. 113–17

Taylor, M. (2007) 'Community participation in the real world: Opportunities and pitfalls in new governance spaces', *Urban Studies*, vol 44 no 2, pp. 297–317

Timko, J. A., Waeber, P. O. and Kozak, R. A. (2010) 'The socio-economic contribution of non-timber forest products to rural livelihoods in Sub-Saharan Africa: Knowledge gaps and new directions', *International Forestry Review*, vol 12, no 30, pp. 284–94

Torfing J., Peters, B. G., Peirre, J. and Sorensen, E. (2012) *Interactive Governance: Advancing the Paradigm*, Oxford University Press, Oxford

Treib, O., Bähr, H. and Falkner, G. (2007) 'Modes of governance: Towards a conceptual clarification', *Journal of European Public Policy*, vol 14, no 1, pp. 1–20

Turner, R. K. and Fisher, B. (2008) 'To the rich man the spoils', *Nature*, vol 451, no 7182, pp. 1,067–8

Turner, S. (2004) 'Community-based natural resource management and rural livelihoods', in C. Fabricius, E. Koch, H. Magome and S. Turner (eds) *Rights, Resources and Rural Development: Community-Based Natural Resource Management in South Africa*, Earthscan, London

Turok, I. and Parnell, S. (2009) 'Reshaping cities, rebuilding nations: The role of national urban policies', *Urban Forum*, vol 20, no 2, pp. 157–74

UNDP (2002) *Human Development Report: Deepening Democracy in a Fragmented World*, United Nations Development Programme, New York, and Oxford University Press, Oxford

UNDP (2005) *A Time for Bold Ambition: Together We Can Cut Poverty in Half*, annual report, United Nations Development Programme, New York

UNEP (2009) 'Environmental change re-draws atlas of Africa', news release, United Nations Environment Programme, 10 June, http://unep.org/documents.multilingual/default.asp?documentid=538&articleid=5834&l=en‡

UN-HABITAT (2008) *The State of African Cities 2008: A Framework for Addressing Urban Challenges in Africa*, United Nations Human Settlements Programme, Nairobi

Walters, W. (2004) 'Some critical notes on "governance"', *Studies in Political Economy*, vol 73, pp. 27–46

Weiss, T. G. (2000) 'Governance, good governance and global governance: Conceptual and actual challenges', *Third World Quarterly*, vol 21, no 5, pp. 795–814

Wicomb, W. and Smith, H. (2011) 'Customary communities as "peoples" and their customary tenure as "culture": What we can do with the Endorois decision', *African Journal of Human Rights Law*, vol 11, pp. 422–46

Whiteman, G. and Mamen, K. (2002) *Meaningful Consultation and Participation in the Mining Sector? A Review of the Consultation and Participation of Indigenous Peoples within the International Mining Sector*, The North-South Institute, Ottawa

WHO (2005) *Ecosystems and Human Well-Being: Health Synthesis, Millennium Ecosystem Assessment*, World Health Organization, Geneva

World Bank (2002) *Governance and Development*, World Bank, Washington, DC

Young, O. R. (2005) 'Why is there no unified theory of environmental governance?', in P. Dauvergne (ed) *Handbook of Global Environmental Politics*, Edward Elgar, Cheltenham

2 Theoretical deliberations on understanding natural resource governance

Kabiri Ngeta

Introduction

This chapter outlines a conceptual framework for enhancing understanding about governance in the natural resource sectors in sub-Saharan Africa. It includes several building blocks that outline a definition of governance, delineate its conceptual schema, and specify the key governance elements that constitute a governance structure. It also situates the governance process within the level, or levels, at which actor interactions are taking place.

The framework contributes to a toolkit for understanding natural resource governance systems in Africa. This chapter thus adds to debates on how to specify a governance system, as opposed to the dominant trend, which focuses on how to evaluate a governance system by tabulating indicators of good governance.

Elements of a governance system pertain to the empirical realm, while the notion of good governance and its associated operationalization, in terms of indicators, belongs to the normative realm. While the concept of elements of a governance structure is applicable across natural resources and actors, that of good governance, by virtue of being normative, cannot be generalized and must be sensitive to the context of both sectors and actors.

The chapter begins with a discussion of the background to the popularization of the concept of governance and goes on to examine the elements central to understanding governance. Key indices that constitute a governance system are then suggested, and the conclusion follows.

Governance: Lack of consensus on a shared meaning

Literature on governance suggests that, while the term is used liberally in day-to-day discussion, there is little consensus, if any, among either academics or practitioners regarding its meaning. The popular opinion appears to be that while, at some point, it has been used as synonymous with 'government', the term actually implies more (Bratton and Rothchild 1992; Hyden 1992; Rosenau 1992; Rhodes 1996; Stoker 1998).

Some scholars, however, argue that the difference between the two (governance and government) is one of means, not ends. This is because both seek the same end: ordered affairs and collective action in society (Stoker 1998). A government achieves these outcomes through formal and institutional processes (Stoker 1998). Governance, on the other hand, is presented as operating both through governmental and non-governmental or extra-governmental means, and specifically through the participation of non-state actors (Rhodes 1996; Stoker 1998; Bressers and Kuks 2003). Some scholars have criticized the use of intervention by non-state actors to distinguish governance from government (Walters 2004);[1] nevertheless, this approach to governance, based on the participation of non-state actors, dominates the contemporary notion of governance, and the statement that governance is not just about government is now a common refrain. As the next section explains, the roots of this dominance lie in the neo-liberal phenomenon of the 1990s.

Genesis of the popularity of the notion of governance

'Governance' as a discursive term gained currency in the post-communist era consequent to the Bretton Woods agenda of economic liberalization in developing countries. While development agencies had long partnered with African governments, the dictates of Cold War politics obliged them to tolerate governmental ineffectiveness. With the end of the Cold War, Western-based development agencies raised the alarm about the inability of developing country governments to design policies and discharge functions (Bratton and Rothchild 1992; Hoon and Hyden 2004).

Consequently, a crisis of governance in Africa was pronounced. Western development agencies and universities founded governance programmes to address this crisis (Bratton and Rothchild 1992; Hyden 1992). Proponents of development diagnosed this crisis as emanating from the way political elites were undermining the state through rent-seeking (Bratton and Rothchild 1992). Consequently, governments were losing legitimacy among their citizens, and this undermined government effectiveness (Bratton and Rothchild 1992).

A remedy was sought in the involvement of non-state actors, both in the direct activities of government and in offloading from the state certain roles and dispersing them to private and voluntary actors in new regimes characterized by such elements as a free press, citizen involvement and the rule of law (Bratton and Rothchild 1992). The conduct of state affairs was seen to depart from the traditional, centralized model dominated by a legislature and bureaucracy, to one engaging autonomous private and voluntary actors. The functions hitherto discharged by governments were now presented as being discharged by a shared coalition of public and private actors.

It is this interdependence of partners and the way they are coordinated that has been popularized as 'governance' (Rhodes 1996; Stoker 1998; Pierre 2000; Treib et al 2007). Some scholars find this conceptualization so all-encompassing that it is devoid of any analytic value (Hyden 2002; Walters 2004). The question of what 'governance' means as a concept therefore remains salient if it is to be deployed for analytic usage.

Conceptualizing governance

The deployment of the concept of governance as discussed above led Bratton and Rothchild (1992) to observe that

> the concept of governance is asked to bear the heavy weight of numerous intellectual agendas . . . no single concept can be all things to all people. The attempt to stretch the concept of governance to perform multiple, perhaps contradictory, purposes helps to explain why its definition in the academic literature has so far been slippery.

They further observed that 'definitional questions remain to be solved' (Bratton and Rothchild 1992, p. 280) (see Box 2.1). In the light of such disparities among analysts in the way they think about governance, there is a need for inquiries on governance projects to clarify the way the term is used. Such clarification is important because it would eliminate a situation in which the conversation on the governance of natural resources in Africa is pursued in terms of different concepts and premises, while the disputants assume they are discussing one and the same thing, thereby obscuring more than they illuminate.

From the list of conceptions of governance in Box 2.1, for example, can the outcomes (successful delivery of goods and services as expected by citizens) envisaged in Delapalme (2011) be secured without any of the routes specified in the other conceptualizations? This can certainly happen, but the question then would be whether those subscribing to the view of networks advanced by Hysing (2009) would consider that process to be part of the governance phenomenon. Moreover, there is the question of whether analyses of the governance of the natural resource sector based on the two perspectives would yield compatible conclusions.

What all this implies is the need to structure the conceptualization of the term 'governance' in such a way that the variances in understanding are narrowed and the various inquiries speak to the same phenomenon.

Box 2.1 Conceptions of governance

Mayntz (2003, pp. 27–8) argues that 'governance' is used in two ways that are distinct from political guidance or steering. There is, she says, 'a new mode of governing that is distinct from the hierarchical control model, a more cooperative mode where state and non-state actors participate in mixed public/private networks'. In the other, 'new', usage, 'governance means different modes of coordinating individual actions, or basic forms of social order'.

Hyden et al (2004, p. 16): 'Governance refers to the formation and stewardship of the formal and informal rules that regulate the public realm, the arena in which state as well as economic and social actors interact to make decisions'.

Kooiman and Bavinck (2005, p. 17): 'Governance is the whole of public as well as private interactions taken to solve societal problems and create societal opportunities. It includes the formulation and application of principles guiding those interactions and care for institutions that enable them'. (See also Kooiman et al 2008, pp. 1–3.)

Hysing (2009, p. 647), citing Sørensen (2006, p. 99): 'The story line tells us that we are witnessing a transformation from hierarchical governing by nationally organized political institutions (i.e. *government*) to modes of governing in which a multitude of public and private actors from different policy levels govern society through networks and soft policy instruments – in other words, *governance*'.

Delapalme (2011, p. 2), citing the Mo Ibrahim Foundation's definition of good governance: 'the successful delivery of those public goods and services

Drawing from the myriad usages of the term, it is possible to outline, as below, a description capturing the broad meaning of the concept of governance and specifying certain features which can be used to identify the phenomenon of governance. It appears that a definition of governance has to be structured around a system of rules that allow decision-making and implementation by social actors in a coordinated way, with specific objectives as the end, within the context of the public realm (Hyden 1992; Mayntz 2003; Hoon and Hyden 2004; Treib et al 2007; Kooiman et al 2008). This conceptualization of governance is elaborated upon by describing what its elements could be.

Broadly, governance has two aspects: content and activity. 'Content' refers to rules and goals. The process of generating these rules must itself be seen as part of governance, while the other part is implementing them. 'Activity' refers to both the production of content and its implementation (Hyden 2002; Hyden et al 2004). Rules are meant to guide public affairs, a process referred to in the literature as the 'steering' or 'coordination' of social systems (Hyden and Court 2002; Hyden et al 2004). If rules are produced but not implemented, then there is no steering, and hence no governance is achieved. To this extent, then, governance is not the same as law or policy-making. Rather, rule-making must be seen as part of a broader governance continuum.[2]

The interplay between rules and coordination implies an existence of goals to be achieved. The rules spell out these goals. Consequently, governance has to be concerned with outcomes (Rosenau 1992; Rhodes 1996; Pierre 2000; Jreisat 2002). In terms of the public realm, this outcome is the social order that entails the management of relations between state and citizens in various sectors within the social, political and economic life of society (Stoker 1998; Hyden et al 2004; Kooiman and Bavinck 2005; Kooiman 2008). While it is not clear to what extent this management of relations includes those among citizens, it can be assumed that matters which are deemed contractual among citizens fall under this management.

The management of relations implies that there is a terrain between rules and goals. This is the implementation sphere (Stoker 1998; Mayntz, 2003; Béné and Neiland 2006). Implementation is about the activities that translate intentions into outcomes. This is largely what is considered the dominant meeting place of the actors involved in the governing system. If, however, it were taken to be what governance was all about, that would be limiting the conceptualization of governance. As discussed above, the coordination of social systems, or steering, is only part of the governance story.

A steering process has to be backed by either formal or informal authority to ensure the desired outcomes. Thus a governance system must be denoted by a sanctioning mechanism.[3] If, as Rhodes (1996, p. 652) concedes, governance, even when non-hierarchical ('hierarchical' is a subtext for state domination), involves the 'authoritative' allocation of resources and exercising control, it becomes difficult to comprehend a controlling authority that is devoid of a sanctioning mechanism.[4] This is especially so when we consider the arena in which the act of governing takes place: the public realm.

The activities that constitute governance take place in the public realm (Hyden 2000; Hyden 2002; Hyden et al 2004). The 'public realm' is defined by Hyden et al (2004, p. 16) as the 'arena in which state as well as economic and societal actors interact to make decisions'. The public realm is to be conceptualized in contradistinction to the private or family arena, unless matters supposedly falling within this latter domain are brought out for adjudication in the public arena.

From the above conceptual schema, a governance structure can then be described. The way the resultant structure plays out in practice determines the form of governance (hierarchical, market, local, self-governance or networks/intermediaries[5]) which ensues.

Governance structures for natural resources

Certain structures or elements of what can constitute a governance edifice may be delineated in broad categories. The dynamics of these structures, either singly or in combination, have direct ramifications for the governance of natural resources.

Actors, world views and preference diversity

Actors in natural resource sectors come in various forms. They include individual as well as corporate, formal, informal, state and non-state actors. Broadly, the tendency is to categorize some actors as government, while others are grouped under both market and civil society. In a case such as that of a river catchment area, the actors can span all three of these broad categories. In the Ruaha River in Tanzania, for example, the Great Ruaha River Catchment stakeholders include the Ruaha National Park, communities downstream (and those in the areas where streams feeding the Ruaha River come from), a hydro-generating agency, non-governmental organizations (NGOs) (e.g. the World Wide Fund for Nature), local government authorities, the Rufiji Basin Water Board and Tanzania's national Ministry of Water.

The complexity of the actor constellation tends to increase as one moves up the governance ladder from the local to the global. This is in part because, at higher levels, even when lower-level actors may not be physically present or even formally represented, they shadow the actors because most natural resources exist at the level where these lower-level actors are to be found. Their response is thus a factor in the activities at these higher levels, and they can therefore be portrayed as a shadow presence at those levels. The reverse need not be the case, where the issue at the local level is not so momentous as to have national or international ramifications.

Actors hold a view or image of the world with respect to a specific issue area: that of what is to be governed (Bressers and Kuks 2003; Kooiman and Bavinck 2005; Jentoft 2007). The issue area in the case of the natural resource sector in Africa could include forests, wildlife, land and specifically rangelands – whose governance involves both wildlife and livestock – and aquatic ecosystems: watersheds, fisheries, coral reefs, mangrove swamps and wetlands. Some sectors are single-resource sectors, while others are multiple-resource sectors, such as wetlands and rivers. The African wetlands are utilized for such purposes as fisheries, pasture, wildlife, veld products, water and land for irrigation (see, for example, Kgathi et al 2011 and Haller 2010 on wetlands and Homewood et al 2009 on rangelands). Natural resource sectors may be valued differently by different actors, with high value resources attracting more complexity and intense forms of governance.

The image of the world that the actors bring to either of these issue areas defines what they perceive as the importance of the sector and how it should be governed to achieve its objectives. Both of these considerations are informed by the actors' assumptions about the world and the information that they have about that area (whether factual or misperceived).

Actors intervening in the natural resource sectors in Africa have been driven by several considerations. These considerations include economic interests, social differentiation, cultural practices, spiritual values, nationalist sentiments and social justice.

The view that natural resources should serve economic purposes has been dominant since pre-colonial times. It intensified and was formalized with the rise of modern economies organized around the nation-state, when sectoral policies and strategies were formulated. According to the policy and strategy of the fisheries sector in Tanzania, for example, the main goal of the sector is to promote the conservation, development and sustainable management of fisheries resources for the benefit of present and future generations. The benefit referred to here is largely economic development. This is clear from the list of objectives of the policy, which include: 'To put into efficient use available resources in order to increase fish production so as to improve fish availability as well as contribute to the growth of the economy' (Ministry of Natural Resources and Tourism [MNRT] 1997). The economic consideration also permeates other images, such as the expression of cultural-spiritual and nationalist sentiments.

This view of certain natural resource sectors in terms of the national heritage took root with the rise of the new states after independence. Foremost among those to give expression to this was Tanzania's President Julius Nyerere. In one of the earliest conferences on the future of African wildlife, Nyerere and others asserted:

> [T]he survival of our wildlife is a matter of grave concern to all of us in Africa. These wild creatures amid the wild places they inhabit are not only important as a source of wonder and inspiration but are an integral part of our natural resources and of our future livelihood and well-being . . . we solemnly declare that we will do everything in our power to make sure that our children's grandchildren will be able to enjoy this rich and precious inheritance.
>
> (International Union for Conservation of Nature [IUCN] 1963, p. 13)

But even as these views on national heritage were being asserted, other actors held that such considerations in the natural resource sector would have to dovetail with the economic imperatives. During the very symposium at which President Nyerere made the above observations, his Minister of Land and Surveys, T.S. Tewa, associated the conservation motive with economic considerations:

> [I]f the mass of my countrymen are to be enlisted in the ranks of conservationists, they will expect to see that what they are told is 'their heritage,' however valuable it may be as a cultural asset, can still be made to earn its keep. I believe it can – through tourism.
>
> (IUCN 1963, p. 21)

Another view that has dominated the natural resource sector is one that conceives it as a theatre of social differentiation. Again, this has been evident since the pre-colonial era, but whereas state power was used to drive this imperative during the pre-colonial and colonial eras, the objectives of social differentiation have been achieved in the post-colonial era by the market, which has been used to price certain classes of society out of these sectors. In the pre-colonial era, centralized societies that had royalty could establish royal hunting grounds, and in the colonial period national parks were established as colonists' recreation havens, from which native populations were largely excluded (Brooks 2005).

The basis for pursuing such exclusionary agendas was, as Anderson and Grove (1987, p. 4) observed, the tendency among 'Europeans to impose their image of Africa upon the reality of the African landscape', as there existed 'a wish to protect the natural environment as a special kind of "Eden"'. In modern times, such thinking has persisted, albeit in more subtle ways. Recently the Tanzanian Minister for Natural Resources and Tourism, Ezekiel Maige, told communities in one of the protected areas that was also a United Nations Educational, Scientific and Cultural Organization World Heritage Site that they would not be allowed to farm in that area because it had been accorded heritage status partly because it was 'natural and free from human activities such as farming' (Philemon 2011).

Of relevance to this topic is that societal conceptions of what should constitute the goals or objectives of natural resource governance (e.g. environmental sustainability, economic development, cultural heritage and social justice) will inform how the actors define the outcomes they seek to achieve through the governance process. Given that the societal conception discussed above is not necessarily homogeneous, the process of defining the outcomes is therefore driven by the diversity of actor preferences. Preferences diversity is bound to be reflected not only in what is done with certain natural resources, but also in how it is done. With the advent of the formal conservation movement during the colonial era, for example, although Europeans and Africans agreed on the need to conserve natural resources, there seemed to be sharp differences about the way forward. While Europeans favoured a route that separated humans and nature, Africans thought that conservation could be achieved by keeping human beings in harmony with nature.

These differences have been reproduced in the contemporary period in the cleavages between preservation and conservation (see, for example, Hutton et al 2005; Goldman 2011). With respect to the question of national heritage referred to above, for example, we see the minister introducing the rider that the national heritage has to pay for itself, which raises the question: what if it is not able to, should resources be shifted from other sectors to sustain the wildlife heritage? Moreover, there are now some who question the nature of the balance between conservation and economic development where such natural resources are not perceived to contribute to the formal economy. These critics question whether large tracts of land should be set aside exclusively for natural resources instead of being opened up to active economic activities in light of the demographic pressures on food supply and shelter (Jones 2006).

The implication here is that it is critical, in delineating the actors, to determine the diversity of their preferences. Actors without a diversity of preference – in whatever dimension: what, how, who and when – would not face a social dilemma and would therefore have no need for a mechanism to aggregate preferences for which institutions become relevant.

Institutional frameworks

Institutional frameworks have to do with both the rules of engagement and the organizations charged with overseeing those rules, or providing the terrain through which the opportunity offered by the rules is accessed. To achieve their objectives, actors need to operate either individually or in collectives, and must have rules which structure and adjudicate their interactions. In a context of preference diversity, as discussed above, rules serve to constrain behaviour so that actors pursue their preferences within limits that recognize the interests of other actors. Rules secure the interests of the various actors similarly or differentially. Of special interest here should be the way rules are formulated (Hyden 1992; Hoon and Hyden 2004), because that will have direct implications for the content of these rules.

In terms of formal institutions, the rules that structure interactions in the various natural resource sectors can be categorized broadly as national constitutions, sectoral policies, sectoral Acts of Parliament and subsidiary legislation. In certain contexts, subnational legislation in the form of local government by-laws that regulate natural resources in those jurisdictions may exist. In other cases, there are customary rules as part of customary governance systems that comprise a coherent set of rules and customary institutions that govern activities and the social activities of the community (Bennett 2008). There may also be sectoral compacts entered into by actors in private or communal conservancies, either among themselves or between themselves and the national or local government authorities. Beyond the nation-state, some sectors are embedded in regional and international instruments.

The constitution is the supreme law from which specific Acts of parliament are derived, sometimes under the umbrella of a policy that outlines the national vision for a specific sector. Such an Act may confer power on the minister in charge of a resource sector to implement subsidiary legislation to further the broad objectives of the Act. Subnational jurisdictions can, within the context of the laws under which they are established, also make by-laws that could be part of the sector's governance. Communal and private conservancies may also have their own rules, some of which are recognized by national legislation and thus have the force of law (for example, communal conservancies in Namibia and Wildlife Management Areas [WMA], in Tanzania), while others are local arrangements without official recognition (see, for example, Rihoy et al 2010, regarding the Communal Areas Management Programme for Indigenous Resources [CAMPFIRE] in Zimbabwe, and the Mahenye community, because the legal unit was the rural district council [RDC]). Customary governance systems have and continue to play a central role in the lives of millions of people worldwide (Jentoft et al. 2003). Although customary legal systems were generally subordinate to the introduced colonial legal systems, in many countries these customary laws continued to be the de facto legal system followed at the local community level and included institutions for rule-making, decision-taking and conflict resolution (Techera 2008; Wicomb and Smith 2011). These customary institutions continued to exist despite being disregarded by formal, largely Eurocentric, legal systems of the colonial governments. Over the past three decades, there has been an increasing trend in the constitutions of many countries in Africa to recognise the existence of customary law as an independent and equal source of law (Wicomb and Smith 2011). This has been largely influenced by various court rulings in favour of the rights of indigenous peoples as well as the emergence of various international instruments that call for the recognition and respect of the rights of customary communities to own, use, develop and manage their own land and resources.

National legislation operates in concert with international instruments ratified by the state. Some of the key instruments are the Convention on International Trade in Endangered Species of Wild Fauna and Flora (CITES), the Convention on Wetlands of International Importance (Ramsar Convention), the Convention on Biological Diversity, the United Nations Convention on the Law of the Sea, the Lusaka Agreement on Cooperative Enforcement Operations Directed at Illegal Trade in Wild Fauna and Flora, and the various protocols of the Southern African Development Community (SADC) on fisheries, forestry and transfrontier conservation areas in southern Africa.

In the international context, there are also initiatives that are not binding, but nevertheless influence natural resource governance. These include the voluntary guidelines of the Food and Agriculture Organization (FAO) on fisheries and land transactions. The formal rules discussed above are shadowed by informal rules and informal networks, in the case of the organizations discussed below, which influence governance in Africa (Bratton and Van de Walle 1997). These formal regulations are executed within an organizational web that spans international and local levels. As mentioned above, these organizations ought to be understood alongside informal governance systems and networks that form part of the governance of natural resources.

In terms of formal organizations, the national ministry in charge of a sector is the apex of that sector's governance. Within the ministry there is a department that manages the sector. That department may have representation through the structures of government down to the local level. In some countries, there is a donor development group that brings together donors to all natural resources in general.

For those sectors with formalized international rules, the international organizational dimension is most evident in the conference of parties of the convention concerned. Some international organizations, such as the FAO, exert their influence through non-binding guidelines such as the ones on responsible fisheries, planted forests and the responsible governance of land tenure.

At the regional level, organizations may have various origins, but most of those in Africa currently are rooted in two regional economic blocks, SADC and the East African Community (EAC), as well as, to a certain extent, the Common Market for Eastern and Southern Africa (COMESA). Such organizations deal primarily, though not exclusively, with transboundary natural resources. Examples are the Lake Victoria Basin Commission, which was formed by five countries in the Lake Victoria catchment area to coordinate fisheries activities in that area. The Lusaka Agreement Task Force, which is in charge of enforcing the Lusaka Agreement, was sponsored by COMESA, while the SADC secretariat normally oversees the protocols that it sponsors.

At the subnational level, local authorities constitute organizations in that they enact by-laws on land, forests, wildlife, wetlands and other natural resources. Zimbabwe's rural district councils (RDC), Tanzania's district councils and Kenya's local and county authorities are examples.

At the community level, communal areas may have local associations that may, in turn, form umbrella bodies at the national level. Such institutions could be a mix of both customary institutions (such as chiefs or councils) as well as modern associations triggered by state law. Examples of such structures are CAMPFIRE in Zimbabwe, community resource boards in Zambia, and the Namibian Association of Community-Based Natural Resource Management Support Organisations. In Tanzania, the WMAs are governed through local-level authorized associations of all the villages forming a single WMA.

As a variation on the communal associations, big landowners form private conservancies but retain individual title to their own land and hence are not 'communal' within the meaning of 'communal landownership'. In Namibia, private white landowners in the 1990s formed freehold wildlife conservancies (Barnes and Jones

2008). Conservation NGOs, as well as local, national and international private firms, also form part of the institutional framework.

All these organizations provide the theatre in which the governance of natural resources in Africa is played out, with actors pursuing their preferred outcomes through the action they take, based on their resources and strategies.

Resources, strategies and action

The interaction of actors is a function of their resources and strategies. Resources are both human and non-human and are deployed to deliver the desired outcome. Human resources include expert personnel who bring technical know-how in the issue area. This is particularly determinative if the matter of concern requires scientific exposition – as, for example, in the case of fisheries. Actors who do not have an adequate command of the relevant information on the natural resources as well as technical competencies are likely to lose out in the contestations over the modalities of utilization and conservation in the sector.

Social and political capital is also a significant resource, especially in matters of governance, where the balance of power in society is critical in tilting the outcome one way or another. This was notably the case in Africa's earlier political dispensations, in which neo-patrimonial tendencies implicitly allowed those with proximity to state apparatus to secure their preferences, as opposed to those without such state connections (see, for example, the case of wildlife politics in Zambia under Kenneth Kaunda [Gibson 1999]). As the democratic space opens, the balance of power may tilt towards those who can marshal the rank and file, because electoral politics tends to drive public policy. This human resource dimension will be strongly felt in sectors such as forestry, fisheries, wildlife and land use, where access to resources is not necessarily capital-intensive, and thus attracts ordinary actors, who constitute the bulk of the electorate.

In terms of non-human resources, the biggest asset has been funding: those able to mobilize funding can command substantial influence in their sectors of interest. Nevertheless, to deploy these resources, actors require a strategy that would translate their visions into action (Bressers and Kuks 2003; Kooiman and Bavinck 2005).

One of the strategies carried over from the colonial era is that of giving one way of knowing about nature a higher status than other ways. Modern formal science is favoured and local and indigenous knowledge underplayed, with the result that certain actors are effectively sidelined (Goldman 2011). In addition, there is a tendency to denigrate certain utilization practices in an effort to depict certain actors as agents of resource degradation, who should therefore be objects of governance, not governance agents. This has particularly been the case among pastoralist, as opposed to sedentary communities. The former are portrayed as degrading the range (Little 1996; Beinart and McGregor 2003; Pollini 2010), even though they are known to contribute to the sustainability of biodiversity in the rangelands (Little 1996).

There have also been direct political economy strategies, such as the state laying claim to all natural resource sectors and thus making other actors subsidiary to it.

Both the colonial and postcolonial states utilized this strategy. Germany's colonial forestry policies in Tanzania, for example, which declared the forests to be ownerless, uninhabited and not for peasant use, were intended to exclude the indigenous peoples from the forest sector so that concessionaire companies could dominate the sector, with the indigenous peoples becoming a pool of cheap labour or taking up cash-crop production, on which they could be taxed (Sunseri 2003).

Modes of governance

Courses of action can be framed between two extremes: coercion and voluntarism (Treib et al 2007). The course taken is largely a function of the existing mode of governance. The central claim defining contemporary conceptualization about governance, as stated earlier, is that governance is not just about government; it also includes other actors. The force of this conceptualization notwithstanding, if either governance or government has to be thought of in terms of a process of enabling interactions among actors who seek to produce a certain outcome through collective action (Rhodes 1996), then that process can take place in various modes.

These modes can be either hierarchical or horizontal. The former is the dominant characteristic in the presumed old way of thinking about government. Some commentaries on governance present it as a top-down style of interaction between a state and its citizens in which the notion of control by one actor is a key concept (Kooiman and Bavinck 2005). In the case of natural resource governance in Africa, this style is best represented by the protectionist approach, whose institutional embodiment is the protected areas model (see, for example, Wilshusen et al 2002; Hutton et al 2005).

The horizontal approach, on the other hand, is associated with supposedly new patterns of interaction. This approach has two major variants: interactive/network modes and the market (Dixon and Dogan 2002). The interactive/network modes have been described in many ways, for instance self-regulation and/or co-regulation, public-private partnerships, cooperative management and joint entrepreneurial ventures. A general characteristic of all variants in this mode is that the societal actors come together for a common purpose under circumstances in which actor autonomy is observed and the actors interact in a horizontal way. The actors define their own rules, act interdependently and are self-organizing. There is an underlying assumption in this narrative of interactive/network modes of governance that such interactions take place outside the purview of the state (Rhodes 1996; Ostrom 1999; Dixon and Dogan 2002; Kooiman and Bavinck 2005; Kooiman et al 2008).

Some of these claims can present difficulties, especially in relation to natural resource governance. This is because the natural resource sector is largely embedded in state policies and laws, so that whatever actors outside the state do is to some degree conditioned by the statutory imperatives that loom in the background, even as the actors' operations adopt a veneer of autonomy. An elaborate argument to this effect has been advanced with respect to the market as the other variant of horizontal modes of governance.

The market has been presented as one of the institutions of governance, along with the state and civil society (Coase 1960; Sbragia 2000; Dixon and Dogan 2002; Kooiman 2008; Hysing 2009). In the context of natural resources, the market certainly does play a coordinating role among actors in such sectors as forestry, as exemplified by the Forest Stewardship Council, a forest certification scheme that seeks to use the market to leverage the proper management of forests and the responsible production of wood products. In the wildlife sector, the market is increasingly playing a direct role in governance, particularly in the emerging privately managed public protected areas in certain parts of Africa (Fearnhead 2008). In addition, the endeavours by such institutions as the World Bank and Food and Agriculture Organization (FAO) to intervene in the land market and issue voluntary guidelines on private sector transactions in land matters arguably constitute an attempt to shore up the role of the market in the governance of land, including that under customary tenure (FAO 2012).

Nevertheless, it is important to note that the market itself also has such variants as command economy and free market. (In the case of natural resources, the liberal approach is best represented by free-market environmentalism.) While the former is directly under the state, this is, as both Walters (2004) and Sbragia (2000) contend, not to say that the latter is wholly independent of the state. Sbragia (2000) argues that the market does not displace public authority; rather, the state looms large in these supposedly non-state manifestations. This is because the market requires the state for it to function.

> [T]he 'rule of law' is critical for a market economy, and the absence of well-defined legal and judicial mechanisms for resolving conflicts among parties is an almost insuperable barrier to the construction of a functioning economy. . . . Markets, just like the welfare state, are absolutely dependent on public authority.
>
> (Sbragia 2000, p. 245)

Thus, interactive/network governance, rather than being seen in terms of the absence of government, should rather be presented as involving interactions among civil, public and private actors (Kooiman et al 2008). The dynamics of these interactions are likely to be amplified by the prevailing resource tenure regime,[6] as well as the level at which they take place.

Levels of interaction

Levels of interaction can be horizontal, vertical or both. The focus on the vertical level needs to be flexible enough to allow for the possibility of an interaction either at or through the local, national, regional and global levels. Governance takes place at all these levels. They are not, however, independent of one another; some or all of them do interact, and what happens at one level often has ramifications at other levels. This is particularly so with respect to the national and international levels, which impact on the subnational and local levels.

At the local level, governance may be organized around individual villages or even households and this could either be customary governance systems or one based on modern statutory provisions. In certain cases, the national legal regime may sanction institutional arrangements for the governance of national resources at the local level. This can be fixed, or may be flexible to allow for local particularities. In other cases, such institutions could be endogenous and reflect a situation in which large families and big landlords drive the governance process (Bandiaky 2008; United Nations Economic and Social Commission for Asia and the Pacific [UNESCAP] n.d.).

Natural resource governance at the subnational level brings together several local jurisdictions that may share a common ecosystem or problem, which therefore has to be governed by a broad range of actors in the region. As stated earlier with respect to river catchment areas, the scale includes not just the immediate vicinity of the river, but also the areas where streams which feed the river originate. The same case applies to rangelands where migratory species roam over a wide territory. This makes it necessary to govern the range as one unit rather than as a simple locale, because wildlife may be at one locale at a certain time of year and elsewhere at other times.

Governance at the national level is concerned with policy-making and the enactment of legislation on natural resources. The policy sets out the main objectives of the sector in terms of how the sector should be managed and for what ends, and it supposedly informs the legislation in the sector, but that does not always happen. The terrain of translating policy into legislation is often a site of actor contestation. This is mostly because, in terms of securing actor preferences, it is the legislation that holds sway, not the policy.

The national level provides the contact point between the state and regional and international initiatives in natural resource governance. There are various regional initiatives involved in the governing of natural resource sectors. In southern Africa, for example, SADC has formulated protocols on the governance of such sectors as fisheries, forestry and wildlife. In eastern Africa, the EAC created the Lake Victoria Basin Commission and the Lake Victoria Fisheries Organization for the governance of the shared lake basin. In addition, there are other initiatives targeting transboundary shared natural resources, and the development of a regional climate-change regulatory framework has been undertaken. In addition to these initiatives by formalized regional bodies, individual states may undertake joint efforts for the governance of shared resources. Such efforts include the Lake Tanganyika Authority, which was formed by the countries sharing the lake to coordinate fishery activities.

At the global level, the situation is rather fluid:

> [T]here is no single model or form of global governance, nor is there a single structure or set of structures. It is a broad, dynamic, complex process of interactive decision-making that is constantly evolving and responding to changing circumstances.
>
> (Commission on Global Governance 1996)

The global governance of natural resources is largely associated with the intergovernmental structures that are constituted as the UN General Assembly and the UN's agencies. The most visible global role is seen in the environmental agreements that deal with the global governance of various natural resource sectors. A multitude of other, non-state actors that participate in the process through their attendance at UN-organized meetings, principally as accredited observers, are able, depending on their organizational abilities, to influence the decision-making process. There are also multilateral institutions, such as the World Bank and the World Trade Organization, that are associated with the UN system but enjoy a measure of autonomy from the UN General Assembly.

Outside the UN framework, global natural resource governance is also driven by multinational corporations, international NGOs and community-based organizations (CBOs), global media and private interests. Thus, as the Commission on Global Governance observed, global governance is multifaceted and involves a network of public, private and mixed institutions and norms that is still developing (Abbott and Snidal 2010). In the realm of natural resource governance, global governance is mostly approached sector by sector.

Outcome of the governance process: Effectiveness or good governance?

Given that every governance system is put in place to generate particular outcomes, it follows that once the steering process has taken place, the resultant outcome should be evaluated for its congruence or otherwise with the intentions of that process. This should in turn disclose the extent of the effectiveness of the governance system. Effectiveness is empirical, unlike good governance, which is normative. Effectiveness is about the extent to which a process generates the outcome it was intended to secure. The securing of that outcome may be achieved even when both the means and the ends cannot be characterized as good governance. In a rent-seeking society, for example, the dominant actors may structure the governance system to be a failure in order to cover up their activities. In such a case, the governance system would be delivering the outcome intended by its designers, even though, from a broader societal perspective, it fails the test of good governance.

A similar case can be made of a governance regime beset by competing visions. Until such competing visions are harmonized, it can be expected that different actors will return a different verdict on what they adjudge to be good governance in the sector being governed. Thus the concept of good governance is better conceived as contextual to both the sector and actors' preferences, such that, in a given sector, competing notions of good governance can be expected, depending on the convergence or divergence of actor preferences.

Perhaps the best example here is to be seen in global financial institutions such as the World Bank and International Monetary Fund, which are proponents of good governance (in the economic sphere), yet have critics who condemn the outcomes that result from the path these institutions consider good governance

(Global Exchange 2000; see also Imrie and Racot 1999, where openness and accountability are presented as liabilities to the pursuit of efficiency, yet their worth is taken as axiomatic by proponents of good governance).

Conclusion

This chapter has outlined some of the conceptual underpinnings for understanding the governance of natural resource sectors in Africa. It has described the popularity of the contemporary notion of governance, especially with reference to Africa, and the need to clarify the concept with regard to natural resources, given the absence of a consensus on the meaning of governance. The chapter has offered a series of building blocks to enhance understanding of governance of natural resource sectors, including analysis of the various concepts of governance and the elements that constitute a governance structure or system.

The chapter has also identified possible resultant modes of governance, and has argued that these need not be conceptualized in terms of mutual opposition; an interaction among them could rather be envisaged, contrary to the popular notion that tends to see governance as not being about government.

Also highlighted in this chapter has been the need to situate the level or levels at which actor interactions take place during the governance process, as well as the need to focus on the effectiveness of the governance process in contradistinction to the popular concern with good governance. The latter, it is advanced, should be contextualized with respect to actors and sectors. The key thrust of this proposition is based on the fact that governance involves actors who may espouse divergent preferences, which implies that they are guided by different world views, a fact that affects either the ends they pursue or the ways they pursue these ends, or both.

Arguments about good governance, and especially the indices often used to mark good governance, are mostly located within this space (that of ends and means to those ends), and different actors can obviously be expected to have different evaluations of either the processes or outcomes of governance. The question of effectiveness, however, can be evaluated on the basis of the stated objectives of the governance process (such as equity and sustainability) and the discernible outcomes, irrespective of whether the actors consider these outcomes as good or not.

Notes

1 Walters (2004) contends that governments have never monopolized control over territories, and that the intervention of non-state actors in governmental processes has always been a feature of society.

2 The dichotomy between rules and steering that Hyden and Court (2002) allude to does not, therefore, arise. They contend that in terms of the substantive content of governance, there are differences between those who believe it is concerned with rules of how to conduct public affairs, and others who see it as steering or controlling public affairs. The two are essentially inseparable, since rules are made to guide the steering in order to make it predictable.

3 There is a notion in the literature suggesting that there is a form of governance not backed by formal authority. This proposition is problematic if it is meant to suggest that the actors involved are engaged in a *laissez-faire* interaction.
4 A sanctioning mechanism may be weak or fragile, and therefore may not realize its mandate, but this does not mean that the governance regime is designed to be without a sanctioning mechanism. The use of the notion of a sanctioning mechanism as opposed to police powers is more robust because the notion of police powers can give the impression that the absence of a police phenomenon implies the absence of a government, and hence the commonly held notion that governance takes place without a formal authority, and hence no government.
5 This could also involve governance under the shadow of hierarchy or in areas of limited statehood (Borzel and Risse 2010).
6 Such tenure regimes include state, community, customary, private, state-communal and state-private ownership. Some resources, even when owned by the state, are de facto open-access, though some actors may decide to govern them if they find them useful for their purposes.

Bibliography

Abbott, K. and Snidal, D. (2010) 'International regulation without international government: Improving IO performance through orchestration', *Review of International Organizations*, vol 5, no 3, pp. 315–44

Anderson, D. and Grove, R. (eds) (1987) *Conservation in Africa: People, Policies and Practice*, Cambridge University Press, Cambridge

Bandiaky, S. (2008) 'Gender inequality in Malidino biodiversity community-based reserve, Senegal: Political parties and the "village approach"', *Conservation and Society*, vol 6, no 1, pp. 62–73

Barnes, J. and Jones, B. (2008) 'Game ranching in Namibia', in H. Suich, B. Child and A. Spenceley (eds) *Evolution and Innovation in Wildlife Conservation: Parks and Game Ranches to Transfrontier Conservation Areas*, Earthscan, London

Beinart, W. and McGregor, J. (eds) (2003) *Social History and African Environments*, James Currey, Oxford

Béné, C., and Neiland, A. E. (2006) 'From participation to governance: A critical review of the concepts of governance, co-management and participation, and their implementation in small-scale inland fisheries in developing countries', *WorldFish Center Studies and Reviews 29*. WorldFish Center, Penang, Malaysia, and CGIAR Challenge Program on Water and Food, Colombo, Sri Lanka

Bennett, T. (2008) "Official' vs 'living' customary law: Dilemmas of description and recognition', in A. Claassens and B. Cousins (eds) *Land, Power and Custom: Controversies Generated by South Africa's Communal Land Rights Act*, Legal Resources Centre and University of Cape Town Press, Cape Town, pp. 138–53

Borzel, T. and Risse, T. (2010) 'Governance without a state: Can it work?', *Governance and Regulation*, vol 4, no 2, pp. 113–34

Bratton, M. and Rothchild, D. (1992) 'The institutional bases of governance in Africa', in G. Hyden and M. Bratton (eds) *Governance and Politics in Africa*, Lynne Rienner, Boulder, CO

Bratton, M. and Van de Walle, N. (1997) *Democratic Experiments in Africa: Regime Transitions in Comparative Perspective*, Cambridge University Press, Cambridge

Bressers, H. T. A. and Kuks, S. M. M. (2003) 'What does "governance" mean? From conception to elaboration', in H. T. A. Bressers and W. A. Rosenbaum (eds) *Achieving Sustainable Development: The Challenge of Governance Across Social Scales*, Praeger, Westport, CT, pp. 65–88

Brooks, S. (2005) 'Images of "wild Africa": Nature tourism and the (re)creation of Hluhluwe Game Reserve, 1930–1945', *Journal of Historical Geography*, vol 31, pp. 220–49

Coase, R. H. (1960) 'The problem of social cost', *Journal of Law and Economics*, vol 3, pp. 1–44

CGG (1996) 'Our global neighborhood: Report of the Commission on Global Governance', http://gdrc.org/u-gov/global-neighbourhood/index.htm, accessed 18 April 2013

Delapalme, N. (2011) 'African governance: The importance of more and better data', *Governance: An International Journal of Policy and Administration*, vol 24, no 1, pp. 1–3

Dixon, J. and Dogan, R. (2002) 'Hierarchies, networks and markets: Responses to societal governance failure', *Administrative Theory & Praxis*, vol 24, no 1, pp. 175–96

FAO, 2012. *Voluntary Guidelines on the Responsible Governance of Tenure of Land, Fisheries and Forests in the Context of National Food Security*, Food and Agriculture Organization, Rome

Fearnhead, P. (2008) 'Privately managed protected areas', in H. Suich, B. Child and A. Spenceley (eds) *Evolution and Innovation in Wildlife Conservation: Parks and Game Ranches to Transfrontier Conservation Areas*, Earthscan, London, pp. 409–24

Gibson, C. (1999) *Politicians and Poachers: The Political Economy of Wildlife Policy in Africa*, Cambridge University Press, Cambridge

Global Exchange (2000) 'Top ten reasons to oppose the IMF', http://globalexchange.org/campaigns/wbimf/TopTenIMF.html, accessed 18 April 2013

Goldman, M. (2011) 'Strangers in their own land: Maasai and wildlife conservation in Tanzania', *Conservation and Society*, vol 9, no 1, pp. 65–79

Haller, T. (ed) (2010) *Disputing the Floodplains: Institutional Change and the Politics of Resource Management in African Wetlands*, Brill, Leiden, Netherlands

Homewood, K., Kristjanson, P., and Trench, P. C. (eds) (2009) *Staying Maasai? Livelihoods, Conservation and Development in East African Rangelands*, Springer Science, New York

Hoon, P. and Hyden, G. (2004) 'Governance and sustainable livelihoods', in H. T. A. Bressers and W. A. Rosenbaum (eds) *Achieving Sustainable Development: The Challenge of Governance across Social Scales*, Praeger, Westport, CT, pp. 43–63

Hutton, J., Adams, W. M. and Murombedzi, J. C. (2005) 'Back to the barriers? Changing narratives in biodiversity conservation', *Forum for Development Studies*, vol 32, no 2, pp. 341–70

Hyden, G. (1992) 'Governance and the study of politics', in G. Hyden and M. Bratton (eds) *Governance and Politics in Africa*, Lynne Rienner, Boulder, CO

Hyden, G. (2000) 'The governance challenge in Africa', in G. Hyden, D. Olowu and H. O. Ogendo (eds) *African Perspectives on Governance*, Africa World Press, Trenton, NJ, pp. 5–31

Hyden, G. (2002) 'Operationalizing governance for sustainable development', in J. E. Jreisat (ed.) *Governance and Developing Countries*, Brill, London, pp. 13–31

Hyden, G. and Court, J. (2002) 'Governance and development', World Governance Survey Discussion Paper 1, United Nations University

Hyden, G., Court, J. and Mease, K. (2004) *Making Sense of Governance: Empirical Evidence from 16 Developing Countries*, Lynne Rienner, Boulder, CO

Hysing, E. (2009) 'From government to governance? A comparison of environmental governing in Swedish forestry and transport', *Governance: An International Journal of Policy, Administration, and Institutions*, vol 22, no 4, pp. 647–72

Imrie, R. and Racot, M. (1999) 'How new is the new local governance? Lessons from the United Kingdom', *Transactions of the Institute of British Geographers*, vol 24, pp. 45–63

IUCN (1963) 'Conservation of nature and natural resources in modern African states: Report of a symposium organized by CCTA and IUCN and held under the auspices of FAO and UNESCO at Arusha, Tanganyika, September 1961', International Union for the Conservation of Nature and Natural Resources, Morges, Switzerland, http://data.iucn.org/dbtw-wpd/edocs/NS-001.pdf, accessed 19 April 2013

Jentoft, S. (2007) 'The limits of governability: Institutional implications for fisheries and coastal governance', *Marine Policy*, vol 31, pp. 360–70

Jentoft, S., Minde, H. and Nilsen, R. (2003) *Indigenous peoples: Resource management and global rights*, Eburon, Delft

Jones, S. (2006) 'A political ecology of wildlife conservation in Africa', *Review of African Political Economy*, vol 33, no 109, pp. 483–95

Jreisat, J. E. (2002) 'Governance and developing societies: Introduction', in J. E. Jreisat (ed) *Governance and Developing Countries*, Brill, London, pp. 1–12

Kgathi, D. L., Ngwenya, B. N. and Darkoh, M. B. K. (eds) (2011) *Rural Livelihoods, Risk and Political Economy of Access to Natural Resources in the Okavango Delta, Botswana*, Nova Science, New York

Kooiman, J. (2008) 'Exploring the concept of governability', *Journal of Comparative Policy Analysis: Research and Practice*, vol 10, no 2, pp. 171–90

Kooiman, J. and Bavinck, M. (2005) 'The governance perspective', in J. Kooiman, S. Jentoft, R. Pullin and M. Bavinck (eds) *Fish for Life: Interactive Governance for Fisheries*, Amsterdam University Press, Amsterdam

Kooiman, J., Bavinck, M., Chuenpagdee, R., Mahon, R. and Pullin, R. (2008) 'Interactive governance and governability: An introduction', *Journal of Transdisciplinary Environmental Studies*, vol 7, no 1, pp. 1–11

Little, P. D. (1996) 'Pastoralism, biodiversity, and the shaping of savanna landscapes in East Africa', *Africa*, vol 66, no 1, pp. 37–51

Mayntz, R. 2003. 'New challenges to governance theory', in H. P. Bang (ed.) *Governance as social and Political Communication*, Manchester University Press, Manchester, pp. 27–40

MNRT (1997) 'National fisheries sector policy and strategy statement', Ministry of Natural Resources and Tourism, United Republic of Tanzania, Dar es Salaam, Tanzania

Ostrom, E. (1999) 'Self-governance and forest resources', *CIFOR Occasional Paper No 20*, Center for International Forestry Research, Bogor, Indonesia

Philemon, L. (2011) 'Ban on farming in Ngorongoro crater rim intact, says minister', http://ippmedia.com/frontend/index.php?l=31097, accessed 19 April 2013

Pierre, J. (2000) 'Introduction: Understanding governance', in J. Pierre (ed.) *Debating Governance: Authority, Steering and Democracy*, Oxford University Press, Oxford, pp. 1–12

Pollini, J. (2010) 'Environmental degradation narratives in Madagascar: From colonial hegemonies to humanist revisionism', *Geoforum*, vol 41, no 5, pp. 711–722

Rhodes, R. A. W. (1996) 'The new governance: Governing without government', *Political Studies*, vol 44, no 4, pp. 653–67

Rihoy, L. Chirozva, C. and Anstey, S. (2010) 'People are not happy: Crisis, adaptation and resilience in Zimbabwe's CAMPFIRE programme', in F. Nelson (ed.) *Community Rights, Conservation and Contested Land: The Politics of Natural Resource Governance in Africa*, Earthscan, London, pp. 174–201

Rosenau, J. N. (1992) 'Governance, order, and change in world politics', in J. N. Rosenau and E. Czempiel (eds) *Governance Without Government: Order and Change in World Politics*, Cambridge University Press, Cambridge, pp. 1–29

Sbragia, A. M. (2000) 'Governance, the state, and the market: What is going on?' *Governance: An International Journal of Policy and Administration*, vol 13, no 2, pp. 243–50

Stoker, G. (1998) 'Governance as theory: Five propositions', *International Social Science Journal*, vol 50, no 155, pp. 17–28

Sunseri, T. (2003) 'Reinterpreting a colonial rebellion: Forestry and social control in German East Africa, 1874–1915', in *Environmental History*, vol 8, no 3, pp. 430–45

Techera, E. (2008) 'Supporting the role of customary law in community-based conservation', Macquarie University School of Law, Sydney

Treib, O., Bähr, H. and Falkner, G. (2007) 'Modes of governance: Towards a conceptual clarification', *Journal of European Public Policy*, vol 14, no 1, pp. 1–20

UNESCAP (n.d.) 'What is good governance?', United Nations Economic and Social Commission for Asia and the Pacific, http://unescap.org/pdd/prs/ProjectActivities/Ongoing/gg/governance.asp, accessed 19 April 2013

Walters, W. (2004) 'Some critical notes on "governance"', *Studies in Political Economy*, vol 73, pp. 27–46

Wicomb, W. and Smith, H. (2011) 'Customary communities as "peoples" and their customary tenure as "culture": What we can do with the Endorois decision', *African Journal of Human Rights Law*, vol 2, pp. 422–46

Wilshusen, P. R., Brechin, S. R., Fortwangler, C. L. and West, P. C. (2002) 'Reinventing a square wheel: A critique of a resurgent "protection paradigm" in international biodiversity conservation', *Society and Natural Resources*, vol 15, pp. 17–40

3 Localizing global environmental governance norms

Implications for justice

Melissa Hansen, Vasna Ramasar and Kent Buchanan

Introduction

With the increasing scale and complexity of environmental and social problems, there has been a trend of elevating responses to the global level. The call to 'think global and act local' means that a global perspective on what is required for sustainability strongly influences actions at national and subnational levels (United Nations [UN] 1992). Global environmental governance is underpinned by a set of normative discourses which hold assumptions on the causes of unsustainability and the corresponding solutions to these challenges (Kates et al 2001; Clark and Dickson 2003; Ostrom 2010). Such normative discourses are important in providing a coherent frame for international strategies in response to large-scale problems such as biodiversity loss, climate change and desertification.

Normative discourses on sustainability, as well as social goals and values are deeply embedded in the minds of agents, the structures of institutions and rules of policy. When global sustainability goals are introduced at the national and local levels, they meet and are influenced by traditional normative objectives that nations aspire to, such as democracy, human rights and economic growth. This chapter explores the intersection of global normative sustainability discourses with others of social and economic development. This is done in order to understand how, at a local level, these normative discourses may correspond or clash, and what the outcomes are for social justice of resulting management decisions. The analysis is based on a case study of the iSimangaliso Wetland Park (IWP) undertaken in 2011 and 2012 in South Africa.

A theoretical discussion on the role of normative discourses in governance follows immediately below, after which the case study is introduced and contextualized. The third section presents the actors involved in the IWP's governance, tracing their primary legislative mandates and the underlying normative discourses that direct their management decisions. The tensions that arise through the application of differing normative discourses at all levels are then described – tensions most often related to a conflict between the divergent goals of the conservation of World Heritage sites and national goals of social and economic development. Evidence from the case study is used to explicate these conflicts in South Africa. This is followed by a discussion of the overall findings and some final conclusions.

Theoretical background

Norms and normative discourses in governance

In our definition, 'governance' refers to ordered rules and collective action in society, where a system of rules around decision-making is implemented by social actors in a coordinated way (Hydén 2001). A system of rules has a distinctive normative foundation, and this is true at all levels. Norms have an influence on the governance regime of the IWP through various initiatives at the global and regional levels, as well as through national legislation and policy frameworks. The Constitution of the Republic of South Africa is based on a cooperative government approach – and this is integral to the governance framework of the IWP[1] (Republic of South Africa 1996: 1267–9). Here it is important to recognize that different agents involved in governance have their own particular mandates and normative approaches.

The *Oxford Dictionary* describes a 'norm' as 'a standard or pattern, especially of social behavior, that is typical or expected' (*Oxford Dictionary* 2012). Norms are embedded in our thinking and often not explicit. They are, however, extremely powerful, in that they represent a prevailing view on a topic and therefore often assume consensus. To better understand what we mean by a 'norm' here, we follow Hydén and Svensson's (2008) ontological analysis, founded on the Aristotelian ideas of 'essence' and 'accident'. A distinction is thus made between the 'essential' and 'accidental' attributes of norms. Three essential attributes of norms are that they are behavioural imperatives, they are socially reproduced and they are the individual's understanding of surrounding expectations regarding their own behaviour (Hydén and Svensson 2008). Accidental attributes of norms include the presence of sanctions, the origin of the norm, the context or arena in which the norm is socially reproduced, whether the norm is system-oriented or value-oriented, the internal function of the norm and the purpose of the norm (Hydén and Svensson 2008).

A legal norm is a compulsory rule of conduct established by the state. For example, South Africa's National Environmental Management Act (Act 107 of 1998) directs the state to 'respect, protect, promote and fulfil the social, economic and environmental rights of everyone and strive to meet the basic needs of previously disadvantaged communities' (RSA 1998a: 2). Legal normative discourses here are those around human rights and restorative justice. A social norm is a habitual rule that governs behaviour in groups and societies. A social normative discourse in the post-apartheid South African context may favour the embracing of a vibrant multiculturalism (Sonnichsen 2009). A technical norm may relate to rules of conservation, for example the necessity of a fence to protect 'sensitive' ecological areas from human impact (United Nations Educational, Scientific and Cultural Organization [UNESCO] World Heritage Centre 2011, pp. 1, 11). An economic normative discourse may presume, for example, that job creation is integral to economic growth, or that gross domestic product is an indicator for economic growth (RSA 2010). A bureaucratic normative discourse may be that of transparency or of democratic participation (RSA 2000).

Normative discourses prescribe what ought to be done. They are the rules that are implicitly followed in management decisions. This chapter delineates the normative discourses underlying legislative and policy documents at different levels (global and regional, national and local). This has important implications, in that conflicting normative discourses may lead to the precedence of global, regional or national priorities and values over those at the local level, or vice versa, leading to important outcomes for social justice. This chapter will look specifically at contradictions arising through normative discourses directing the divergent goals of the conservation of World Heritage, and social and economic development, influencing the management decisions of actors at different levels.

Normative discourses of sustainable development and the iSimangaliso Wetland Park

The IWP is a UNESCO World Heritage Site. Our analysis is based upon the argument that, as a site of ecological world heritage, the governance framework of the IWP is partly founded upon normative discourses of global conservation and the public interest. Through the World Heritage Convention Act (Act 49 of 1999), which incorporates the World Heritage Convention into South African legislation, a global commitment to the conservation of areas of 'outstanding universal value' has received national legislative support and, in this case, been given effect through the establishment of the IWP. Thus global norms of conservation become entrenched in national and local levels of government.

At the national level, although normative discourses around economic development are present, intersectional justice, human rights and social development also come strongly into play, in light of South Africa's post-apartheid priorities. The Bill of Rights in the Constitution of South Africa states: 'Everyone has the right to an environment that is not harmful to his or her health or well-being' (RSA 1996). Even the World Heritage Convention Act includes a strong emphasis on intersectional justice. Among the fundamental principles of the Act are that 'participation by vulnerable and historically disadvantaged persons must be ensured' and that 'decisions must take into account the interests, needs and values of all interested and affected parties' (RSA 1999).

At a southern African regional level, the IWP comprises a major node of the Lubombo Spatial Development Initiative (LSDI) and of the Lubombo Transfrontier Conservation and Resource Area (LTFCA),[2] two collaborative development projects of the governments of Mozambique, South Africa and Swaziland. Jourdan (1998) has argued that spatial development initiatives are guided by growth-based definitions of development. Among the priorities for the LSDI are to generate economic growth by making maximum use of the inherent, but underutilized, potential of the area; to maximize private-sector involvement and create an attractive and stable climate for investors to operate in; and to maximize job creation by ensuring that the new industries being stimulated are competitive and have a long-term future in the region (National Department of Environmental Affairs [NDEA] n.d.). At the signing of the Trilateral Protocol for the LTFCA on 22 June 2000,

Mohammed Valli Moosa, South Africa's Minister of Environmental Affairs and Tourism at the time, stated that the LTFCA 'aims to consolidate conditions for the development of Lubombo's considerable tourism potential that underpins the promise of a revitalized regional economy' (Moosa 2000). The role of the iSimangaliso Wetland Park Authority (IWPA) (then the St Lucia Wetland Park Authority) would be to 'accelerate development, generate sustainable jobs and create conditions for the establishment of an internationally competitive tourism destination' (Moosa 2000). This shows that there is a strong emphasis on economic development goals at the southern African regional level. Here the policy view is that the IWP will facilitate socio-economic development through tourism.

Introduction to the case study

The study area

The IWP in northern KwaZulu-Natal was listed as South Africa's first UNESCO World Heritage Site in December 1999, in recognition of its superlative natural beauty and unique global values (UNESCO World Heritage Centre [WHC] 2000). Specifically, three of the ten criteria of UNESCO were met (UNESCO WHC 2000). First, the IWP is an example representing ongoing ecological and biological processes in the evolution and development of ecosystems and communities of plants and animals. Second, it contains superlative natural phenomena or areas of exceptional natural beauty and aesthetic importance. Lastly, it contains the most important and significant natural habitats for in-situ conservation of biological diversity, including those containing threatened species of outstanding universal value from the point of view of science or conservation. The IWP also contains four wetlands of international importance under the Ramsar Convention (Department of Environmental Affairs and Tourism [DEAT] 2009).

The IWP was elevated to the status of an icon in the history of the environmental struggle in South Africa (DEAT 2009). In the late 1980s and early 1990s, a proposal by a multinational company to mine the dunes on the eastern shores of Lake St Lucia for titanium and other heavy metals was met with vehement and polarized public debate (Bainbridge 1993/1994). An extensive environmental impact assessment (EIA) process was characterized by a clash between two contrasting views: economic benefits versus aesthetic value and sense of place (Drogin 1994; Kruger et al 1997; Lyman 1993).

An independent review panel, chaired by Justice Ramon Leon, was appointed to review the EIA process and the final reports, to assess public opinion and to submit a recommendation to the cabinet as to which land use for the area was considered to be the most appropriate (Bainbridge 1993/1994). The panel put forward a proposal that future development be based on ecotourism as the primary land use option (Dominy 1993/1994). This proposal was influenced by an overwhelming national and international outcry about the loss of a global biodiversity hotspot, representing an explicit connection of the IWP to global conservation goals. Significant normative discourses here were those around global biodiversity conservation and the public interest.

Informed by the recommendations of the Leon Commission, South Africa's new democratic government ruled that dune mining on the eastern shores of Lake St Lucia be prohibited and the area's fragile beauty and sense of place protected for future generations (DEAT 2009). This was a landmark decision for the South African environmental movement, contributing to a change in the way conservation areas were thought of in South Africa, and creating the opportunity to view ecotourism as a viable alternative to primary-sector natural-resource extraction as an economic growth strategy (Dominy, 1993/1994; Chellan and Khan 2008; Walker 2008).

Saved from dune mining, the IWP (then known as the Greater St Lucia Wetland Park) was proclaimed in 2000, in terms of regulations published under the World Heritage Convention Act. At the same time the IWPA was set up to manage the park on behalf of the state. The IWP effectively consolidated 16 parcels of previously fragmented land – a patchwork of former proclamations (the earliest going back to 1895), state-owned land, commercial forests and former military sites[3] – to create an integrated park for the first time (DEAT 2009; IWPA 2009).

The IWP covers more than 330,000 hectares, stretching 220 kilometres from Kosi Bay, just below the Mozambican border in the north, to Maphelane, south of the St Lucia estuary (DEAT 2009) (Figure 3.1). Its eastern boundary is the Indian

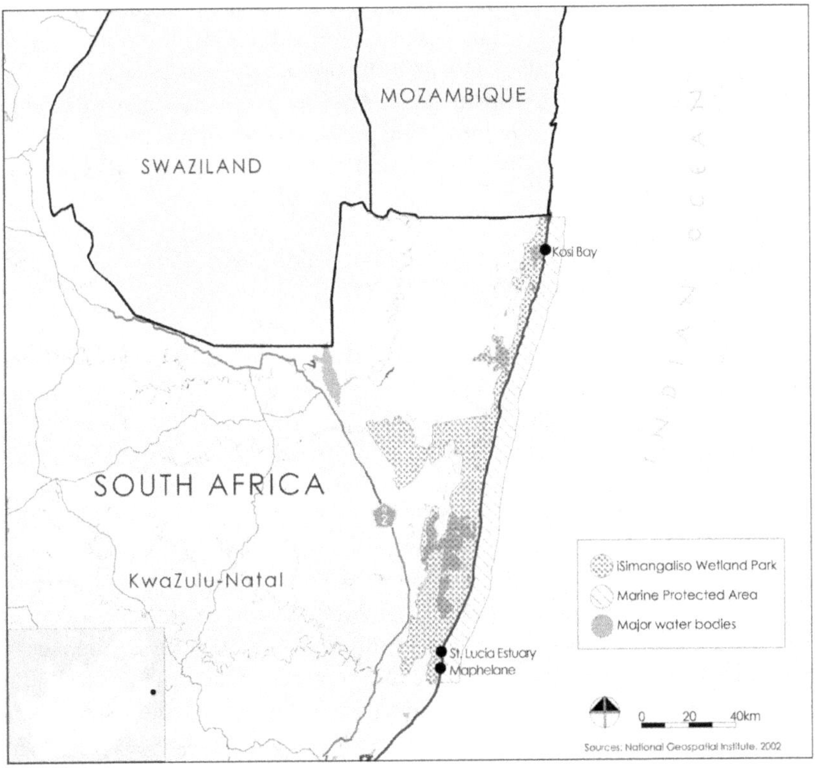

Figure 3.1 Geographical location of the iSimangaliso Wetland Park.

Ocean and its western boundary is irregular, incorporating the entire Kosi, Sibaya and St Lucia lake systems, as well as the uMkhuze Game Reserve. It encompasses one-third of the KwaZulu-Natal coastline and 9 per cent of the entire coastline of South Africa (DEAT 2009).

Socio-economic context

The IWP falls almost entirely within the boundaries of the uMkhanyakude District Municipality (IWPA 2008). Some of the most impoverished communities in South Africa are found in the uMkhanyakude region, and areas neighbouring the IWP in particular (IWPA 2008). As of 2007, 614,046 people live in the uMkhanyakude council area, translating into 114,973 households (Community Survey 2007, cited in uMkhanyakude District Municipality 2011/2012). Within the Coastal Forest Reserve section of the park, there are six small townships[4] whose combined population in 1998 comprised approximately 200 families (KwaZulu-Natal Nature Conservation Services [NCS] 1998). The area is characterized by high levels of underdevelopment, unemployment and poverty, along with some of the highest rates of HIV/AIDS in the country (uMkhanyakude District Municipality 2011/2012). Significantly, many people rely on natural resources for their livelihoods. For example, our fieldwork identified the common use of ncema reeds (*Juncus krausii*) for mat-making and the building of traditional Zulu structures, ilala palm (*Hyphaene coriacea*) for the production of ilala wine, fish from the coastal area and grazing land for cattle. The integrated management plan for the IWP raises a concern about increasing pressure on such resources inside the park, through the depletion and degradation of natural resources in communal areas (IWPA 2008).

Data collection

Field research was carried out in 2011 and 2012, with numerous visits to communities residing both in and adjacent to the IWP, as well as interviews with local tribal authorities, municipal officials from the Big Five False Bay Local Municipality (a subdivision of the uMkhanyakude District Municipality) and several non-profit organizations working in the Big Five False Bay local municipal area. Data collection consisted primarily of semi-structured interviews and direct observations (Kvale 1996; Brockington and Sullivan 2003), as well as focus groups (Bryman 2008) in KwaDapha, a small community residing within the boundaries of the IWP. In addition, household surveys were conducted with around half of the 49 households belonging to this community, in order to gain information about the socio-economic context of the area and perceptual data on the impacts of the IWP on everyday life and the nature of community relations with the IWPA. Relevant legislative and policy documents were also collected and analysed and newspaper articles relating to the management of the IWP were also reviewed. To improve the credibility of results, most of the findings were verified through triangulation.

Five interviews of approximately one hour each were undertaken with officials from the IWPA in February and March 2011 and August 2012.

The governance framework of the iSimangaliso Wetland Park

This section discusses the actors constituting the governance framework of the IWP, their enabling legal framework and the normative discourses that guide their management decisions.

The iSimangaliso Wetland Park Authority

The IWPA is the management authority for the IWP. Its major objective is to ensure the implementation of the proposal put forward by the Leon Commission – that the development of the park be based on ecotourism as the primary land use option, integrating both the conservation of World Heritage and local economic development. The authority reports directly to the national Department of Environmental Affairs, from which it receives its core funding[5] (DEAT 2009). It has a board of nine members, including the CEO, who represent business, traditional councils, land claimants, the provincial government (specifically Ezemvelo KZN Wildlife), the national government (specifically the Department of Environmental Affairs) and local government (DEAT 2009).

The goal of the IWPA is 'develop to conserve' (IWPA 2008, p. 2). The authority specifically strives to balance conservation and sustainable development, taking into account the 'pressing social development priorities of the region' and aiming to end the 'paradox of poverty amongst natural plenty' (IWPA 2008, p. 2). Conservation objectives however, are foremost, in order to ensure that World Heritage values are not compromised (IWPA 2008). Although the goal of intersectional justice is of fundamental importance in the IWP's policy framework, it is overshadowed by that of global conservation.

In addition to conservation as its primary aim, the mandate of the IWP includes the facilitation of optimal tourism-based development in the park (IWPA 2008). The integrated management plan for the IWP aims to achieve this through creating an enabling environment for tourism development. It clearly recognizes the role of the private sector as the primary actor in the development of tourism (IWPA 2008). The plan explicitly strives to balance conservation, tourism development and the local economic development of historically disadvantaged communities in and adjacent to the IWP (IWPA 2008). The latter is expected to be achieved through equity partnerships between the private sector and mandatory community partners (IWPA 2008). An example is the Thonga Beach Lodge and Mabibi community campsite – cited as benchmarks for the development of ecotourism partnerships between the private sector and communities (Sunde and Isaacs 2008). Here, normative discourses centre strongly on human rights and restorative justice.

The IWP is also conceptualized as a 'commercial asset that has the potential to help drive the economic revival of a region that was systematically underdeveloped in the past' (IWPA 2008, p. 3). This is typical of an ecological modernisation discourse, which views nature as an instrumental resource (Cock 2007, cited in Walker 2008).

Ezemvelo KZN Wildlife

The IWPA has contracted the provincial conservation agency, Ezemvelo KZN Wildlife, as its nature conservation agent (IWPA 2008). The organization consists of the KwaZulu-Natal Nature Conservation Board and the KwaZulu-Natal Nature Conservation Service. The board is a public entity reporting to the KwaZulu-Natal provincial Department of Agriculture and Environmental Affairs. The service carries out the day-to-day operation of nature conservation in KwaZulu-Natal and is accountable to the board. In line with its statutory mandate, Ezemvelo KZN Wildlife is responsible for the management of nature conservation within the province of KwaZulu-Natal, and the development and promotion of ecotourism facilities within protected areas (EKZNW 2009). Most tourist facilities within the IWP, such as camping grounds and rustic cottages, are managed by Ezemvelo KZN Wildlife, although there are some community-owned and private facilities within the park, for example the Thonga Beach Lodge and Mabibi community campsite (interview, 14 February 2010).

The mission of Ezemvelo KZN Wildlife is 'to ensure effective conservation and sustainable use of KwaZulu-Natal's biodiversity in collaboration with stakeholders for the benefit of present and future generations' (EKZNW 2009). Since democracy, the organization has invested in repositioning ecotourism so that it contributes more substantially to provincial growth and development (EKZNW 2009). The Ezemvelo KZN Wildlife strategy for the period 2009–14 states that responsible management of biodiversity conservation is recognized worldwide as being a critical factor in the success of sustained economic development, and that often protected areas are a catalyst for economic development (EKZNW 2009). Objectives guiding Ezemvelo KZN Wildlife involve stakeholder participation, as well as ecotourism and conservation as means for achieving economic growth.

uMkhanyakude district and local municipalities

Until mid-2011, the IWP was a district management area[6] falling almost entirely in the uMkhanyakude District Municipality and contiguous to all five of its local municipalities. Since mid-2011, however, district management areas have formed part of their adjacent municipalities (uMkhanyakude District Municipality 2011/2012). This means that the IWP is now geographically split among three local municipalities, the Big Five False Bay, Mtubatuba and uMhlabuyalingana (uMkhanyakude District Municipality 2011/2012). Local municipalities represent a subdivision of district municipalities, usually in rural areas, with district municipalities offering coordination and support to those local municipalities under their respective jurisdictions (Frödin 2009).

The South African Constitution declares that the South African government is founded on three distinct but interdependent and interrelated spheres, thus making local government not just a subordinate level of government, but a significant sphere in its own right (Frödin 2008). Local government has considerable autonomy and the responsibility to promote social and economic development, in addition to providing water, sanitation, roads, stormwater drainage, electricity and municipal health services (Cameron 2003, cited in Frödin 2008). Municipalities also provide the linkages to the provincial and national departments that are responsible for other services, such as health care and education (RSA 1998b).

Tribal authorities

An important intricacy of the South African Constitution is the provision it makes for traditional government, acting through customary law, to function within the local governance sphere (RSA 2003). This manifestation of traditional government is commonly referred to as a tribal authority, which comprises an inkosi, or chief, and indunas, or headmen, who oversee the community. The inkosi is entitled to the position through his bloodline, while the indunas are usually appointed by the inkosi.

Formally, the role of the tribal authority is to work with the municipality while promoting functions under customary law (RSA 2003, section 4). In practice this parallel governance is characterized by unclear roles and questionable jurisdictions over the people, necessitating the negotiation of memorandums of understanding between the municipalities and tribal authorities (Buchanan 2011). There is thus some uncertainty as to how much power and influence they have within municipal structures (Buchanan 2011). Nevertheless, data collected from our interviews and observations show that the tribal authorities were consistently considered the legitimate representatives of local communities in and adjacent to the IWP, by both the IWPA and the communities themselves. Tribal authorities functioned as the communication link between the IWPA and local people. Tribal authorities also oversee much of their community's affairs, including social rules and regulations.

Governance through tribal authorities comes with its own set of legal, social, economic, technical and bureaucratic norms. Though tribal authorities showed respect for the ecological system managed by the IWPA, the importance of social and economic development – and guarantees to uphold their traditions, including the medicinal, spiritual, nutritional and economic uses of natural resources – were observed to be their main priorities (Buchanan 2011).

Tensions between normative discourses and resulting management decisions around the conservation of World Heritage, and social and economic development

Various actors play a role in the governance of the IWP, mandated through policy and legislation at all levels – global and regional, national and local. Management decisions taken by these actors are in turn guided by various normative discourses. As these discourses can be contradictory or ambiguous, management decisions often lead to conflict. This chapter highlights three areas where tensions are visible: the imposition of restrictions on everyday life for people residing within the IWP, the construction of physical conservation measures such as fences and the conditions and channels for participatory governance.

Restrictions on everyday life as a source of conflict between the public interest and local social and economic development

The consolidation of the IWP has led to the imposition of new rules of governance, which constrain the economic and social development activities and opportunities of local people falling within its boundaries. Our research in KwaDapha showed that negative perceptions of these restrictions are widespread. For example, one interviewee stated that: 'after iSimangaliso came in 1999 they put sanctions on us. Life was better before. Now there are sanctions even in the lake. People can't renovate their houses, can't fish on the lake' (interview, 6 September 2012).

One example of local perceptions of limitations to social development is the accusation that the IWPA was opposed to the building of the KwaDapha Primary School (interview, 6 September 2011). This is, however, in contradiction with the stated position of the IWPA, which aimed to ensure that the school was constructed in an ecologically sensitive manner (interview, 3 August 2012). It has also been stated that community members have been stopped from renovating the (Methodist) church in KwaDapha (interview, 6 September 2012).

Another example of perceptions of restrictions on economic activities and opportunities is related to illegal tourism development on the part of individuals in the Coastal Forest Reserve section of the IWP. As of 2 August 2011, there had been at least three concluded civil cases and one concluded criminal case, and there remained one outstanding criminal case, against local people and their partners (Savides 2011). The applicants in these cases – the Minister of Water and Environmental Affairs, the IWPA and Ezemvelo KZN Wildlife – feared that the IWP would suffer irreparable damage, that it might lose its status as a World Heritage Site and that the communities which could benefit through controlled management of the park might suffer hardship, unless unlawful occupiers were stopped and evicted before it was too late (Kuppan 2009). The IWPA likened these tourism development initiatives to 'ecological theft' (Kuppan 2009). Nevertheless, two interviewees involved in these initiatives in KwaDapha stated that they believed they had gone through the necessary channels for authorisation – receiving the go-ahead from the local induna and the owners of the land, the iNgonyama Trust (interview, 7 September 2011). One interviewee, quoted in a newspaper report, stated that 'we believe that the court was wrong to rule against us. We followed all the relevant channels before we started building' (South African Press Association [SAPA] 2009). In addition to obtaining permission from the local induna, they had also submitted their plans to the magistrate at iNgwavuma who allowed them to build (ibid.). It was additionally reported that developers would mobilize the community against the IWPA (ibid.). During our household surveys and interviews in the area, it was clear that many community members held significant feelings of anger and resentment towards the IWPA because of the actions taken against tourism development initiatives in the area. The local induna stated in reference to this, that 'we are not free in this area' (interview, 11 September 2011).

Our interviews, focus group meetings and household surveys in KwaDapha also show that many community members had negative attitudes towards the IWPA as a result of a perceived lack of jobs in KwaDapha, due to these restrictions on tourism development. One interviewee stated in reference to this: 'employment gives money. Money gives food' (interview, 11 September 2011). Another interviewee stated that,

> we want the government to intervene to build big lodges for people to have jobs at KwaDapha. People won't then have a problem with permits or sanctions. . . . We have submitted an application to develop a 4-star diving lodge where the community tented camp currently is and to upgrade the Kosi Bay Beach Camp. Then iSimangaliso will find it easier to work with communities. If iSimangaliso doesn't stop development, they will find it easier to work with the community.
>
> (focus group meeting, 6 September 2012)

The above examples reveal a tension between the IWPA and local people's perceptions of their socio-economic development opportunities. There seems to be a conflict between sustainability norms of intergenerational justice and the conservation of World Heritage, on the one hand, and those of intersectional justice on the other. Although perceptions from local people of the restrictions placed on their livelihoods and social and economic development opportunities might be anecdotal, it is important to note the lacking or insufficient communication between the IWPA and local people (discussed in more detail in the section on democratic participation in the IWP). Negative perceptions of the impact of the IWP on everyday life is further complicated in the case of KwaDapha, as the area falls under the Coastal Forest land claim, which is yet to be settled. In practice this means that there has been limited benefit flow from the designation of the IWP as a World Heritage Site to local people. Walker (2008) highlights the issue of the public interest in protected areas in national environmental legislation and policy frameworks. She argues that the current orthodoxy for settling land claims in protected areas in South Africa, premised on commitment to social justice for those who were dispossessed of their rights in the past, tends to downplay or disregard the interests of other constituencies who are not claimants. Nevertheless, although we recognize the overall benefits of the IWP in terms of the public interest, we argue that where local livelihood and socio-economic opportunities are constrained, in light of South Africa's post-apartheid priorities that include intersectional justice and human rights, but also in terms of the policy framework for the IWP, some measure of recompense is necessary.

Fencing as a source of conflict between conservation and livelihoods

The IWP is considered a natural asset of global significance that must be conserved for the people of the region, the country and the world (IWPA 2008, p. 2). In order to conserve this asset, conservation and preservation measures have been instituted. In many instances, these measures have entailed restrictions on the livelihood activities of local communities, resulting in tensions.

A vivid example of such tensions between conservation and livelihoods is the construction of a fence between the IWP and adjacent areas. The IWP managers use the fence to preserve ecological integrity and endemism at the site in accordance with technical conservation norms (UNESCO WHC 2011). For adjacent communities however, access to natural resources in the IWP (for example land for grazing and agriculture) has been an important social and economic norm. Tribal authority leaders of the Mbila, Makhasa, Nibela, and Mnqobokazi communities adjacent to the IWP have all criticized the construction of a fence as potentially limiting their access to natural resources that are considered important for traditional use, economic use, health and food. Even where gates allow access, the communities are not confident that they will be allowed in.

Representatives of three of these four communities have refused to allow a fence. The other community has permitted the erection of a fence, even though the residents knowingly ignore the IWPA's rules for access to the park. For instance, a tribal authority representative of the Mnqobokazi community explained that the tribal authority was not complying with the IWPA's requests to restrict cattle grazing in the park, because the authority was not 'listening to them' (interview, 15 March 2011). Fences between the IWP and adjacent communities have even been cut down at various times and locations, according to the tribal authority leaders interviewed. The Mbila tribal authority representatives confirmed that such a fence cutting event occurred at the time of our fieldwork in 2011 (interview, 20 February 2011).

In a media statement released on 4 November 2009, the committee representing the Bhangazi, Dukuduku, Western Shores, Sokhulu, Mbila, Mdletsheni, KwaJobe and Triangle communities expressed their 'wish to bring to the attention of the world and government', the concern that their 'rights to access land for grazing, cropping and hunting are severely curtailed' and that 'community members who are trying to access the land to support themselves and their families are being subjected to all sorts of injustices' (Savides 2009).

Fences have also been a problem for communities residing within the boundaries of the IWP. One interviewee recounted an instance when they had been denied access to the park at its Coastal Forest Reserve access gate, after returning to KwaDapha on foot from KwaNgwanase, late at night. The gate was burned down in 2009 (interview, 24 September 2011).

In our surveys, most households reported that they had experienced difficulties with what they called 'nature's problem'. Older respondents in particular explained that they had detected an increase in forest cover over the preceding ten or so years. They viewed this in a negative light, as they were not allowed to cut the trees down for fuel wood, and because hippopotami (*Hippopotamus amphibious*) and vervet monkeys (*Chlorocebus pygerythrus*) frequently destroyed household subsistence gardens.

Depending on one's perspective, a fence is either a progressive tool for ecological conservation or a stumbling block to local social and economic development. The tensions between the elements of sustainability and different normative approaches to sustainable development become clear in any attempt to constitute a governance structure that brings these together.

Democratic participation in the IWP

'Democratic participation' is a normative discourse that appears frequently at all levels of governance. The South African Constitution emphasizes 'cooperative government' and 'participatory democracy'. This democratic norm has established channels and conditions requiring governance to be conducted with consideration of the voices of stakeholders, including local residents. In practice, it is important to consider the depth of stakeholder engagement or public participation. If participation lacks depth, either purposely or accidentally, voices from the community and/or other actors are excluded. This can result in a bias towards some normative discourses in preference to others, leaving social justice and democracy in question.

The history of nature conservation in southern Africa is complex, with conservation more often than not in conflict with democratic values of participation (Fabricius 2004). In addition, the demarcation of conservation areas has often resulted in forced evictions and exclusion from natural resource use (Fabricius 2004; Sunde and Isaacs 2008). The IWP is no exception in this regard: there have been a total of 14 land claims within the park (interview, 14 February 2011). Three of these were settled in 1998 and 2002, six in 2007, and five remained to be settled at the time of writing (IWPA 2010). Land claims within the IWP have been settled through co-management agreements. The co-management process includes representatives of IWPA and the land claims committee, usually made up of tribal authority members in a given community.

At the People and Parks National Conference in 2010, communities presented a list of persistent problems and challenges they faced. A major concern related to co-management agreements. Many communities represented at the conference felt that co-management arrangements were not being implemented in a way that allowed communities to participate as much as they would like in local, regional and national decision-making processes, and that the government was failing to involve them adequately in the management of protected areas (NDEA 2010).

In the case of the IWP, the relationships between the IWPA and land claims committees vary from community to community, as our fieldwork revealed. They range from cooperative to obstructionist, with co-management agreements ignored in the latter case. In an interview with the Nibela tribal authority, the relationship with the IWPA was described as good, because both sides respected the co-management agreement (interview, 25 February 2011). In the case of Mnqobokazi, however, the community was not observing the rules and regulations established in their co-management agreement, which had not yet been formally signed, and the IWPA was not meeting the community's expectations. For example, according to a Mnqobokazi tribal authority representative, the community is not restricting access to the park, while the IWPA is withholding the gate fee payout promised to the Mqnobokazi tribal authority (interview, 15 March 2011).

In addition, we find the depth of participation in planning and development activities related to the IWP to be lacking. According to various community members and tribal authority representatives in the Nibela and Mbila communities, as well as officials of the Big Five False Bay Municipality, the views of many community members are not always heard, since attendance at stakeholder and public meetings can be low. Nzama (2009) has found that despite the fact that regular outreach workshops are held to foster communication between the IWPA and local communities, participation in planning and development activities is still limited. One reason elicited by our research for low attendance at stakeholder meetings, is that community members know that instead of attending decision-making meetings, they can go to a second, and shorter, informational meeting the next day. The result is that those community members have no voice in decision-making.

Another reason is that participation through tribal leaders may favour community members close to those leaders and exclude others. Sunde and Isaacs (2008) report that the Mabibi community, who reside within the IWP, are adamant that they are not able to participate in the management of the IWP – and that the community is unaware of the potential benefits flowing to them from the Thonga Beach Lodge and the Mabibi community campsite. The fact that some members of the community are unaware of this demonstrates the lack of adequate and effective participation in the planning process (Sunde and Isaacs 2008). One of the reasons Sunde and Isaacs (2008) give is the hierarchical structure of the local tribal authority.

The depth of participation is further limited by conflicting rationales, including sustainability norms and political interests. One such limiting factor observed during research was the lack of interaction between the IWPA and the municipalities. This relationship is limited at best. Municipal officials from the Big Five False Bay Local Municipality stated that they had never had contact with the IWPA (interviews, 11 and 22 February 2011). The IWPA, for its part, stated that the roles of the authority and the municipalities were 'separate and unrelated' (interview, 14 February 2011). This suggests that differing normative perspectives of sustainability – the municipality striving for social development and the IWPA mainly for the conservation of World Heritage – can limit participation. Nevertheless, the municipalities hold information about the local residents, such as details of the areas needing social development more urgently (Big 5 False Bay Municipality, 2010/2011) that could well serve the socio-economic development objectives of the IWPA.

In addition, observations during our fieldwork showed that elected officials have been known to steer the municipalities' social development activities in directions likely to win themselves political favour. For example, both Operation Upgrade and the Makhasa Adventist Group, non-profit organizations working in the area, stated that the mayor of the Big Five False Bay Municipality took credit for their work, without publicly acknowledging them in any way.

Discussion: The iSimangaliso Wetland Park as a place for inclusion and exclusion

The IWP's status as a UNESCO World Heritage Site indicates that normative discourses around the conservation of World Heritage and the public interest play a significant role in the management decisions taken by the IWPA, as well as its very existence. As the IWP's integrated management plan affirms, the World Heritage concept implies that some sites on earth are important to all peoples of the world, irrespective of where they are located (IWPA 2008). In these terms, the IWP is a site for global conservation, its declaration as a World Heritage Site is introduced and justified in part by UNESCO, a global actor. World Heritage status places extra responsibilities on member states and site managers. In this context, questions arise about who should bear the costs of the IWP's designation, particularly relating to communities living within and adjacent to the park.

The case study presents a local space where we see global, regional, national and local normative discourses intersecting (Cox 1998). Norms from all levels and all aspects of sustainability (ecological, social and economic) influence the management of the IWP. This is a space where different agents exert their normative beliefs about what is required to achieve sustainability. At the same time, norms are embedded in the policies, institutions and management decisions related to the IWP. In the context of the IWP, there are tensions between the different goals of sustainability, the conservation of World Heritage, and social and economic development and livelihoods. The achievement of all of these goals requires some trade-offs in decision-making. For example, the fence surrounding the IWP represents a strategy to conserve biodiversity, but can also be seen as an instrument reducing the livelihood options of local communities.

This raises questions of justice in the governance of the IWP. Arguably, South Africa's political and conservation history demonstrates the need to think about reducing the injustices of the past, more than aspiring to achieve an ideally just society in the present. The theoretical approach to justice we take in this chapter is thus realization-focused and comparative, following Sen's work presented in *The Idea of Justice* (2009). This approach focuses on the pressing need to remove identifiable injustices in the world, and is concerned with social realizations resulting from actual institutions, actual behaviours and other influences (Sen 2009). Specific questions asked around justice in this chapter focus on 'Who gains what?' and 'At the expense of whom?' The divergent goals of the conservation of World Heritage, and social and economic development, are admirable and essential for sustainable development. This is reflected clearly in the numerous policies that speak to society's conception of sustainability that demands intersectional justice, in addition to the maintenance of ecological integrity.

Achieving environmental governance for social justice is, however, a formidable task. Part of the challenge stems from the interaction of the underlying norms that guide different actors in trying to achieve this goal. Different normative approaches may be in conflict, leading to tensions that result in social injustices.

In the case of the IWP, management choices, decision-making structures, and policies, in support of conserving a World Heritage Site, are guided by a normative discourse around the global conservation. Although local needs are acknowledged, the impetus to 'think global' means that some choices have been made beyond the bounds of the local area. A question of justice arises not from simply looking at the local level, but rather through understanding the interplay caused by the intersection of several levels – in this case, the implementation of different policies and initiatives relating to the IWP. Nancy Fraser suggests that injustices at intersecting scales can lead to the social exclusion of the global poor (Fraser 2010).

The IWP's identification as a UNESCO World Heritage Site raises it above the status of the local territory and, sometimes, beyond the decision-making authority of local people. There is no prioritization in explicit policy of the global conservation need over local social and economic development needs. Nevertheless, the strength of the global impetus is reflected in the fact that World Heritage status privileges certain actors and goals over others. This could lead to injustices of 'misframing', in which some issues are framed as being primarily of local importance, yet obliged to compete for resources with issues that are considered to be of international or national importance.

The construction of physical conservation measures can also be regarded as a matter of distributive justice. According to technical norms of conservation, the fence is a logical choice to manage the movement of people and animals in and out of the IWP. Legally, the IWPA is well within its rights to erect such a structure. The tension arises when we consider the notion of access. People living in and around the IWP have had access to the park's natural resources for decades. Although the question of ownership and property rights is still a contested one and will not be addressed here, another aspect of access refers to the 'right to benefit from things' (Ribot and Peluso 2003). In this case, it is a question about not simply a bundle of rights, but a bundle of powers. Clearly, the construction of a fence removes the power of the local residents to manage their movement in and out of the park. Management of people and animals is at the discretion of the IWPA. Building a fence implicitly suggests that the local residents will not self-manage their use of natural resources sustainably and must be managed instead. This limits the powers of the local community to play an active role in governing natural resources sustainably.

Finally, channels for participation and cooperative governance can also be used as means of inclusion and exclusion. A wide range of local, national and international actors operate in the area, and this raises questions of legitimacy and accountability. It is not clear whether cooperative governance actually does foster democratic consolidation, or whether the multitude of approaches and actors produces fragmented forms of authority, where the involvement and voice of individuals, and some groups, are lost (Ribot et al 2008). Local leaders are involved in specific decisions that are deemed to pertain to them, but management of the

IWP, for the most part, is left to the conservation 'experts'. Local people are recognized as important constituents of the heritage site, but more as beneficiaries than as conservators. In this light, socio-economic development is limited to a model based on ecotourism. Benefits for local people are in the form of material gains.

There are indeed present and potential injustices around the IWP, but these clearly do not result from explicit strategies to exclude local people. The pervasive nature of norms means that there are some fundamental beliefs about how protected areas such as the IWP should be managed. The World Heritage Convention Act and the IWPA espouse the goal of recognizing intergenerational equity with a level of priority that is not the same for municipalities or tribal authorities. Also, the means of achieving intergenerational equity may differ greatly based on the normative perspective. For example, to one person intergenerational equity is best achieved through conserving biodiversity, while to another person this may be done by improving the living conditions of the current generation (Sen 2009).

In this way, norms pervade decisions and actions about the IWP. In the process, different groups are recognized and have access differently to resources, decisions and benefits. In the end, the way people are included or excluded has implications for justice.

Conclusion

The analysis has revealed significant tensions between varying normative discourses of sustainability and sustainable development at different levels: global and regional, national and local. First, the conservation of World Heritage is in conflict with the IWP as a regional political economy and development project through ecotourism development. Second, both of these have an impact upon social justice at a local level. These impacts must be addressed if the governance of the IWP as a World Heritage site is to be truly sustainable.

In effect, what this case study demonstrates is that management decisions arising from normative discourses around the conservation of World Heritage, as well as social and economic development, and livelihoods, are often in conflict. Effective governance cannot be negotiated without taking this into account. The study also illustrates the complex nature of conservation in the southern African context, and highlights important tensions that exist in realizing admirable concepts such as World Heritage at a local level, where competing goals exist.

Negotiating the environment-development nexus demands a deeper examination of diverging values and normative discourses. In this regard, analyses that take into account the normative concerns of multiple actors at varying levels (global, regional, national and local) are integral to a reflexive and adaptive governance regime.

Notes

1 The Constitution of the Republic of South Africa states: 'All spheres of government and all organs of state within each sphere must exercise their powers and perform their

functions in a manner that does not encroach on the geographical, functional or institutional integrity of government in another sphere; and cooperate with one another in mutual trust and good faith' (RSA 1996, section 41(1)(g) and (h)).

2 Transfrontier conservation areas (TFCAs) are conservation areas that straddle the boundaries of two or more states, aiming to promote regional cooperation and development.

3 Cape Vidal State Forest, Dukuduku State Forest, Eastern Shores State Forest, False Bay Park, Makasa State Forest, Maphelane Nature Reserve, Maputaland Marine Reserve, uMkhuze Game Reserve, Nyalazi State Forest, Sodwana Bay National Park, Sodwana State Forest (Ozabeni), St Lucia Game Reserve, St Lucia Marine Reserve, St Lucia Park, Coastal Forest Reserve and Lake Sibaya Freshwater Reserve (IWPA 2008).

4 Enkovukeni, KwaDapha, Mqobela, Mbila, Shazibe and Hlabezimhlope.

5 Additional funding for specific projects has been received from the Global Environment Facility through the World Bank, among other sources.

6 The Local Government: Municipal Structures Act (Act 117 of 1998) defines a district management area as part of a district municipality which has no local municipality and is governed by that district municipality alone (RSA 1998b).

Bibliography

Bainbridge, B. (1993/1994) 'Lake St Lucia and the Eastern Shores: The Natal Parks Board's role in the Environmental Impact Report', *Natalia*, vol 23/24, pp. 45–60

Big 5 False Bay Municipality (2010/2011) 'Final IDP review: 2010/2011', http://mfma.treasury.gov.za/Documents/01.%20Integrated%20Development%20Plans/2010-11/02.%20Local%20Municipalities/KZN273%20The%20Big%205%20False%20Bay/KZN273%20Big%205%20Falsebay%20-%20IDP%20-%201011.pdf, accessed 22 November 2011

Brockington, D. and Sullivan, S. (2003) 'Qualitative research', in R. Scheyvens and D. Storey (eds) *Development Fieldwork: A Practical Guide*, Sage Publications, London, pp. 57–74

Bryman, A. (2008) *Social Research Methods*, Oxford University Press, Oxford

Buchanan, K. (2011) 'Disjoined action in a conjoined world: An analysis of human development governance in rural KwaZulu-Natal, South Africa', MSc thesis, Lund University, Sweden

Chellan, N. and Khan, S. (2008) 'Contesting eco-tourism development in the iSimangaliso Wetland Park in KwaZulu-Natal', *Alternation*, vol 15, no 1, pp. 268–89

Clark, W. C. and Dickson, N. M. (2003) 'Sustainability Science: The Emerging Research Program', *Proceedings of the National Academy of Sciences of the United States of America*, vol 100, pp. 8,059–61

Cox, K. R. (1998) 'Spaces of dependence, spaces of engagement and the politics of scale, or: Looking for local politics', *Political Geography*, vol 17, no 1, pp. 1–23

DEAT (2009) *Fifteen Years: A Review of the Department of Environmental Affairs and Tourism*, Chief Directorate: Communications, Department of Environmental Affairs and Tourism

Dominy, G. (1993/1994) 'Editorial', *Natalia*, vol 23/24, pp. 5–6

Drogin, B. (1994) 'Environment: South Africa makes waves over mine plan: 300,000 have signed petitions to protect wetlands, energising the nation's tiny' "green" movement', *Los Angeles Times*, January 11

EKZNW (2009) 'Ezemvelo KwaZulu-Natal Wildlife Five Year Strategic Plan and Performance Plan 2009–2014', Ezemvelo KwaZulu-Natal Wildlife, Queen Elizabeth Park, Pietermaritzburg

Fabricius, C. (2004). *Rights, Resources and Rural Development: Community-Based Natural Resource Management in Southern Africa*, Earthscan, London

Fraser, N. (2010) 'Injustices at intersecting scales: On "social exclusion" and the "global poor"', *European Journal of Social Theory*, vol 13, no 3, pp. 363–71

Frödin, O. (2008) 'Synergies, stalemates and social dilemmas: Governance in South Africa in comparative perspective', PhD monograph, Lund University, Sweden

Frödin, O. (2009) 'Generalised and particularistic thinking in policy analysis and practice: The case of governance reform in South Africa', *Development Policy Review*, vol 27, no 3, pp. 287–306

Hydén, G. (2001) 'Operationalizing governance for sustainable development', *Journal of Development Studies*, vol 17, no 2, pp. 13–31

Hydén, H. and Svensson, M. (2008) 'The concept of norms in sociology of law', *Scandinavian Studies in Law, Law and Society*, vol 53, pp. 15–32

IWPA (2008) 'Integrated Management Plan', iSimangaliso Wetland Park Authority, The Dredger Harbour, St Lucia, South Africa

IWPA (2009) 'iSimangaliso Authority', http://isimangaliso.com/index.php?view_page+1907, accessed 10 October 2012

IWPA (2010) Presentation by the iSimangaliso Wetland Park Authority at the Fourth People and Parks National Conference, 29 August–1 September, Cape Vidal, iSimangaliso Wetland Park, South Africa

Jourdan, P. (1998) 'Spatial development initiatives (SDIs): The official view', *Development Southern Africa*, vol 15, no 5, pp. 717–25

Kates, R. W., Clark, W. C., Corell, R., Hall, J. M., Jaeger, C. C., Lowe, I., McCarthy, J. J., Schellnhuber, H. J., Bolin, B., Dickson, N. M., Faucheux, S., Gallopin, G. C., Grübler, A., Huntley, B., Jäger, J., Jodha, N. S., Kasperson, R. E., MAbogunje, A., Matson, P., Mooney, H., III, B. M., O'Riordan, T. and Svedin, U. (2001) 'Sustainability science', *Science*, vol 292, pp. 641–2

Kruger, F. J., van Wilgen, B. W., Weaver, A. V. B. and Greyling, T. (1997) 'Sustainable development and the environment: Lessons from the St Lucia environmental impact assessment, South Africa', *Journal of Science*, vol 93, pp. 23–33

Kuppan, I. (2009) 'Tear it down – or pay', *The Daily News* (Durban), 24 November

Kvale, S. (1996) *Interviews: An Introduction to Qualitative Research Interviewing*, Sage Publications, Thousand Oaks, CA

KZN NCS (1998) 'Nomination proposal for the Greater St Lucia Wetland Park to be listed as a World Heritage Site', submitted to the UNESCO World Heritage Centre, KwaZulu-Natal Nature Conservation Service

Lyman, R. (1993) 'Royal snub where mining is king: fate of game reserve', *The Inquirer*, December 28

Moosa, M. V. (2000) 'Lubombo TFCA protocol: Development and investment', address by Minister Valli Moosa at the signing of the LTFCA protocol, Durban, 22 June, https://www.info.gov.za/speeches/2000/000713110p1005.htm, accessed 15 June 2012

NDEA (2010) *4th National People and Parks Conference, Conference Report 2010*, Directorate: Protected Areas Planning and Development, Department of Environmental Affairs, Republic of South Africa

NDEA (n.d.) 'Spatial Development Initiatives: the Lubombo Spatial Development Initiative', https://www.environment.gov.za/ProjProg/SDIs/SDI_contents.htm, accessed 16 April 2012

Nzama, A. T. (2009) 'The nexus between sustainable livelihoods and ecological management of the World Heritage Sites: Lessons from iSimangaliso World Heritage Park, South Africa', *Inkanyiso, The Journal of Humanities and Social Sciences*, vol 1, no 1, pp. 34–42

Ostrom, E. (2010) 'Beyond markets and states: polycentric governance of complex economic systems', *American Economic Review*, vol 100, pp. 641–72

Oxford Dictionary (2012) 'Norm', Oxford University Press, http://oxforddictionaries.com/definition/norm, accessed 29 May 2012

Ribot, J. C. and Peluso, N. L. (2003) 'A theory of access', *Rural Sociology*, vol 68, no 2, pp. 153–81

Ribot, J. C., Chhatre, A. and Lankina, T. (2008) 'Institutional choice and recognition in the formation and consolidation of local democracy', *Conservation and Society*, vol 6, no 1, pp. 1–11

RSA (1996) 'Constitution of the Republic of South Africa (Act 108 of 1996)', *Government Gazette*, vol 378, no 17,678, Cape Town

RSA (1998a) 'National Environmental Management Act (Act 107 of 1998)', *Government Gazette*, vol 401, no 19,519, Cape Town

RSA (1998b) 'Local Government: Municipal Structures Act (Act 117 of 1998)', *Government Gazette*, vol 402, no 19,614, Cape Town

RSA (1999) 'World Heritage Convention Act (Act 49 of 1999)', *Government Gazette*, vol 414, no 20,717, Cape Town

RSA (2000) 'Local Government: Municipal Systems Act (Act 32 of 2000)', *Government Gazette*, vol 425, no 21,776, Cape Town

RSA (2003) 'Traditional Leadership and Governance Framework Act (Act 41 of 2003)', *Government Gazette*, vol 462, no 25,855, Cape Town

RSA (2010) *The New Growth Path: Framework*, Economic Development Department, Cape Town and Pretoria, www.info.gov.za/view/DownloadFileAction?id=135748, accessed 15 June 2012

SAPA (2009) 'Builders brush off order', *iafrica.com*, 29 December 2009, http://news.iafrica.com/sa/544585.html, accessed 14 June 2012

Savides, D. (2009) 'Threat to "shut down tourism"', *Zululand Observer*, 10 November 2009

Savides, D. (2011) 'Park development stopped in court', *Zululand Observer*, 2 August 2011

Sen, A. (2009) *The Idea of Justice*, Belknap Press of Harvard University Press, Cambridge, MA

Sonnichsen, A. (2009) 'Unmaking "the people": Unity and division in post-apartheid South African politics', PhD dissertation, Department of Political Science, University of Copenhagen

Sunde, J. and Isaacs, M. (2008) 'Marine conservation and coastal communities: Who carries the costs? A study of marine protected areas and their impact on traditional small-scale fishing communities in South Africa', SAMUDRA monograph, International Collective in Support of Fishworkers, Chennai, India

uMkhanyakude District Municipality (2011/2012) *Integrated Development Plan (IDP) Annual Review for 2011/12: Final*, IDP/PMS Unit, Office of the Municipal Manager, uMkhanyakude District Municipality, Mkhuze, South Africa

UN (1992) *Agenda 21: The United Nations Programme of Action from Rio*, United Nations Commission for Sustainable Development

UNESCO WHC (2000) *Report of the 23rd session of the committee*, WHC-99/CONF.209/22, 2 March, UNESCO World Heritage Centre, Paris

UNESCO WHC (2011) *Operational Guidelines for the Implementation of the World Heritage Convention, WHC 11/01*, November, UNESCO World Heritage Centre, Paris.

Walker, C. (2008) 'Land claims, land conservation and the public interest in protected areas', *South African Review of Sociology*, vol 39, no 2, pp. 232–44

4 Governing ancestral land in Madagascar

Have policy reforms contributed to social justice?

Barry Ferguson, Charlie J. Gardner, Mijasoa M. Andriamarovolonona, Tim Healy, Frank Muttenzer, Shirley M. Smith, Neal Hockley and Mathilde Gingembre

Introduction

Madagascar has long been renowned for its unique and threatened biodiversity, as well as the rich and diverse cultures of the Malagasy people and the severe poverty and exposure to natural disaster that they endure. The island has frequently appeared in the media, however, as a result of political turmoil and natural resource governance dynamics. The infamous Daewoo land deal, subsequent political protests and the ousting of the country's president, Marc Ravalomanana, in 2009 were followed by international media attention directed at the illegal rosewood and ebony trade, controversial oil sands and iron extraction deals and the negative social and environmental impacts of high-profile multinational mining projects (see Harbinson 2007a, 2007b; Andrew Lees Trust 2009; Global Witness and Environmental Investigation Agency 2009; Ballet et al 2010; Draper 2010; Gingembre et al 2010; Vinciguerra, 2010; Andrianirina-Ratsialonana et al 2011; Burnod et al 2011).

Beyond these recent high-profile cases, Madagascar has, for the past decade and more, been the site of numerous mining ventures, agricultural land acquisitions and new protected areas. The social and environmental impacts of many of these initiatives have been questioned in both popular media and academic literature, and many projects still operate beyond the radar of agencies responsible for these matters and do not even feature in regional development planning.[1]

This burgeoning interest in Madagascar's resources has not occurred in a policy vacuum. Indeed, since the late 1980s, the country has seen numerous public policy and institutional reforms concerned with the management of natural resources and the governance of revenues gained from them (Fauroux and Blanc-Pamard 2004; Rochegude and Plançon 2009; Teyssier et al 2009; Evers et al 2011). However, many of these reforms, as described in this chapter, have been observed to suffer serious setbacks and failings when it comes to their operationalization.

The rapid growth of interest in Madagascar – from the private sector in the island's land and minerals, and from global conservationists focused on its biodiversity – has without doubt led to cases of social injustice against the rural Malagasy people, whose customary and ancestral lands have been caught up in a modern natural resource rush (Fritz-Vietta et al 2011). These incidents are also, however, an indication of the potential which exists for more extensive and serious impacts in future. Both the actual trends and the potential for future impacts have led to the emergence of a number of civil society platforms, lobby groups and communication networks that have been established because of the rising concern about environmental destruction and social ills that neoliberal forces are articulating across the island.[2] This has been accompanied by an additional focus on Madagascar by a number of international environmental and social justice campaigning groups,[3] as well as significant institutional reviews of the challenges of natural resource governance by international agencies and donors including GIZ (formerly GTZ), the German technical cooperation agency (Ballet et al 2010), the United Nations Children's Fund (Juttersonke and Cartas 2010), the World Bank (Carret et al 2010; Plangemann et al 2010) and the United States Agency for International Development (USAID) (Freudenberger 2010). Typical of this domain, much additional activism on these issues is also carried out by individuals or unnamed groups and coalitions.

This chapter will address issues of social justice in environmental governance through its focus on the dynamics of 'participation' versus 'exclusion'. Participation has been important in international development thinking for decades (e.g. Chambers 1983) and more recently has begun to pervade discourse and policy in natural resource management (Randrianalijaona et al 2011). However, the term itself is somewhat nebulous, as pointedly shown in Sherry Arnstein's (1969) 'ladder of participation'. Participation has the potential to genuinely empower local actors to take control of a given issue, but the term can also be applied to approaches that are little more than tools for placating or even manipulating less powerful participants, who become subject to outcomes predetermined by the instigators of 'participation'. These less desirable manifestations of participation have been called tyrannical (Cooke and Kothari 2001). Nevertheless, participation offers a powerful lens through which to investigate social justice issues (Berghofer and Berghofer 2006).

The first four sections of this chapter explore four policy areas – agriculture, mining, forests and protected areas – which have had, and continue to have, a profound impact on the governance of ancestral lands in Madagascar. Each case is introduced with a brief description of policy reforms, after which the role and depth of participation in the design and implementation of reforms, and the consequences of this for social justice, are investigated. The fifth section reflects on conceptions of ancestral land, rights and power and their implications for environmental governance. Finally, certain conclusions are drawn about the overall social justice impacts of three decades of policy reform.

Land tenure reform and agribusiness

Madagascar has attracted a large number of investors in the global land rush (Burnod et al 2011), partly spurred by reports claiming that the island boasts large tracts of 'idle' land (World Bank 2010). According to the Malagasy Ministry of Agriculture, a total of eight million hectares of arable land is available on the island (Ministère de l'Agriculture, de l'Elevage et de la Pêche [Ministry of Agriculture, Livestock and Fisheries] [MEAP] 2009, cited in Burnod et al 2011) while, according to the Food and Agriculture Organization (FAO) (2007, cited in Burnod et al 2011), the true extent could even be 15 million or 20 million hectares. Investors were also enticed by favourable investment policies under the Ravalomanana regime, which prevailed from 2002 to 2009, when, for the first time in Malagasy history, foreigners were allowed to buy land.[4] They were also offered a facilitating structure, the Economic Development Board of Madagascar, a one-stop service provider for investors, which ranked agribusiness among its six priorities and was empowered to authorize land prospecting and streamline bureaucratic processes to catalyse investment.

From 2005 to 2010, more than 50 agricultural projects were recorded, each involving 1,000 hectares or more (Andrianirina-Ratsialonana et al 2011). The largest was led by the South Korean company Daewoo, which sought to acquire 1.3 million hectares, and then came the Indian company Varun with 230,000 hectares. Both were ultimately suspended because of international resistance and through the political instrumentalization of the threats these projects represented to the local landholders (Gingembre et al 2010; Vinciguerra 2010).[5] Of the 50 or so original projects, only one-quarter are still under way at the time of writing, and the total area of land concerned has decreased from three million to 130,000 hectares. More than two-thirds of the interested operators are foreigners, most focused on leasing rather than buying land. Contrary to what the Daewoo case may have suggested, most ongoing investments aim to produce biofuel, rather than address food security concerns (Andrianirina-Ratsialonana et al 2011).

Madagascar boasts a legal framework that embodies most of the principles advocated by the international community in the Voluntary Guidelines on the Responsible Governance of Tenure of Land, Fisheries and Forests in the Context of National Food Security[6] agreed upon in 2012 (Anseeuw et al 2012), and is therefore an important site for the study of these corporate land-related investments. Indeed, the country's legal framework, in theory, supports best practice in commercial farming, including land securitization, the transfer of authority for land matters to local governments, and local participation in natural resource management (Teyssier 2010). Madagascar is one of the few southern countries where untitled land is no longer considered state-owned (the presumption of state-ownership having been abolished in 2005).[7] Moreover, populations are provided with affordable tools to secure their land property rights officially, an issue that is discussed further later in this chapter. However, initial empirical studies have revealed flaws in the procedures for lease issuance, infringements of people's rights under the regime of 'untitled private property' and a generally low degree of understanding among the rural populations of what their rights are with regard to these procedures[8] (Andrianirina-Ratsialonana et al 2011).

Agricultural land tenure and reform

Malagasy legislation requires investors to fulfil a certain number of obligations to ensure that their projects do not present any environmental and social risks to the localities concerned. Besides providing a business plan, investors need to obtain an environmental permit from the National Office for the Environment if their project affects 1,000 hectares or more. According to the 1999 decree *Mise en Compatibilité des Investissements avec l'Environnement* (MECIE), which includes provisions on environment impact assessment,[9] such a permit is only granted after the validation of an impact assessment that considers environmental and socio-economic issues, while also setting out in a management plan how impacts will be monitored and mitigated. In December 2010, the central government issued a new letter of instruction to the administration, which imposed a strict and lengthy procedure for investors looking to acquire or lease tracts of land equal to or greater than 2,500 hectares.[10] Although this instruction stated that no land should be leased or bought without the population and local government having been consulted, it should be noted that only one of the 18 procedural steps investors are required to follow involves non-state actors.

In terms of customary property rights protection, the first attempt towards official recognition came with a 1998 decree dealing with relative land tenure security,[11] which was designed for local communities to whom the state had transferred management of natural resources through a 1996 law regulating the local management of renewable natural resources.[12] These communities were encouraged to demarcate their village land (*terroir*) and start formally registering their rights over it. The procedure remained traditional (boundary marking and area demarcation), but opened the possibility of having collective titles to land (Rochegude and Plançon 2009). Although the land tenure decree represented progress in that it allowed communities to exclude colonizing immigrants from their lands, implementation was weak and, in any case, did not provide recognition or security for individual customary land.

The next major change occurred in 2005, when a thorough land reform process established that untitled land would no longer be considered state property, and would instead be deemed private property over which current dwellers and/or users should have a say.[13] This was backed by a decentralization of authority for land matters to local communes and by the creation of *guichets fonciers* (local land offices) to whom anyone able to prove an ongoing de facto use or occupation of a tract of land within the sovereignty of these institutions could apply for a land certificate. The intention was to simplify the procedure by consolidating the process in one place, thus reducing the number of agents involved, the time taken and the cost, while providing the holder with the same rights and prerogatives as land titles (Teyssier et al 2009).

Foreign investment, land reform and social justice

Despite these legal constraints, livelihoods and environmental diversity remain poorly protected when confronted with major agribusiness investments. In states such as Madagascar, where power is personalized and regime turnover is high, changes in laws and policies are frequent. Natural resource management policies are therefore superseded by new policies or preferences before having been properly rolled out, and responsibilities are regularly transferred from one institution to another. In addition to this, land titling procedures in Madagascar are extremely long and drawn-out, with an average period of more than six years to obtain a title (Malagasy Land Observatory 2008). In practice, this results in complex situations, incomplete rights (e.g. where an area has been demarcated but is not titled), blocked status and competing rights or claims (Evers et al 2011).

In addition, an examination of land access negotiation procedures shows that the law-making has not taken sufficient consideration of political structures, socio-economic conditions, culturally embedded power dynamics and natural resource management norms. For instance, the obligation to obtain the mayor's consent before undertaking any land deal does not ensure that the local users' rights will be adequately defended. A review of ongoing land acquisition projects in the country demonstrated that no mayors had ever raised objections; this was attributed to the hierarchical 'political and economic pressures' that constrain the freedom of mayors to make real decisions (Andrianirina-Ratsialonana et al 2011). In Madagascar, local leaders have very limited autonomy indeed. Some major local officials, such as heads of region and sometimes mayors, are directly appointed by central authorities,[14] while others, thanks to the domination of the ruling presidential party, are indebted and therefore loyal to higher-ranked officials.

More importantly, for municipalities with underfunded budgets, private investments are often a highly welcome source of revenue (in the form of land taxes and rental fees) and providers of social service infrastructure (through compensation mechanisms). Mayors and other local officials involved may therefore be tempted to obfuscate certain land uses and rights in order to ease the way for investment (Burnod et al 2013).

The implementation of land reform is also hampered by insufficient communication and information. The law provides that, before untitled land can be allocated to an investor, a commission has to go into the field and check whether there are already established occupants. Observation of this recognition procedure being carried out in recent investment projects confirms that little consideration is given to the status of untitled private property or to the protection of local rights (Burnod et al 2011), and that the category of 'untitled land' is frequently confused with that of 'state land'. All commissions so far have concluded that there is 'nothing to declare', even where grazing and/or agricultural activities are clearly evident. Even property rights that were formalized through the establishment of land certificates or use rights recognized through the GELOSE (Gestion Locale Securisée) procedure have, on occasion, been ignored.

Since the maps used by state land titling services do not yet feature these certificates: the only institutions that can attest to their existence in land claims procedures are the *guichets fonciers*. However, their teams are not always included in the land recognition procedure (Burnod et al 2011). Needless to say, in places where there still are no *guichets fonciers*, including many of the prominent mining areas and the hinterlands of many new protected areas, the private property presumption of untitled land is even more remote a prospect.

Legal tools do not always work as intended, and, in certain instances, the rules are flouted. Most investors either lack information or do not see the point in engaging in long, complex procedures before they can be certain that their investment will be fruitful. Often advised by national brokers, they seem to be aware that following the official institutional pathway is of less importance than securing high-level political support. Their targets and ethics, and the advice they get, drive them to adopt different institutional paths: some strive for support from high-ranking officials, whether from the region or from local authorities (Andrianirina-Ratsialonana et al 2011). These contacts are often validated by 'contracts' or 'conventions' which are not legally valid or binding, but can nonetheless help in securing the trust of financial backers and ensuring, through respect for patronage networks, the approval of other public agencies and leaders (Burnod et al 2011).

Observation of Malagasy realities suggests that securing local land rights is a very long and complex process, and that the legal regulation of investment, which is called for by many (Cotula et al 2009; Von Braun and Meinzen-Dick 2009; World Bank 2010), may not be sufficient to ensure that people's customary property rights are not undermined or flouted.

Mining, centralization and political power

Motivated by the desire to attract foreign direct investment and shaped by institutional reforms advocated by the World Bank, Madagascar, like many other developing countries, has implemented structural changes to its governance systems, as part of a liberalization drive towards manufacturing and urbanization (Pegg 2003; Kambhampati 2004). These reforms include reducing taxes, royalties and other fees; eliminating restrictions on the repatriation of profits; and strengthening investor rights (Pegg 2003, 2006). Such economic measures, however, facilitate the transfer of profits by large-scale mining companies from the host country (i.e. where the mineral reserves are) to where those companies are domiciled, resulting in a disconnect from the local economy, since less money is available to stimulate the local economy or for government reinvestment in development projects. Moreover, although companies benefit from attractive fiscal policies, host governments typically receive minimal revenues from mineral extraction (Szablowski 2007).

Policy reforms such as these, and inherently weak governance mechanisms, have enabled large-scale mining to expand into certain vulnerable countries, such as Madagascar, without effective regulatory control (Civil Society Structural Adjustment Participatory Review Initiative [SAPRIN] 2002). Dependence on mineral wealth can have a negative impact on the political stability and economic and social development of a country – the paradox of the resource curse (see Atkinson and Hamilton 2003; Auty 1993; Karl 1997; Pegg 2003, 2006; Sachs and Warner 2001; Stevens and Dietsche 2008). Although large-scale mining operations have been the focus of recent discussion, artisanal mining, particularly for gemstones and gold, are not insignificant and have considerable socio-economic and environmental impacts (Cook and Healy 2012).

Access to land by mining companies – as regulated by mining reforms, land tenure and environmental legislation – and potential conflicts involving ancestral lands are key issues for the industry. Madagascar presents an interesting case study for mining operations, as modern titling coexists with informal or ancestral tenure systems (Jacoby and Minten 2007), though not always harmoniously. Ancestral lands are by far the most significant land ownership category in rural areas and 'belong' to either communities or individual families. Private mining companies need to secure their shareholders' investments with titled land or leases to surface rights. The investment law of 2008 states that registered companies in Madagascar can acquire titled land for business development,[15] however, the process requires the support of existing land authorities. Titled land is, of course, the preference for the developers of large mines, although some potential foreign mining investors may have to be content with a lease, as securing title can be difficult due to the persistent resistance of some authorities to selling land to foreigners.

Mining policy reform

An important restructuring of mining policy began in 1998, with further reforms in 2005. The principal aims of these reforms were to liberalize the mining sector and disengage the state (GLW Conseil 2009). These aims coincided with the goals of encouraging better governance and transparency in mining through collaboration with the Mining Sector Reform Project. The Mining Code also reinforced relations between departments in the Ministry of Mines and the Ministry of Environment and Forests.[16] Meanwhile, the ongoing political crisis has stalled further policy reform, which would integrate mining into regional development planning, environmental protection and good governance in the artisanal mining sector (GLW Conseil 2009). Although previous efforts by the United States Agency for International Development (USAID) and the World Bank did establish the concept of integrated regional planning in the Anosy region (southern Madagascar) through a dynamic mineral resources management scheme and the establishment of a regional stakeholder alliance, this process became dependent upon the stimulus provided by transnational company Rio Tinto and public-private partnerships such as that facilitated by the German technical cooperation agency, GIZ.

However, the focus of reform was skewed to conserving biodiversity, rather than ensuring the well-being of human populations. Mining reform processes in 2007 created environmental units within the regional departments for mines, which are orchestrated by a central unit within the Ministry of Mines. The principal objective of these environmental units should be to evaluate simplified environmental impact assessments (EIAs) and support the National Office for the Environment's technical committees with evaluations of full EIAs. The environmental unit should have an important role working with decentralized authorities, focusing principally upon regional and local governments, as well as civil society organizations and community associations, non-governmental organizations (NGOs) and regional representatives from various ministries, including the Ministry of Lands.

However, the quality of EIAs varies, and they can also be complex or ambiguous. The competence or political bias of associations or civil society groups in the EIA process could be prejudicial to or limit constructive participation in EIAs. Indeed, the decision over the modalities of public participation in the evaluation is left to the National Office for the Environment and barely constrained by law (article 15 of MECIE decree). Therefore it is imperative that communities be visible and proactive throughout the processes.

Meanwhile, the processing of certain EIAs can be fast-tracked where companies are keen to maintain investment momentum. Land office commissions can conduct site assessments under the direction of the specific head of region, who does not systematically inform every person affected by the project, as this process is delegated to mayors and heads of villages (Andrianirina-Ratsialonana et al 2011). These local representatives may not understand the processes involved in securing lands and instead may focus solely on compensatory measures. Furthermore, the *guichets fonciers* are scarce in many regions of the country, meaning that the dissemination of information about, and provision of, land tenure services are limited. However, existing local land offices in some parts of the country are now attempting to adopt the National Tenure Programme approach, although empirical evidence of whether, or how, these have functioned in areas of mining interest is not yet available.

Equally problematic are compensation schemes, since the availability of alternative cultivable land is sometimes limited for persons affected by the project, as was the case in some areas near the Rio Tinto mining project (Andrianirina-Ratsialonana et al 2011). In addition, the National Office for the Environment is centralized in the capital Antananarivo, although it liaises with regional offices of the Ministry of Environment and Forests, and the process of creating decentralized offices has not been completed. The regulations exist to address tensions between mining ventures and community interests, although these are severely constrained in terms of good governance and transparency, which are further compounded by weak political and institutional capacities (World Bank 1998; Sarrasin 2004, 2006).

Madagascar's mining code levies a royalty tax of 2 per cent of a mutually agreed initial price of the mined product. This tax is apportioned at two levels: 1.4 per cent (*ristournes*) to the regional government where the mine is situated and 0.6 per cent (*redevance*) to the state. Regional *ristournes* are further distributed at three administrative levels: commune (60 per cent), region (30 per cent) and province (10 per cent) (ATW 2010). *Ristournes* spending at commune level from the Rio Tinto project, however, does not always provide sustainable assistance to the families most significantly impacted by the mining project (Smith 2012). This observation is in line with evidence from studies in other sub-Saharan countries, which suggest that mining revenues rarely benefit the poorest 20 per cent of the population and may, in fact, worsen their situation (Ross 2001). Thus, as increased mine production releases more money into the regional economy, more equitable *ristournes* distribution at the grassroots will be critical to reducing the risk of an economic gap widening between beneficiary and non-beneficiary communities.

Political power, participation and social justice

Motivated by the revenue associated with mining investments, which reduces its commitment to the regulation of the country's extractive industries, the state centralizes control of Madagascar's large-scale mining agenda in certain ways, including the manipulation of regional and local government (Smith et al 2011). In strategic mineral-rich locations, elections are routinely decided by political patronage more than the preferences of the electorate, and this creates a hierarchy of dependencies. Local government is further weakened by a decentralization strategy in which budgets and authority are not devolved. Moreover, political pressure and funding difficulties militate against the responsibility of the National Office for the Environment and the Ministry of Mines to monitor the impact of mining activities.

Mediation mechanisms, such as forums chaired by local government officials, typically are weak, lack momentum, have few grassroots links and are held in low regard by the communities they are intended to represent. This dichotomy of local government – that it acts as an agent of national government while also being supposed to represent broader society – brings into question the legitimacy of certain civil society representatives. In effect, individuals and communities in rural Madagascar generally have limited access to mechanisms for consultation with, or making representations to, elected politicians. The outcome, in effect, is a general suppression of both opposition and community resistance.

Community disempowerment is exacerbated by disconnects at many levels, partly due to the relative isolation of rural communities, the economic divergence between rural and urban populations and conflict between modern local government and traditional values (Marcus 2008). Lack of collaboration and solidarity among those with common concerns (Fauroux and Blanc-Pamard 2004) weakens communities' abilities to negotiate with mining companies. For example, in the Anosy region, the site of Rio Tinto's operations, displaced communities situated in different neighbouring administrative districts have protested separately, not as a group, against the government and mining company regarding land compensation.

Madagascar's official mechanism of public consultation regarding proposed mining projects inherently limits community input into the design and development of plans. First, consultation typically occurs at an advanced stage of negotiations between the company and government, when permit approval is already virtually assured. The power balance is heavily weighted against the community, and therefore there is little motivation for the government or mining company to invest in expensive community participation processes. The consultation is mostly expert-led, rather than participant-driven, entailing issues and concepts beyond the experience of most rural communities. Pointing to such power relations, Szablowski (2007) also contests the claim that the World Bank's voluntary resettlement policy is participatory. Essentially, communities have no opportunity to oppose resettlement or development plans for mining projects, so while they might be said to have 'participated' in the consultations, in reality they have no chance of influencing the outcome, which leaves them, as manipulated citizens, at the bottom of Arnstein's (1969) ladder of participation.

Aiming to emulate successful societies worldwide, the Malagasy government initiated a network of platforms of civil society organizations to act as regional community representation forums that could also build capacity within and between community-based associations. The intention was to create a framework that would link existing grassroots associations to one another and to regional forums, and subsequently link those regional platforms to one another, thus strengthening civil society.

A key role of these platforms is to represent civil society within the Extractive Industries Transparency Initiative (EITI). The EITI is an externally validated voluntary scheme that advocates financial transparency by governments, and addresses the need to develop public understanding of mining revenues and expenditure to improve expectations of accountability (EITI 2008). This is a key response to concerns about mining companies achieving and retaining a 'social licence' to operate. However, in Madagascar, mainly large players, including the Ambatovy (a nickel and cobalt mine in eastern Madagascar) and Rio Tinto projects, participate at present, while many smaller foreign or national mining investors remain outside the process. In addition, civil society engagement in Madagascar's EITI process has been limited, as stakeholder meetings were cancelled when government commitment to the EITI declined during the 2009 political crisis (Smith et al 2011). Several platforms of civil society organizations also lacked the coherence and motivation to continue functioning autonomously when external funding was withdrawn.

Project finance agreements can require companies to comply with a complex array of economic, social and environmental standards: that is, economic requirements to deliver positive investment returns, social policy objectives regarding people issues, and quantifiable environmental performance indicators and impact assessments. Of these stipulations, social standards are the least evolved and are often based on environmental objectives (Overseas Development Institute [ODI] 2005).

Moreover, although the extractive industries attract significant research interest, little of that research is about the social consequences at the grassroots. However, Rio Tinto's history of environmental and social problems around the world is regularly debated (Martinez-Alier 2001; International Longshore and Warehouse Union 2010) and, thanks to the unique biodiversity in Madagascar, the project there has attracted significant external interest (see e.g. World Wide Fund for Nature [WWF] 2005; Friends of the Earth, 2007; Harbinson, 2007b). Several lobbyists have raised concerns about the quality of the social and environmental impact assessment (WWF 2005), particularly the inadequate consideration of social issues, such as moving traditional burial grounds, health risks, inflation in the cost of living, pressure on already limited natural resources and insufficient focus on capacity-building prior to mining activities commencing.[17]

Nevertheless, since Madagascar is recognized as a biodiversity hotspot, external influences and weak governance have combined to perpetuate the unbalanced focus on environmental and conservation issues rather than socio-economic development (Myers et al 2000; Sarrasin 2006, 2009). For instance, Rio Tinto, the first international large-scale mining company to begin operating in the country, embarked on environmental research at an early stage of exploration with leading international partners such as the Missouri Botanical Garden and Kew Millennium Seed Bank (Ganzhorn et al 2007). Rio Tinto's environmental department, which employs a number of local people as field agents and research assistants, has established strong links at grassroots levels and operates a strong, although non-transparent, multi-stakeholder biodiversity committee. In contrast, the social department – which was established later, has no equivalent social committee and consists of a significantly smaller team – has recognized difficulties in ongoing dialogue and consultation with local communities and other social sector stakeholders (Smith 2012). Weaknesses in Rio Tinto's social strategy, particularly land purchase and resettlement policies, as well as employment practices, are an ongoing source of local discontent, manifesting as protests, blockades and demonstrations in and around the company's mining locations (ClientEarth 2010; Friends of the Earth 2007), which local politicians exploit or incite for their benefits.

Although society expects corporate social responsibility from the industry, investment in high-profile infrastructure such as schools and clinics is more motivated by public relations, aimed at achieving legitimacy, than embedded in firm policies designed to achieve measurable social change and poverty reduction (Smith et al 2011). Such projects are routinely cited in annual reports without consideration also being given to the overall net social impacts of company activities.

The Global Mining Initiative, established by nine transnational mining companies in the 1990s, aimed to explore the industry's approach to sustainable development, and launched the Mining, Minerals and Sustainable Development project for that purpose. In its report, the project conceded that the industry was not addressing the key elements of sustainable development, such as poverty alleviation, employment opportunities, capacity-building, good governance and stakeholder engagement, and, as a consequence, had failed to convince stakeholders of its 'social licence to operate' (International Institute for Environment and Development [IIED] and World Business Council for Sustainable Development [WBCSD] 2002).

In contrast to the fragility of civil society and weak institutional capacity of government, the power and authority of the mining industry and international lending institutes give them strategic influence. As Sarrasin (2009) explains, Rio Tinto's influence over the policies developed under Madagascar's Mining Sector Reform Project and Mineral Resources Governance Project, both backed by the World Bank, is evidenced by the low levels of royalties and taxes being paid by mining companies operating in that country.

In conclusion, civil society's fragility and low expectation of accountability encourages a general acceptance of the lack of enforcement and regulation.

Moreover, voluntary standards, such as the EITI, which advocate stakeholder engagement as a mechanism to achieve good governance, fail to take account of complex power relations between stakeholders and the key motivation behind relationships. Given the fragility and neo-patrimonial character of post-colonial states such as Madagascar, the significant challenge is to identify what kind of pragmatic approaches to improving governance and revenue distribution can be developed.

Community forest management and the free lunch

Renewed interest in Madagascar's biodiversity from the mid-1980s was initially manifested in a phase during which the original protected areas were reinforced and a small number of new ones established. Then, in the mid-1990s, the donor-funded National Environmental Action Plan began a process of transferring the management of forests to communities[18] (*Transfert de Gestion* [TDG]).[19] Community forest management (CFM) was expected to promote both ecological integrity and social justice in Madagascar, as in other countries, but it has been critiqued on both counts (Bertrand 1999; Antona et al 2004; Raik and Decker 2007; Toillier et al 2008; Kull 2004; Dressler et al 2010).

CFM can promote ecological integrity and social justice in two ways. First, it can convert de facto open-access resources (nominally state-controlled) into common property, from which communities gain the legal rights to exclude outsiders and control exploitation levels. Given appropriate social and ecological conditions (Ostrom et al 1999) this may reduce implicit discount rates, shifting harvested populations to new equilibria, both more productive and ecologically robust (Tietenberg 2000; Reynolds et al 2001), solving Hardin's infamous tragedy of the commons (Hardin 1968). Second, CFM may achieve its aims by replacing agents of the state (who may lack the capacity or incentive to protect forest resources) with communities with greater ability and/or incentive to manage the resources sustainably, due to their knowledge of, proximity to and reliance upon these resources. In neither case, however, is it certain that CFM will produce the level or type of forest management desired by the external agencies, including the state and conservation NGOs, which often drive the process forward, as happened in Madagascar. Nor can it be assumed that CFM will necessarily advance social justice.

Conception, design and implementation of forest management transfers

Madagascar's TDG or management transfer policy was the product of several influences, including the rise of CFM internationally and criticisms of earlier conservation paradigms in Madagascar and elsewhere for exacerbating poverty. These factors led to a desire to demonstrate that conservation could benefit the rural poor, whom conservationists nevertheless held responsible for biodiversity loss. More specifically, TDG was part of a process of decentralization and privatization in Madagascar, which sought to make the National Environmental Action

Plan more 'participatory', and conservation 'self-sustaining', by moving the costs of conservation from the state and donor community to others, including communities and the private sector. Finally, TDG was shaped by the considerable influence of the international donor and NGO community in Madagascar, driven to a large extent by biodiversity conservation interests and the desire to increase the area of forest under conservation management (see e.g. Kull 2004; Kaufmann 2008; Corson 2010).

Overall, while no one wanted conservation in general, or TDG in particular, to have a negative effect on social justice, the process was dominated by conservation-oriented organizations, with no comparable representation from community or social justice groups. Thus, while TDG was couched in a rhetoric of dual goals (ecological protection combined with community participation and poverty alleviation), there was in practice an imbalance of intentions and the main focus was on achieving conservation cheaply.

Two important features of Madagascar's TDG policy result from this. First, TDG was not necessarily 'community-based', since forests could be transferred to private operators such as hotels. The emphasis in transferring management was to move it *away* from the state, rather than *towards* communities. TDG assumed the existence of a well-organized 'operator', able to turn the forest to more productive use while also protecting it. Second, it is important to note that *management* was to be transferred, not ownership – which remained with the state.

Management is a responsibility (in other words, it has a cost), and whether changes in management yields benefits to communities depends on the degree to which the rights are transferred. Rights to clear forest could not be granted under TDG, and rights to the commercial exploitation of forest resources were neither automatically nor commonly included, despite being among the central elements of the original TDG concept in Madagascar, as formulated by its proponents in the French cooperation agencies.

Apart from certain well-funded exceptions, the implementation of TDG was often hurried and imperfect, with donors accused of engaging in a 'race for contracts' (Josserand 2001). A key feature was the enormous difference between the 'oral contract' proposed to communities and the written contract signed by communities with the state. In some cases the former threatened communities with the possible privatization of their forests to outsiders if they did not sign a TDG contract, and promised development assistance if they did (Hockley and Andriamarovololona 2007). The contracts actually signed by most communities imposed stringent management and reporting responsibilities on the communities, while granting few, if any, use rights and guaranteeing no assistance with their duties, nor any development aid. In most cases, in effect, communities won the dubious 'right' to carry out the state's management work for it, and were implicitly expected to reach a higher standard than the state had achieved, yet with no external support.

We know very little about the effects of TDG generally in Madagascar because there have been no rigorous, quantitative evaluations of either the environmental or the socio-economic effects, reflecting the situation internationally (Bowler et al

2012). Given the funding allocated to TDG in Madagascar and similar policies elsewhere (raised through public donation and taxation), this is surprising. One must therefore rely on mostly qualitative case studies of TDG, which may or may not be representative of the country as a whole.

What can be said, based on published accounts and work done by the authors of this chapter, is that while there have probably were some success stories, it seems likely that in many cases TDG had limited positive effects on either the condition of the forest or the welfare of local communities. This was not through lack of enthusiasm on the part of communities, who often embraced the concept. Rather it stemmed from an imbalance in the conception and design of TDG. Those stakeholders who drove the development of TDG believed that it would provide a win-win solution: stopping deforestation and improving local livelihoods. This was based on a simplistic view of slash-and-burn agriculture as being primarily a tragedy of the commons, rather than a rational use of forest lands, and overestimation of the potential of non-timber forest products and ecotourism to generate income for communities, without the need to allow logging (which few conservationists wanted to do).

As a result, the state and its donors drove too hard a bargain with communities (who were greatly influenced by the 'oral contract') and tried to extract a free lunch: better, cheaper forest protection and improved rural livelihoods. When the 'oral contract' proved illusory, communities found TDG to provide little value, the written contract became unstable and matters reverted more or less to the pre-TDG situation (Hockley and Andriamarovololona 2007).

Participation and social justice

CFM may improve environmental governance, but good governance will not necessarily pay for itself, besides delivering global public benefits. Where ecological integrity does not simply amount to a tragic scramble for open access resources, CFM alone will not deliver improved ecological integrity without inputs from other stakeholders who stand to benefit: there is no free lunch for conservationists. Nor will it deliver social justice gains, unless it transfers significant rights to communities and allows them to access new, high-value markets for their resources, which may impact ecological integrity (e.g. low-impact logging), or to rely on slowly developing international markets (e.g. ecotourism or carbon credits). Though well intentioned, TDG was destined to perform poorly, suffering as it did from an imbalance of interests in its design and an asymmetry of power and information in its implementation.

While conservationists may regard the recent ravages of Madagascar's forest resources as an unfortunate and unpredictable 'blip' in Madagascar's environmental governance caused by temporary political instability, they should perhaps consider whether the damage was not caused in part by internationally driven environmental programmes that refused to significantly alter the forest tenure regime inherited from colonial times (Brandon and Wells 1992). If communities had held widely recognized, comprehensive and long-term property rights to 'their' forests, the current fire sale of Madagascar's precious timbers might have been more limited or, at the very least, might have generated more value for local communities, rather than for corrupt agents of the state.

Participation in creation of new protected areas

One of the newest areas of environmental governance policy in Madagascar is the country's second generation of protected areas, launched in 2003 at the World Parks Congress in Durban, South Africa. Under strong donor pressure (Duffy 2006; Horning 2008), the government declared its intention to triple the country's protected area coverage, in line with International Union for Conservation of Nature (IUCN) recommendations, to cover 10 per cent of the country's total land area (Mittermeier et al 2005; Norris 2006). Known as the Durban Vision, this ambitious programme required a wholesale rethink of the way in which the country looked at its protected areas. The existing strict protected area models were seen as unsuitable for most new sites because the remaining natural habitats were largely surrounded by rural communities, who were heavily dependent on natural resources for their subsistence and household income. Furthermore, the national parks agency lacked the capacity to oversee the expansion programme, and protected area governance was therefore opened to a range of actors including NGOs, local community associations and private landowners, as well as various forms of collaboration between these groups.

New policy for new protected areas

Madagascar's protected area system (*Système d'Aires Protégées de Madagascar* [SAPM]) comprises two parallel sub-networks, each with its own management objectives, conservation approaches and systems of governance. The country's first protected area was created in 1927, and by 2003 this network had grown to include 46 protected areas in IUCN categories I to IV (Randrianandianina et al 2003). Originally the responsibility of the Waters and Forests Service (now the Minister of Environment and Forests), from the mid-1990s the management of these protected areas was transferred to the parastatal national park agency (formerly the *Association Nationale pour la Gestion des Aires Protégées*, rebranded Madagascar National Parks in 2008), which was created in the first phase of the National Environmental Action Plan.

Many of these protected areas were established in 'wilderness areas' with negligible human populations (Ferguson 2010), and their management objectives focused primarily on the conservation of biodiversity in its natural state, alongside scientific research and recreation. As of 2001, the management and use of protected areas was regulated under national legislation[20] which strictly forbade the collection or destruction of plants in all categories, although exceptions were made for traditional use at some sites.

The implementation of the Durban Vision and the second wave of protected areas entailed a raft of changes in protected areas policy and legislation, particularly with regard to social justice issues. These changes are defined in revised protected area legislation[21] and supporting government guidelines, which specify procedures for the creation and management of new protected areas. For all new protected areas, an environmental and social impact assessment has to be carried out, including a census of all neighbouring populations. This is supposed to lead to the development of a social safeguards plan, in which all people potentially affected by the project (defined as people that depend on natural resources from within the protected area and may suffer a loss as a result of restrictions placed on access) and materially or socially vulnerable groups are identified (Commission SAPM 2007, 2009). A plan should then be developed in which appropriate compensation, usually in the form of development micro-projects and alternative livelihood activities, is determined in consultation with the communities involved. In order to gain temporary protected status, protected area promoters are required to submit the minutes of consultations with all communities potentially affected by the project, in which their agreement to the creation of the protected area and its associated zoning and internal regulations is explicitly stated.

Although Madagascar National Parks is creating and expanding a number of national parks within the Durban Vision framework, the majority of new protected areas are intended to be co-managed by local community associations in conjunction with NGOs (Raik 2007) and are designated or proposed to be in International Union for Conservation of Nature (IUCN) categories III, V and VI, that is to say multiple-use protected areas (Gardner 2011). Zoning of such sites typically includes a core area for the conservation of biodiversity in its natural state, surrounded by a buffer zone which may include sustainable use zones, controlled occupation zones (including the territories of pre-existing settlements within the protected area boundaries) and infrastructure development zones. Within these zones, the exercise of customary use rights (with respect to hunting, fishing, the collection of natural products, and agricultural and forestry activities) is permitted as long as these conform to national legislation and the internal regulations of the protected area, which is certainly not always the case.

These new protected areas may include management transfers (in terms of GELOSE and GCF [Gestion Contractualisée des Forêts]) that either pre-date protected area establishment or are created as part of the protected area establishment process. In such cases, the protected area's internal regulations cannot trump pre-established legislation associated with the management transfers.

With the exception of certain Madagascar National Parks sites and private protected areas, the governance of new protected areas is intended to be largely participatory (Commission SAPM 2006), with governance structures arranged according to spatial scale. At the local level, community associations (including the management committees of management transfers) are responsible for enforcing regulations within their village territories; these associations may be regrouped into a union, forming the participatory management committee of the protected area. This committee is supported by a steering committee operating at the protected area level, composed of representatives of regional authorities, relevant ministries and an inter-commune association which brings together the mayors and councillors of all rural communes affected by the protected area project. The role of the steering committee is focused primarily on strategic matters, the application and enforcement of protected area regulations and conflict resolution. Lastly, a body of supporting organizations (primarily NGOs) provides technical and financial support, and orients biodiversity conservation strategies.

Internal regulations of new protected areas parallel the spatial arrangement of their governance structures, and operate at two levels: at the scale of the village territory and at the inter-commune scale of the protected area. Rules are defined within *dina*, traditionally sets of informal, customary rules (Henkels 1999) which may also be written and ratified to assume the status of local by-laws. Many new protected areas thus have two sets of complementary *dina*: those applying throughout the protected area, developed by an inter-commune association, and specific *dina* developed for each village territory or management transfer. The principles of good governance (see Graham et al 2003) are promoted for adoption at all levels by government guidelines (Commission SAPM 2006).

Participation and social justice in new protected area management

The discourse around the implementation of the Durban Vision suggests clear improvements with regard to protected areas and social justice compared to the previous protected area network, although strict no-go protected areas and inhabited multiple-use protected landscapes are not strictly comparable. This is illustrated by the adoption of a social safeguards policy, the increased participation of local communities in protected area governance and management, the relaxation of legal restrictions on natural resource use and the emphasis on sustainable natural resource use for poverty alleviation and development.

Nevertheless, the rapid expansion of the protected area system has received criticism on a number of grounds, partly as a result of the limited time and resources available to promoting agencies. Freudenberger (2010) reports that levels of community consultation and participation have rarely matched the rhetoric, in some cases amounting to community approval of externally generated plans and regulations rather than truly participatory decision-making. This applies in particular to internal protected area zoning and the definition of *dina*, which often represent externally imposed rules rather than community-developed agreements (Andriamalala and Gardner 2010; Berard 2011; Blanc-Pamard and Ramiarantsoa 2007). Similarly, while many protected areas are nominally managed or co-managed by local community associations with the support of NGOs, the capacity and motivation of local stakeholders remains limited, with the result that the protected area establishment and management process remains largely driven by NGOs.

Furthermore, some protected areas have received criticism for failing to sufficiently recognize and integrate customary land tenure or to fully understand local social-ecological systems, with the result that proposed development alternatives and resource-use restrictions fail to meet the aspirations and needs of local resource users (Marie et al 2009). Critically, protected area legislation prohibits the securing of private property within protected areas, thereby preventing rural communities from securing legal tenure over either individual or collective customary lands within the framework of Madagascar's national land titling programme (*Programme National Foncier*).

Reflections on the governance of ancestral land

Imagine you saw ancestral land with your eyes, without knowing it was ancestral land. You might also see land that your informant says is ancestral, but is not actually ancestral. So whether it is ancestral land or not does not depend on what your informant says; such a claim simply gives you an indication of what is more plausible.[22]

Thought experiments with subjective reality are central to the environmental justice discussion, if only because justice in distribution is relative to social meanings, though not purely and simply so (Walzer 1994, p. 26). Social anthropologists infer explanations of culture, understood as symbolic systems with social meanings embodied in action, from their own observations of everyday life among the people they study, as well as from rules and concepts enacted in rituals or explicitly stated by informants. In Madagascar, ethnographies of ancestral land frequently report such constructions to involve counter-intuitive contradictions of normal understandings of the world. The understanding that the dead differ physically from the living and not yet born is negated by the statement that individuals affiliated to the same tomb are 'one person' (Bloch 1968). The understanding that land physically differs from the people who own it is negated by statements that owners 'are possessed by the land' (Goedefroit 1998, p. 183; Ottino 1998, p. 540; Woolley 2002).

At this point it may not be entirely clear why the understandings reported by anthropologists should be relevant to policy experts, legislators or social critics. After all, sector-specific policy and land use management across sectors must balance local needs and interests with the need for global governance, and must do so within hegemonic frameworks of thought and action which do not include beliefs that living people are consubstantial with material objects or spirits.

What is needed is a method to translate the understandings of rural Malagasy into the normal discourses of social science. Law consists of normative and cognitive conceptions in which validity for certain groups of people is asserted (Benda-Beckmann 2002, p. 48). Differences between transplanted and locally grown law concern not only the rules but also the cognitive postulates which guide the interpretation of the rules (Chiba 1987; Greenhouse 1998). These mental schemes are culturally inflected. Resource claims by rural Malagasy invoking ancestors or nature spirits, for example, mystify the universal concerns of environmental justice which are linked to the objective realities of issues such as poverty, marginalization, environmental degradation and human displacement.

Land tenure is a less value-loaded term to use when discussing these environmental justice issues. It refers to property relations at the economic, political and ideological levels which can be analysed in terms of layered bundles of rights. Rights are enforceable claims, though we should note that the enforcing need not come from the state: other organizations are also capable of sanctioning property rules (Geisler 2006). Legal pluralism as an analytical concept dissociates law from the state; its basic claim is that besides state law there is a whole array of non-state laws (Benda-Beckmann 2001; Griffiths 2002). Property rights bundles may be defined either by state-legal rules or by other legal rules, or by a combination of both. If we accept this claim, we can see that the gap separating policy discourses from implementation is not a gap between rules and practices – the normative 'ought' and the empirical 'is' – but between multiple empirical sets of rules and legitimating ideologies, any of which may attain a degree of compliance in everyday life.

Although Madagascar's 2005 land use policy follows a worldwide tendency to give legal recognition to common property and individual tenure arrangements outside state law, one cannot assume that the presumption of 'untitled private property' – which, as the term indicates, applies to privatized, not communal, ancestral lands (Muttenzer 2010) – simply replaces the domain declaration (presumption of state ownership) and its conceptual divides between public and private ownership, protected areas and agricultural land, nature and culture. Indeed, the sector-specific policy reviews making up the rest of this chapter suggest that these conceptual divisions are now, for the first time in history, influencing the lives of rural people.

The environmental policy transition from preservation to so-called integrated conservation and development provides a nice illustration, at least to the extent that the new protected areas are not merely 'virtual realities' or 'paper parks' (West and Brockington 2006). In spite of common property and legal pluralism having been explicitly taken into account, the new forest and environmental policy reproduces a style of scientific ecosystem management that excludes humans from nature. 'Coercing conservation' has been shown to be a source of social conflict, and the hope for the future has been that community-based natural resource management will restore the voice of non-state actors, as well as address the economic needs of resource-dependent local populations (Dressler et al 2010). The fact that its implementation in Madagascar since 1996 has so far had very limited impact therefore requires an explanation.

Community-based natural resource management differs from coercive conservation in its claim to be responsive to other ways of thinking about nature and society (Goedefroit, 2006). Its policy narratives typically combine a harmony model of society with an equilibrium model of human-environment interactions (Berkes 2004; Blaikie 2006). Ancestral land is viewed mainly from the perspective of compliance with stable rules that are expected to be well adapted to ecological systems and processes. Concomitantly, community-based models associate environmental degradation with the market-induced or state-imposed social change, fluid rules and low compliance that presumably characterize the individualistic strategies of displaced, landless migrants (Aubert et al 2003), as discussed in this chapter in relation to community forest management. The implicit rationale of restoring historically constituted land rights is to restrict access to resource frontiers for settlers who are perceived to act outside the ancestral order.

A similarly distorted view of customary tenure characterizes the 2005 policy of giving legal recognition to 'untitled private property' in non-forest commons. Only individual claims to permanently cultivated fields are acknowledged, whereas lineage or village rights to land on which these fields are located, or rights in the as yet uncultivated commons or fields under long and short periods of fallow, are not. If implemented, this policy would achieve the transition from communal to private property. But such an outcome is unlikely because 'common property' systems are based on a complementary relationship of separable resource claims of individuals or families with political authority claims of village or descent group elders to oversee property transactions (Benda-Beckmann and Benda-Beckmann 2006; Muttenzer 2006).

Under these assumptions, statements that people are 'one person'[23] are intended to legitimize, or correct, distortions in the normal order of inheritance and succession. Statements that people are 'owned by the land'[24] are intended to legitimize or restrain appropriation by living founders or members of other descent groups or communities, as well as contractual transfers of property. Both are obliquely arguments about environmental justice, and 'odd' only in the sense of being meaningless in the conceptual framework of untitled private property.

It is, of course, individual rural people who deforest, settle, appropriate and cultivate land, lend it to others, transmit it to children and so on and who would talk about their actions in the language of everyday life, if indeed they were to talk about them at all (Bloch 1991, 1992). In certain contexts, these individuals would also invoke 'odd' beliefs in ancestors and spirits, for instance during funerals, agricultural cycle ceremonies or dispute settlements, or in the ritualized setting of ethnographic interviews. But these beliefs and *post hoc* rationalizations cannot be considered explanations of property relations. They are abstract concepts to be explained by the trained fieldworker's perceptions of 'what goes without saying' (Bloch 1992).

What is this evidence of? First, it would seem that the strategy of occupation of protected areas and clearing forest for agriculture is not merely a matter of economic needs at the household level (Keller 2008). If it were, compensating for the opportunity costs of slash-and-burn cultivation might be an effective means to reduce deforestation (Pollini 2009). However, clearing forest is also a legitimate strategy to claim new resources and territory, and to become established as the founder and recognized as the master of the land by other settlers and local officials. Second, it appears that such founder claims may refer to ancestral land, meaning places that were settled long ago, while others refer to frontier land whose settlers are living (Muttenzer 2008, 2010). Third, there is evidence that the concepts of being 'one person' and of being 'owned by the land' are shared by prior occupants and later settlers. Immigrants can marry earlier settlers and become (one person with) ancestral founders by female descent, or they can invoke counter-legitimacy by demonstrating that the land possesses them (Deschamps 1959; Goedefroit 2002).

In what ways do these customary legal concepts relate to the implementation of state policies? To what extent are the social organizational outcomes of legal pluralism determined by norms implicit in the customary law? Legal institutions and legitimations are not necessarily framed in terms of legal discourses, whereas other symbolic genres, such as ritual, or symbols contained in instrumental action, do have legal consequences and sanctions (Zerner 2003).

Moreover, villagers interacting with local state officials or with agents of pilot projects selectively borrow state-legal rules and procedures (e.g. illicit official authorizations to occupy land or clear forest, community-based management contracts, certificates of untitled private property), in addition to traditional property rules and legitimations in terms of prior occupancy and descent-based inheritance. In this way, illegal squatters in protected areas and state-legal concessions for carbon sequestration, agriculture or mining claim recognition as ancestral or living founders. Their hybrid claims are 'legal' in the sense that the rules are enforceable, albeit by non-state organizations or diffuse social pressure, and they are 'customary' in the sense that the rules are not codified.

Given that state public policy and legislation may recognize, as well as be recognized by, customary legal rules, there can be no strict organizational boundary between customary law and state law as a result of their interaction. This boundary exists only in the categorical statements of either legal system, and only in the relevant contexts in which such assertions are meaningful.

Accordingly, the local administrative practices before and after the ideological shift in state-legal discourse from coercive to community-based natural resource management have remained relatively stable. This continuity is explained by the previous village-level interactions of local officials of the postcolonial state, who did not, in fact, coerce conservation as intended by repressive legislation, but instead participated in villagers' reinterpretations of coercive policy.

Reports of involuntary displacements from ancestral land and forest frontiers following large-scale investments, protected area establishment and land-grabs call into question the relevance of participation and recognition as effective tools of government or means to enhance environmental justice. Local participation in sector-specific policy and in land reforms across sectors is usually linked to pre-existing local practices, institutions and goals that differ from those envisioned by NGOs and state agencies promoting the 'legal recognition of customary tenure' as a sustainable development and conservation initiative (Muttenzer 2008).

Prior occupants, later immigrants and local officials base their complementary resource claims on an array of customary legal arrangements (combining precedence in occupation, inheritance, membership in associations, informal authorizations and contractual property transfers) whose concepts and rules reflect a local theory of justice. Community-based user associations, together with other state-legal authorization or recognition procedures, are strategically deployed by competing resource users who invoke shared understandings of what it means to be one person, or owned by the land.

It seems unlikely that the recognition of customary law in public policy to govern ancestral land will transform village commons into ancestral enclosures, multiply open-access counter-claims from landless immigrants or generate voluntary concern for the environment among the subjects of global governance.

Conclusions

This chapter has explored the environmental justice impacts of policy reform across four land-use sectors. Certain trends and inconsistencies between rhetoric and reality appear common across sectors, and together result in social justice outcomes that commonly fail to meet the standards that the policies were purportedly designed to ensure. Discrepancies between the desired outcomes of policy reform and realities on the ground can be largely ascribed to the five key factors discussed below.

Good intentions, mixed results

Policy reforms in the natural resources sectors in Madagascar over the past 20 years have sought to increase participation significantly, with the intention of achieving improved environmental governance and social justice, but the outcomes of these reforms have not always lived up to expectations because of flaws in their conception, design and implementation. Some examples of positive outcomes at localized sites do exist, but more often social justice is not achieved. In many cases, seemingly most, the status quo is maintained and policies are largely ignored by local people or not operationalized by government authorities in local contexts. This implies a need for more effective monitoring and independent scrutiny.

Power asymmetries and participation

Numerous obligations exist within the various policy reforms outlined in this chapter to ensure that local people are informed, consulted and organized into associations and committees. However, a critical analysis of what power or say peasant participants have in that process is vital to understanding whether it is a genuinely empowering, decentralizing process, or simply a case of powerful actors placating or manipulating local populations. Inequitable or repressive policy has often proven to be unstable, to the detriment of both environmental governance and social justice.

Capacity to participate

The problems we observe in unsatisfactory social justice outcomes are due in part to limited capacity to participate effectively in new structures. Examples include local mayors' lack of legitimacy or of political capital to sign off land-deal papers. Limitations of technical capacity, literacy and awareness of laws among local actors, as well as the risk and reality of serious politicization, manipulation or corruption of deals, undermines legal frameworks, however well designed they may be. Furthermore, a virtually absent civil society in many natural resource areas compounds the difficulties of tackling these problems.

Fragmentation of governance between sectors and resistance to decentralization

The fragmented nature of the governance and monitoring of natural resources in Madagascar undermines the capacity to improve standards and therefore meet aspirations for social justice in natural resource domains. The regulation of business, including environmental and social impact assessments and monitoring systems, needs to be better coordinated across sectors.

The participation gap

The fact that legal pluralism exists in rural Madagascar – that is, state and customary rules coexist – and that the apparatus of the state, democratic structures and civil society is virtually absent in many rural parts, or certainly dysfunctional, means that regular reality checks are needed of the gap between policy narratives and the implementation of policies on the ground. The local perception of rights and power when participating in environmental governance processes may be very different to the perceptions and techno-narrative of those designing natural resource policy (mining companies, land title cooperation programmes, agribusiness companies and conservationists), and the reality of implementation may differ greatly from intentions. The cultural lens through which local peoples' realities are scrutinized (be it in the form of social anthropology studies or the testimonies of local people) is important to anyone who is trying to understand what the situation is really like.

Notes

1 Useful reviews of international land acquisitions in Madagascar have been carried out (e.g. Andrianirina-Ratsialonana et al 2011), but these reports are not exhaustive and have not included various biofuels projects, such as the castor plantations of Flora Eco-Power and PhileoL in the Amboasary, Ambovombe and Tsiombe districts (Anosy and Androy regions). Similarly for mining, regional development spatial planning includes the high-profile mining projects of multinational companies such as Rio Tinto, but neglects to mention other less well-known mining projects, including the Malaysian sapphire mining company Nan Tin Polychrome and British uranium mining companies Blue Sun Mining, Bekitoly Resources and Vuna (Anosy 2005).

2 *Alliance Voahary Gasy* is a prominent civil society group in this field. Lobbying organizations include the *Collectif pour la Défense des Terres Malgaches* (Collective for the Defence of Malagasy Land), Human Rights in Masoala, the *Comité pour le Droit des Paysans à Madagascar* (Committee for the Rights of Malagasy Peasants), the *Sehatra Iombonana ho an'ny Fananan-tany* or *Solidarité des Intervenants sur le Foncier* (Speakers of Land Solidarity), the *Syndicat des Organisations Agricoles* (Union of Agricultural Organizations), and the Firaisankinan'ny Tantsaha eto Madagasikara (Farmers' Coalition in Madagascar). Communication and information networks include *Justice et Droits de l'Homme à Madagascar* (Justice and Human Rights of the Malagasy People) and the Madagascar Environmental Justice Network.

3 The Global Witness and Environmental Inestigation Agency (2009) and Friends of the Earth have renewed their interest due to the illegal trade in precious wood in the eastern

forests and the high-profile Total and Madagascar Oil tar sands projects in Bemolanga and Tsimiroro.

4 Law No. 2003-028 of 22 August 2003, extended by law 2007-036 of 14 January 2008.

5 In 2009, Andry Rajoelina overthrew President Marc Ravalomanana with the support of a protest movement and a rebellious army unit. His anti-regime propaganda was partly based on various scandals, including the Daewoo land deal, which had caused great public outrage. With civil society groups watching him closely, the new leader was under pressure to deliver on this issue, and one of the first decisions he took upon seizing power in March 2009 was to cancel the Daewoo deal.

6 After a long discussion process praised for its inclusiveness, these guidelines were endorsed by the Committee on World Food Security at its 38th (special) session on 11 May 2012.

7 Law No. 2005-019 of 17 October 2005, determining the principles governing the status of lands.

8 Recent investigations were conducted by the International Land Coalition, the Land Observatory in Madagascar, the *Centre de Coopération Internationale en Recherche Agronomique pour le Développement* and some postgraduate students working with the last-mentioned two bodies. Their findings and analyses can be read in Andrianirina-Ratsialonana et al (2011) and Burnod et al (2011).

9 Decree No. 99-954 of 15 December 1999, to make investments compatible with the environment, modified by decree No. 2004-167.

10 *Circulaire* 321-10/MATD/SG/DGSF, 'Instructions relative to the procedure to follow in cases of large-scale land requests', General Direction of Land Services, 25/10/2010 (authors' translation).

11 Decree No. 98-610 of 13 August 1998, *Sécurisation Foncière Relative* (SFR), regulating the implementation of relative land tenure security.

12 Law No. 96-025 of 30 September 1996, *Gestion Locale Securisée* (GELOSE), regarding the local management of renewable natural resources.

13 Law No. 2005-019 of 17 October 2005, determining the principles governing the status of lands.

14 From the establishment of regions in 2004 until 2010, heads of region were nominated by the council of ministers, and a mayor could be replaced by a 'head of special delegation' (*président de la délégation spéciale*, PDS), chosen by the President of the Republic himself. The Constitution of the Fourth Republic, accepted by referendum on 17 November 2010, stated that heads of region would be elected by the universal suffrage, but no such election has yet been held, and a number of cities have had unelected PDSes filling the place of mayors since then.

15 Law No. 2007-036 of 14 January 2008.

16 Mines-Environment Inter-ministerial Decree of 2000.

17 The relocation of burial grounds is a highly emotive matter owing to the significance of ancestors in Malagasy culture.

18 Decree 2001/122 completing law 96-025.

19 This took place in terms of the GELOSE law, which applied to all land, and GCF (Gestion Contractualisée des Forêts), which applied only to forests, and was intended to be more streamlined than GELOSE, though the tenure offered was more limited.

20 Law No. 2001/05 of 11 February 2003, on the management of protected areas.

21 Due to the ongoing political crisis since 2009, the revised management code for protected areas has yet to be ratified.

22 This formulation of the problem is indebted to thought-provoking discussions with Jacques Pollini, an expert on Malagasy forestry.

23 This statement invokes descent-group rights to make use of land.

24 This statement invokes village rights to make use of land.

Bibliography

Andrew Lees Trust (2009) *Madagascar Voices of Change: Oral Testimony of the Antanosy People*, Andrew Lees Trust and Panos London, London

Andriamalala, G. and Gardner, C. J. (2010) 'L'utilisation du *dina* comme outil de gouvernance des ressources naturelles: Leçons tires de Velondriake, sud-ouest de Madagascar', *Tropical Conservation Science*, vol 3, pp. 447–72

Andrianirina-Ratsialonana, R., Ramarojohn, R. L., Burnod, P. and Teyssier, A. (2011) *After Daewoo? Current Status and Perspectives of Large-Scale Land Acquisition in Madagascar*, Madagascar National Land Observatory and CIRAD (Agricultural Research Centre for International Development), Antananarivo

Anosy (2005) *Region de l'Anosy: Plan regionale de developpement*, version de Février, Ministere de l'Interieur et de la Reforme Administrative

Anseeuw W., Wily A. L., Cotula L. and Taylor, M. (2012) *Land Rights and the Rush for Land: Findings of the Global Commercial Pressures on Land Research Project*, International Land Coalition, Rome

Antona, M., Biénabe, E. M., Salles, J.-M., Péchard, G., Aubert, S. and Ratsimbarison, R. (2004) 'Rights transfers in Madagascar biodiversity policies: Achievements and significance', *Environment and Development Economics*, vol 9, pp. 825–47

Arnstein, S. R. (1969) 'A ladder of citizen participation', *Journal of the American Institute of Planners*, vol 35, pp. 216–24

Atkinson, G. and Hamilton, G. (2003) 'Savings, growth and the resource curse hypothesis', *World Development*, vol 31, pp. 1,793–807

ATW (2010) *Premier rapport de réconciliation des revenues miniers Madagascar 2007–2009*, ATW Consultants Madagascar, Antananarivo

Aubert, S., Razafiarison, S. and Bertrand, A. (eds) (2003) *Déforestation et systèmes agraires à Madagascar: Les dynamiques des tavy sur la côte orientale*, Cirad, Cite, Fofifa, Antananarivo

Auty, R. M. (1993) *Sustaining Development in Mineral Economies: The Resource Curse Thesis*, Routledge, London

Ballet, J., Lopez, P. and Rahaga, N. (2010) 'L'exportation de bois précieux (*Dalbergia* et *Diospyros*) "illégaux" de Madagascar: 2009 et après?', *Madagascar Conservation and Development*, vol 5, no 2, pp. 110–16

Benda-Beckmann, F. von (2002) 'Who's afraid of legal pluralism?', *Journal of Legal Pluralism*, vol 47, pp. 37–80

Benda-Beckmann, F. von, and Benda-Beckmann, K. von (2006) 'How communal is communal and whose communal is it? Lessons from Minangkabau', in K. von Benda-Beckmann, F. von Benda-Beckmann and M. Wiber (eds) *Changing Properties of Property*, Berghahn, New York, pp. 194–217

Benda-Beckmann, K. von (2001) 'Legal pluralism', *Tai Culture*, vol 6, no 1–2, pp. 18–40

Berard, M.-H. (2011) 'Légitimité des normes environnementales dans la gestion locale de la forêt à Madagascar', *Canadian Journal of Law and Society*, vol 26, no 1, pp. 89–111

Berghofer, U. and Berghofer, A. (2006) '"Participation" in development thinking: Coming to grips with a truism and its critiques', in S. Stoll-Kleeman and M. Welp (eds) *Stakeholder Dialogues in Natural Resources Management: Theory and Practice*, Springer, Berlin, pp. 79–116

Berkes, F. (2004) 'Rethinking community-based conservation', *Conservation Biology*, vol 18, no 3, pp. 621–30

Bertrand, A. (1999) 'La gestion contractuelle, pluraliste et subsidiaire des ressources renouvables à Madagascar (1994–1998)', *African Studies Quarterly*, vol 3, no 2, pp. 75–81

Blaikie, P. (2006) 'Is small really beautiful? Community-based natural resource management in Malawi and Botswana', *World Development*, vol 34, no 11, pp. 1,942–57

Blanc-Pamard, C. and Ramiarantsoa, H. R. (2007) 'Normes environnementales, transferts de gestion et recompositions territoriales en pays betsileo (Madagascar): La gestion contractualisée des forêts', *Nature Sciences Sociétés*, vol 15, pp. 253–68

Bloch, M. (1968) Tombs and conservatism among the Merina of Madagascar, *Man*, New Series, vol 3, no 1, pp. 94–104

Bloch, M. (1991) 'Language, anthropology, and cognitive science', *Man, New Series*, vol 26, no 2, pp. 183–98

Bloch, M. (1992) 'What goes without saying: The conceptualization of Zafimaniry society', in A. Kuper (ed) *Conceptualizing Society*, Routledge, London and New York, pp. 127–46

Bowler, D. E., Buyung-Ali, L. M., Healey, J. R., Jones, J. P. G., Knight, T. M. and Pullin, A. S. (2012) 'Does community forest management provide global environmental benefit and improve local welfare?', *Frontiers in Ecology and the Environment*, vol 10, no 1, pp. 29–36

Brandon, K. E. and Wells, M. (1992) 'Planning for people and parks', *World Development*, vol 20, no 4, pp. 557–70

Burnod, P., Gingembre, M. and Andrianirina-Ratsialonana, R. (2013) 'Competition over authority and access: International land deals in Madagascar', *Development and Change* vol 44, no 2, pp. 357–79

Burnod, P., Gingembre, M., Andrianirina-Ratsialonana, R. and Ratovoarinony, R. (2011) *From International Land Deals to Local Informal Agreements: Regulations of and Local Reactions to Agricultural Investments in Madagascar*, International Conference on Global Land Grabbing, 6–8 April 2011, Land Deals Politics Initiative (LDPI) in collaboration with the Journal of Peasant Studies, Future Agricultures Consortium at the Institute of Development Studies, University of Sussex

Carret, J. C., Rajaonson, B., Feno, J. P. and Brand, J. (2010) 'L'environnement à Madagascar: Un atout à préserver, des enjeux à maitriser', in World Bank (ed.) *Madagascar: Vers un agenda de relance économique*, World Bank Madagascar Office, Antananarivo

Chambers, R. (1983) *Rural Development: Putting the Last First*, Longman, Harlow, UK

Chiba, M. (1987) 'Three dichotomies of law: An analytical scheme of legal culture', *Tokai Law Review*, vol 1, pp. 279–90

ClientEarth (2010) *Referral to the Financial Reporting Review Panel Re: The Rio Tinto Group Annual Report 2008*, ClientEarth, London

Commission SAPM (2006) 'Système d'aires protégées de Madagascar: Orientations générales sur les catégories et les types de gouvernance des aires protégées', Commission SAPM orientation document, July, Antananarivo

Commission SAPM (2007) 'Cadre fonctionnel de procédures: Création et gestion d'aires protégées; mesures de sauvegardes; mécanisme de financement', Commission SAPM orientation document, April, Antananarivo

Commission SAPM (2009) 'Création d'aires protégées: Mesures de sauvegarde, cadre de procédure', Antananarivo

Cooke, B. and Kothari, U. (2001) *Participation: The New Tyranny?*, Zed Books, London

Cook, R. and Healy, T. M. (2012) *Madagascar Case Study: Artisanal Mining Rushes in Protected Areas and a Response Toolkit*, World Wide Fund for Nature and Estelle Levin, Ltd

Corson, C. (2010) 'Shifting environmental governance in a neoliberal world: US AID for conservation', *Antipode*, vol 42, no 2, pp. 576–602

Cotula, L., Vermeulen, S., Leonard, L. and Keeley, J. (2009) *Land Grab or Development Opportunity? Agricultural Investment and International Land Deals in Africa*, International Institute for

Environment and Development, Food and Agriculture Organization and International Fund for Agricultural Development, London

Deschamps, H. (1959) *Les migrations intérieures passées et présentes à Madagascar*, Berger-Levrault, Paris

Draper, R. (2010) 'The pierced heart of Madagascar', *National Geographic*, vol 218, pp. 80–109

Dressler, W., Buscher, B., Schoon, M., Brockington, D., Hayes, T., Kull, C. A., McCarthy, J. and Shrestha, K. (2010) 'From hope to crisis and back again? A critical history of the global CBNRM narrative', *Environmental Conservation*, vol 37, no 1, pp. 5–15

Duffy, R. (2006) 'Non-governmental organisations and governance states: The impact of transnational environmental management networks in Madagascar,' *Environmental Politics*, vol 15, no 5, pp. 731–49

Evers, S., Burnod, P., Teyssier, A., Andrianirina-Ratsialonana, R. (2011) 'Foreign land acquisitions in Madagascar: Competing jurisdictions of access claims', in T. Dietz, K. Havnevik, M. Kaag and T. Oestigaard (eds) *African Engagements: On Whose Terms? Africa Negotiating an Emerging Multi-Polar World*, Africa-Europe Group for Interdisciplinary Studies 5, Brill NV, Leiden, The Netherlands, pp. 112–30

EITI (2008) 'Secretariat workplan 2008', Extractive Industries Transparency Initiative Secretariat, Oslo

Fauroux, E. and Blanc-Pamard, C. (2004) 'L'illusion participative: Exemples ouest-malgache', *Autrepart*, vol 31, pp. 3–19

Ferguson, H. B. (2010) 'REDD in Madagascar: An overview of progress', in O. Springate-Baginski and E. Wollenberg (eds) *REDD, Forest Governance and Rural Livelihoods: The Emerging Agenda*, Centre for International Forestry Research, Bogor, Indonesia, pp. 135–72

Freudenberger, K. S. (2010) *Paradise Lost? Lessons from 25 Years of USAID Environment Programs in Madagascar*, International Resources Group, Washington, DC

Friends of the Earth (2007) 'Mining madagascar: Forests, communities and Rio Tinto's white wash', media briefing, Friends of the Earth, London, http://foe.co.uk/resource/media_briefing/mining_madagascar.pdf

Fritz-Vietta, N. V. M., Ferguson, H. B., Stoll-Kleemann, S. and Ganzhorn, J. U. (2011) 'Conservation in a biodiversity hotspot: Insights from cultural and community perspectives in Madagascar', in F. E. Zachos and J. C. Habel (eds) *Biodiversity Hotspots: Distribution and Protection of Conservation Priority Areas*, Springer, Heidelberg and Berlin, pp. 209–33

Ganzhorn, J. U., Goodman, S. M. and Vincelette, M. (2007) *Biodiversity, Ecology and Conservation of Littoral Ecosystems in Southeastern Madagascar, Tolagnaro (Fort Dauphin)*, Smithsonian Institution, Washington, DC

Gardner, C. J. (2011) 'Application of IUCN protected area management categories to new, multiple-use protected areas in Madagascar', *Oryx*, vol 45, no 3, pp. 336–46

Geisler, C. (2006) 'Ownership in stateless places', in K. von Benda-Beckmann, F. von Benda-Beckmann and M. Wiber (eds) *Changing Properties of Property*, Berghahn, New York pp. 40–57

Gingembre, M., Bouhey, A., Ratrimoarivony, M. and Randrianarimanana, P. (2010) 'L'affaire Daewoo à Madagascar', in Allaverdian, C. (ed) *Agricultures familiales et sociétés civiles face aux investissements massifs dans les terres: Six études de cas*, Commission C2A (Agriculture et Alimentation) Coordination Sud, Paris

Global Witness and Environmental Investigation Agency (2009) *Investigation into the illegal felling, transport and export of precious wood in SAVA region, Madagascar*, Global Witness and Environmental Investigation Agency, Washington, DC, and London

GLW Conseil (2009) 'Actualisation du plan de gestion environnemental du programme et gouvernance des ressources minérales', Ministère de Mines, Antananarivo

Goedefroit, S. (1998) 'L'identité de terre et de territoire chez les Sakalava du Menabe (Madagascar)', in G. Dominique, M. Seysset and A. Walter (eds) *Hommage à Joël Bonnemaison: Le voyage inachevé*, IRD Eds and Prodig, Paris, pp. 179–84

Goedefroit, S. (2002) 'Stratégies patrimoniales au paradis de la nature: Conservation de la biodiversité, développement et revendications locales à Madagascar', in M. C. Cormier-Salem, D. Juhe-Beaulaton, J. Boutrais and B. Roussel (eds) *Patrimonialiser la nature tropicale: Dynamiques locales, enjeux internationaux*, Colloques et séminaires, IRD éditions, Paris, pp. 125–66

Goedefroit, S. (2006) 'La restitution du droit à la parole', *Etudes rurales*, vol 178, no 2, pp. 39–64

Graham, J., Amos, B. and Plumtre, T. (2003) 'Governance principles for protected areas in the 21st century', paper presented at the Fifth World Parks Congress, Durban, South Africa, Institute on Governance, Ontario, Canada

Greenhouse, C. (1998) 'Legal pluralism and cultural difference: What is the difference? A response to Professor Woodman', *Journal of Legal Pluralism*, vol 42, pp. 61–71

Griffiths, A. (2002) 'Legal pluralism', in R. Banakar and M. Travers (eds) *An Introduction to Law and Social Theory*, Hart, Portland, OR, pp. 289–310

Harbinson, R. (2007a) *Development Recast? A Review of the Impact of the Rio Tinto Ilmenite Mine in Southern Madagascar*, report for Friends of the Earth, Panos London, London

Harbinson, R. (2007b) *A Mine of Information? Improving Communication around the Rio Tinto Ilmenite Mine in Madagascar*, Panos, London

Hardin, G. (1968) 'Tragedy of the commons', *Science*, vol 162, no 3,859, pp. 1,243–48

Henkels, D. M. (1999) 'Une vue de près du droit de l'environnement malgache', *African Studies Quarterly*, vol 3, no 2, pp. 39–59

Hockley, N. and Andriamarovololona, M. (2007) 'The economics of community forest management in Madagascar: Is there a free lunch? An analysis of "Transfert de Gestion"', report to USAID, Madagascar, http://pdf.usaid.gov/pdf_docs/PNADL290.pdf

Horning, N. R. (2008) 'Strong support for weak performance: Donor competition in Madagascar', *African Affairs*, vol 107, no 428, pp. 405–31

IIED and WBCSD (2002) *Breaking New Ground: Mining, Minerals and Sustainable Development. The Report of the MMSD Project*, Earthscan, London, for International Institute for Environment and Development and World Business Council for Sustainable Development

International Longshore and Warehouse Union (2010) 'Rio Tinto: A shameful history of human and labour rights abuses and environmental degradation around the globe', 20 April, London Mining Network, http://londonminingnetwork.org/2010/04/rio-tinto-a-shameful-history-of-human-and-labour-rights-abuses-and-environmental-degradation-around-the-globe/

Jacoby, H. G. and Minten, B. (2007) 'Is land titling in sub-Saharan Africa cost-effective? Evidence from Madagascar', *World Bank Economic Review*, vol 21, no 3, pp. 461–85

Josserand, H. P. (2001) *Community Based Natural Resource Management in Madagascar*, Programme de Transition Eco-régional, USAID Madagascar

Juttersonke, O. and Cartas, M. (2010) *Peace and Conflict Impact Assessment (PCIA): Madagascar*, Centre on Conflict, Development and Peacebuilding, Graduate Institute of International and Development Studies, Geneva

Kambhampati, U. S. (2004) *Development and the Developing World*. Polity Press, Cambridge

Karl, T. L. (1997) *The Paradox of Plenty: Oil Booms and Petro-States*, University of California Press, Berkeley, CA

Kaufmann, J. C. (2008) *Greening the Great Red Island: Madagascar in Nature and Culture*, Africa Institute of South Africa, Pretoria

Keller, E. (2008) 'The banana plant and the moon: Conservation and the Malagasy ethos of life in Masoala, Madagascar', *American Ethnologist*, vol 35, no 4, pp. 650–64

Kull, C. A. (2004) *Isle of Fire: The Political Ecology of Landscape Burning in Madagascar*. University of Chicago Press, Chicago, IL

Malagasy Land Observatory (2008) 'Décentralisation de la gestion foncière: Un premier bilan synthéthique', April, Observatoire du Foncier, Antananarivo, www.observatoire-foncier.mg/get-file.php?id=72

Marcus, R. R. (2008) 'Tòkana: The collapse of the rural Malagasy community', *African Studies Review*, vol 51, no 1, pp. 85–104

Marie, C. N., Sibelet, N., Dulcire, M., Rafalimaro, M., Danthu, P. and Carrière, S. M. (2009) 'Taking into account local practices and indigenous knowledge in an emergency conservation context in Madagascar', *Biodiversity and Conservation*, vol 18, no 10, pp. 2,759–77

Martinez-Alier, J. (2001) 'Mining conflicts, environmental justice and valuation', *Journal of Hazardous Materials*, vol 86, nos 1–3, pp. 153–70

Mittermeier, R. A., Hawkins, A. F. A., Rajaobelina, S. and Langrand, O. (2005) 'Wilderness conservation in a biodiversity hotspot', *International Journal of Wilderness*, vol 11, no 3, pp. 42–5

Muttenzer, F. (2006) 'The folk conceptualization of property and forest-related going concerns in Madagascar', in K. von Benda-Beckmann, F. von Benda-Beckmann and M. Wiber (eds) *Changing Properties of Property*, Berghahn, New York, pp. 269–92

Muttenzer, F. D. (2008) '"Integrated fortress conservation" in the buffer zone of Ankarafantiska National Park: Malagasy narratives of conservation, participation and livelihoods', in M. Galvin and T. Haller (eds) *People, Protected Areas and Global Change: Participatory Conservation in Latin America, Africa, Asia and Europe*, perspectives of the Swiss National Centre of Competence in Research (NCCR) North-South, University of Bern, vol 3, Geographica Bernensia, Bern, Switzerland, pp. 253–86

Muttenzer, F. D. (2010) 'Politiques foncières et pluralisme juridique à Madagascar', *Taloha*, vol 19, www.taloha.info/document.php?id=755

Myers, N., Mittermeier, R. A., Mittermeier, C. G., da Fonesca, G. A. B. and Kent, J. (2000) 'Biodiversity hotspots for conservation priorities', *Nature*, vol 403, pp. 853–58

Norris, S. (2006) 'Madagascar defiant', *BioScience*, vol 56, no 12, pp. 960–65

ODI (2005) *Meeting the Social and Local Performance Standards of International Project Finance Institutions: Guidance for Energy Companies*, Overseas Development Institute, London

Ostrom, E., Burger, J., Field, C. B., Norgaard, R. B. and Policansky, D. (1999) 'Sustainability – revisiting the commons: Local lessons, global challenges', *Science*, vol 284, pp. 278–82

Ottino, P. (1998) *Les champs de l'ancestralité à Madagascar: Parenté, alliance et patrimoine*, Karthala/ORSTOM, Paris

Pegg, S. (2003) *Poverty Reduction or Poverty Exacerbation: World Bank Group Support for Extractive Industries in Africa*, Oxfam America, Friends of the Earth-US, Environmental Defense, Catholic Relief Services and the Bank Information Center

Pegg, S. (2006) 'Mining and poverty reduction: Transforming rhetoric into reality', *Journal of Cleaner Production*, vol 14, no 3–4, pp. 376–87

Plangemann, K. A., Lefebvre, A.-L., Randriamanampisoa, D., Keener, S., Rajaonson, B., Marcus, R. R., Blum, J. and Poole, A. (2010) 'Governance and development effectiveness review: A political economy analysis of governance in Madagascar', World Bank Report No 54277-MG

Pollini, J. (2009) 'Carbon sequestration for linking conservation and rural development in Madagascar: The case of the Vohidrazana-Mantadia corridor restoration and conservation carbon project', *Journal of Sustainable Forestry*, vol 28, nos 3–5, pp. 322–42

Raik, D. B. (2007) 'Forest management in Madagascar: An historical overview', *Madagascar Conservation & Development*, vol 2, no 1, pp. 5–10

Raik, D. B. and Decker, D. J. (2007) 'A multisector framework for assessing community-based forest management: Lessons from Madagascar', *Ecology and Society*, vol 12, no 14, http://ecologyandsociety.org/vol12/iss1/art14/

Randrianalijaona, M., Ballet, J., Koffi, J. M. and Koména, K. B (2011) *Comment préserver les ressources naturelles?*, Editions rue d'Ulm, Sciences Durables, Paris

Randrianandianina, B. N., Andriamahaly, L. R., Harisoa, F. M. and Nicoll, M. E. (2003) 'The role of protected areas in the management of the island's biodiversity', in S. M. Goodman and J. P. Benstead (eds) *The Natural History of Madagascar*, University of Chicago Press, Chicago, IL, pp. 1,423–32

Reynolds, J. D., Mace, G. M., Redford, K. H. and Robinson, J. G. (2001) *Conservation of Exploited Species*, Cambridge University Press, Cambridge, UK

Rochegude, A. and Plançon, C. (2009) *Décentralisation, foncier et acteurs locaux: Mise en perspective juridique des textes sur la décentralisation et le foncier en Afrique de l'Ouest et du Centre*, Fiche Madagascar, PDM Cotonou, Benin, Coopération Française, Paris

Ross, M. (2001) *Extractive Sectors and the Poor*, Oxfam America, Washington, DC

Sachs, J. D. and Warner, A.M. (2001) 'Natural resources and economic development: The curse of natural resources', *European Economic Review*, vol 45, pp. 827–38

SAPRIN (2002) *The Policy Roots of Economic Crisis and Poverty: A Multi-Country Participatory Assessment of Structural Adjustment*, joint World Bank/Civil Society Structural Adjustment Participatory Review Initiative (SAPRIN) and the Citizens' Assessment of Structural Adjustment (CASA), Washington, DC

Sarrasin, B. (2004) 'Madagascar: A mining industry caught between environment and development', in B. Campbell (ed.) *Regulating Mining in Africa: For Whose Benefit?*, Coll. Discussion Paper 26, Nordic Africa Institute, Uppsala, Sweden, pp. 53–66

Sarrasin, B. (2006) 'The mining industry and the regulatory framework in Madagascar: Some developmental and environmental issues', *Journal of Cleaner Production*, vol 14, no 3–4, pp. 388–96

Sarrasin, B. (2009) 'Mining and protection of the environment in Madagascar', in B. Campbell (ed.) *Mining in Africa: Regulation and Development* Pluto Books, London, pp. 150–86

Smith, S. M. (2012) 'Large-scale mining in Madagascar: Stakeholder engagement and community representation', doctoral thesis, School of Agriculture, Policy and Development, University of Reading, Reading, UK

Smith, S. M., Shepherd, D. D. and Dorward, P. T. (2011) 'Perspectives on community representation within the Extractive Industries Transparency Initiative: Experiences from south-east Madagascar', *Resources Policy*, vol 37, no 2, pp. 241–50

Stevens, P. and Dietsche, E. (2008) 'Resource curse: An analysis of causes, experiences and possible ways forward', *Energy Policy*, vol 36, no 1, pp. 56–65

Szablowski, D. (2007) *Transnational Law and Local Struggles: Mining Communities and the World Bank*, Hart, Oxford, UK, and Portland, OR

Teyssier, A. (2010) 'Decentralising land management: The experience of Madagascar', Perspective (policy brief) no 4, CIRAD, Paris

Teyssier, A., Andrianiriana-Ratsialonana, R., Razafindralambo, R. and Razafindrakoto, Y. (2009) 'Décentralisation de la gestion des terres à Madagascar: Processus, enjeux et perspectives d'une nouvelle politique foncière', in J. P. Colin, P. Y. Le Meur and E.

Lénorad E. (eds) *Les politiques de reconnaissance des droits fonciers: Du cadre légal aux pratiques locales*, IRD Karthala, Paris, pp. 273–97

Tietenberg, T. (2000) *Environmental and Resource Economics*, 5th ed, Addison-Wesley, Reading, MA

Toillier, A., Lardon, S. and Herve, D. (2008) 'An environmental governance support tool: Community-based forest management contracts (Madagascar)', *International Journal of Sustainable Development*, vol 11, no 2–4, pp. 187–205

Vinciguerra, V. (2010) 'How the Daewoo attempted land acquisition contributed to Madagascar's political crisis in 2009', paper from the conference Africa for Sale: Analysing and Theorizing Foreign Land Claims and Acquisitions, Groningen, Netherlands, 28–29 October

Von Braun, J. and Meinzen-Dick, R. (2009) '"Land Grabbing" by Foreign Investors in Developing Countries: Risks and Opportunities', IFPRI Policy Brief 13, April, International Food Policy Research Institute, Washington, DC

Walzer, M. (1994) *Thick and Thin: Moral Argument at Home and Abroad*, University of Notre Dame Press, Notre Dame, IN

West, P. and Brockington, D. (2006) 'An anthropological perspective on some unexpected consequences of protected areas', *Conservation Biology*, vol 20, no 3, pp. 609–16

Woolley, O. (2002) 'The earth shakers of Madagascar: An anthropological study of authority, fertility and creation', *LSE Monographs on Social Anthropology*, vol 73, Continuum, London

World Bank (1998) 'Project appraisal document for a mining sector reform project', Report no 17788-MAG, 2 June, World Bank, Washington, DC

World Bank (2010) 'Rising global interest in farmland: Can it yield sustainable and equitable benefits?', World Bank, Washington, DC

WWF (2005) *An Update on the QMM Mining Project in the Anosy Region, Southeastern Madagascar*, 2005 WWF Review, WWF

Zerner, C. (2003) *Culture and the Question of Rights: Forests, Coasts and Seas in Southeast Asia*, Duke University Press, Durham, NC

5 Exclusionary spaces

Power, poverty and community-based natural resource management in West Africa

Marlène Buchy and Roy Maconachie

Introduction

Over the past three decades, participatory natural resource management has undergone a significant transformation. Originally conceived as an innovative, groundbreaking philosophy and methodology (Chambers 1997), it has more recently been subject to accusations of tyrannizing those it is supposed to empower (Cooke and Kothari 2001), despite being mainstreamed in the world of international development. Indeed, no well-intentioned development project today would be given the green light without having some participatory component in its programming (Cornwall and Brook 2005). In situations where natural resource management is a key aspect of project activities, this rise to institutionalization has also taken place in parallel with genuine policy changes in resource governance at global, national and local levels.

The devolution of natural resource management structures and the decentralization of decision-making around resources have also come of age as community-based natural resource management (CBNRM) has emerged in its modern form in South Asia – that is, in India and Nepal since the 1970s in the forestry sector – and spread to other parts of the world more recently (Larson 2005; Ribot et al 2006). While CBNRM is an approach to resource management that is designed to give communities the legal rights, local institutions and economic incentives to take substantial responsibility for the sustained use of their resources, local people have of course, in many cases, been managing their resources communally and 'sustainably' for centuries. However, colonial regimes of centralized and extractive management have often disrupted these pre-existing systems of community stewardship. Today, these same local-level processes continue to be shaped by wider power structures, most notably the commodification of natural resources through global market exchange and the impact that this has on state-level decision-making. In this chapter, that is the context in which we are concerned by the paradoxes contained within the more recent advent of 'invited spaces' (Gaventa 2006) of CBNRM – those created in-country by governments to hand over the responsibility for sustainable resource management to local communities.

The benefits and limitations of participatory natural resource management have been well documented elsewhere (Agrawal and Gibson 1999; Cleaver 1999;

Khadka 2009) and much has been written about the need to enhance popular participation and empowerment in resource planning and decision-making. However, it is an unfortunate reality that apart from a few flagship projects (such as the Communal Areas Management Programme for Indigenous Resources [CAMPFIRE] initiative in Zimbabwe (Alexander and McGregor 2000), CBNRM has fallen short of delivering significant poverty alleviation, social and gender equity or sustainable management (Buchy and Rai 2008). Even the most widely hailed success stories, including CAMPFIRE, have recently come under attack from critics (e.g. see Alexander and McGregor 2000; Logan and Mosely 2002; Dzingirai 2003).

Drawing on case studies in the mining and forestry sectors in West and Central Africa, we argue that it is the systematic ignorance of power relationships and how they shape social structures that compromises the chances of promoting social change and poverty reduction in resource-rich communities.

A framework for power relations

To understand how power relations are inextricably linked at different levels and places, but are also multifaceted, we draw upon the concept of the power cube developed by the Participation Group at the Institute of Development Studies in Sussex. Power, according to Gaventa (2006), is three-dimensional: it operates at different scalar levels (international, national and local) and within different spaces (closed, invited and claimed), and it can be of different dimensions (visible, hidden or invisible – i.e. internalized). In this chapter, we explore the notion that CBNRM initiatives are invited spaces: within a country, they are initiatives designed and launched by the government, and often managed by the state bureaucracy (the forest department or the ministry of mines), functioning according to non-negotiable rules. Communities are 'invited' to participate in the initiatives, but the modality for their participation has already been defined and legislated by more powerful actors.

Within such spaces, all three types of power are present. *Visible* power is situated within the explicit rules and regulations and the local and state institutions represented in the CBNRM initiative. *Hidden* power is embedded within the tacitly observed rules of socialization (for example, women knowing not to speak in public in front of their husbands, and especially not to contradict them) which govern social relationships in the community. A seasonal labourer who depends on a landlord for employment may not be willing to lobby for decisions that would threaten the landlord's status at community level. *Invisible* power is the internalization of defined social orders, often prevalent where one social group has historically dominated another. For example, many Dalits (formerly known as 'untouchables') in India have internalized and accepted the norm that they are destined to be at the bottom of the social hierarchy and thus do not claim the rights granted to them by the Indian constitution and law.

Such hidden and invisible forms of power are usually separated by a very fine line. For example, in some cultures, many women have internalized that men are superior and should rightly control them (invisible power), and also understand that in certain contexts they need a man to access livelihood resources, such as land, to ensure household survival (hidden power). This is despite the fact that women may have more rights *de jure* (visible power). Where women can legally inherit land from their fathers, in practice they may choose not to claim their share. This is because in the case of divorce or widowhood, they would need to rely on their kinship for survival, which would be difficult if they had dismantled the family assets. In this case, refusal to claim the land that a woman is entitled to is partly an insurance policy, but also an expression of the de facto power imbalance.

The motivation to analyse and understand power structures is linked to the premise that development interventions and especially CBNRM aim to facilitate some form of social change which will ensure more sustainable futures. However, the belief that community-based approaches will deliver more equitable outcomes for the poor and the marginalized has confused origins and has often been an afterthought rather than the initial driving factor. The history of wildlife conservation practices in southern Africa is a case in point. Management policies that initially excluded local populations in conservation practices have gradually elevated those originally labelled as 'poachers' and 'encroachers' by the colonial administration to the status of co-managers (Carruthers 1985).

Similarly, Aboriginal people in Australia were initially excluded from the management of Kakadu National Park because their traditional use of fire as a management tool was deemed destructive. However, many Aboriginal groups have subsequently become co-managers, largely because scientists now realize that fire has been used for hundreds of years to maintain the unique Kakadu landscape (Baker et al 2001). Such examples of the promotion of local communities to co-management thus originate in a functional approach rather than an adherence to rights principles.

Deeply rooted in conservation policy is the proposition that local people who depend on natural resources for their livelihoods will become 'poachers' and 'encroachers' if they do not somehow benefit from that policy. As many developing countries have limited resources to control vast areas of land effectively, the tactic of involving local people in management is seen as a win-win situation. The substantial international aid contributions – in the form of loans, grants and project resources – that support the development of these models also partly explain their popularity. In Nepal and, to a lesser extent, India – two countries with a very long history of modern community forestry management – the move to involve local communities was initially driven by donors (see Khadka 2009).

It can also be argued that the policy shift toward more decentralized governance is a response to inefficient and centralized systems of state governance and is aimed at improving resource conservation and local livelihood benefits. This idea hinges on the belief that if local people have an economic incentive to protect their resources, they will manage the resources more effectively than the state, and at a much lower cost.

There are many different models of CBNRM around the world, but typically a CBNRM approach is based on the legally binding handing over of a resource (e.g.

a patch of forest, a section of a mine or an area of national park) by the state to a formal community group composed of local residents and governed by a board of elected committee members. The board usually deals with the day-to-day running of the operation, while the general assembly of members meets less frequently to deliberate and vote on strategic decisions. The relevant state authority – for example, the forest department or the ministry of mineral resources – continues to be involved in various ways, as discussed below.

Coupled with the argument that community groups would be better managers than many 'outsiders' is the view that poverty is the root cause of environmental degradation and that participatory management would more effectively benefit the poor. The nexus between poverty and environmental degradation has been debated for many years (Eckholm 1975; Hobley 2007), but in this chapter we take the position that poverty is fundamentally an outcome of social exclusion and unequal power relationships which have excluded the poor from equitable access to resources. We therefore contend that policy changes in resource management which aim to alleviate poverty must first tackle the root causes of unequal power relationships.

However, addressing power differentials is not a simple issue. Pearce (2006, p. 2) reminds us that while acknowledging the need for change, we also need to understand that 'knowing when to talk to the powerful, when to challenge the powerful and when to confront the powerful, requires a strong understanding of the kind of power in particular spaces and structures'. Accordingly, Haugaard (2003, p. 109) has argued that there are seven ways of creating power. Here, we would like to consider power created by social order ('causal predictability created through the reproduction of meaning; theorized as structuration and confirming structuration') and power created by system bias ('order [that] precludes certain actions: destructuration') to demonstrate how, in conjunction with visible, hidden and invisible power, CBNRM as an invited space cannot deliver empowerment and social change and thus is destined to fall short of its poverty alleviation target.

Diamonds and ebony: two resources, one paradox

Diamonds, power and community-led decision-making in Sierra Leone

Concerns for citizen voice and community participation in decision-making processes over key mineral resources are today particularly acute in policy circles in Sierra Leone. This is especially so owing to the country's unstable past and its association with violent conflict around diamond resources. While there is considerable difference of opinion on the specific motivations and causes of the country's decade-long civil war in the 1990s, there does appear to be consensus that diamonds were a central feature of the conflict, allowing the various fighting factions to fund their warring activities. Many commentators also appear to agree that poor governance, the overcentralization of power and the creation of a socially excluded 'underclass' were prime factors responsible for fomenting the preconditions for the war. It thus seems that in the case of Sierra Leone, unequal power relationships around diamond management played a central role in the country's turbulent past.

One well-known recent governance initiative in the diamond sector, the Diamond Area Community Development Fund (DACDF), aims to redress the most visible unequal power relationships within the diamond industry and make local-level decision-making over resources more equitable. Building on the popularity of previous CBNRM experiments elsewhere in Africa, the DACDF, an invited space of participatory governance, was formally approved by Sierra Leone's Ministry of Mineral Resources in December 2001, as part of a broader reform programme for the diamond sector following the end of the war. Because it returns a percentage of mining revenue to the producing chiefdoms, the fund has been widely hailed as providing a considerable incentive for miners and diamond-rich chiefdoms to engage in legal diamond mining activities and revenue reporting, thereby enhancing the Kimberley Process Certification Scheme for Rough Diamonds. A small portion of the government's three per cent diamond export tax (amounting to 0.75 per cent of the total export value) is now put into the DACDF for community-managed, small-scale development projects in diamondiferous regions.

Since the initiation of the DACDF, an unprecedented amount of diamond revenue has been returned to diamond-mining communities. There is also much evidence to suggest that the fund has been utilized wisely by some chiefdoms and councils for financing community infrastructure, education, health and vocational skills training centres (Temple 2005). On the other hand, several constraints, particularly in the early days of the initiative, have made the implementation of the fund a great challenge.

Foremost among these challenges is the inadequacy of the DACDF in circumventing the entrenched, unequal power relationships in the diamond industry. While the fund aims to make local-level decision-making over resources more equitable, it has failed to address many of the underlying invisible power issues. This remains a significant threat to prospects for peace-building and post-conflict reconstruction in the country, particularly because pre-war patrimonial relationships endure and continue to subvert fair access to, and control of, the nation's diamond resources (Maconachie 2012). Although the government of Sierra Leone acknowledges that the fund has not produced the development results anticipated, to the extent that the Ministry of Mineral Resources has, in fact, recently developed a new set of operational guidelines to address its shortcomings, it is still not clear whether this will remedy the underlying issue of how community members participate in natural-resource decision-making (Maconachie 2010).

Participatory forestry in Cameroon

The Cameroonian economy is well diversified, with the service industry accounting for 44 per cent of 2009 gross domestic product (GDP), agriculture and manufacturing for 19 per cent each and oil and mining for 7 per cent (World Bank 2011). Cameroon's forests also play an important role in the national economy, with 160,467 kilometres squared of dense lowland forest covering the 466,326 kilometres squared total land area of the country. The forestry sector generates 6 per cent of GDP and employs 13,000 in the industrial sector, but estimates put the number of people involved in the informal sector at ten times that of the formal sector (Karsenty 2007, p. 21).

The forest estate is divided into permanent and non-permanent domains and exploited through a number of logging titles. The extractive model of forestry in Cameroon, where the state and foreign logging companies hold the majority of key concessions and take the lion's share of the benefits, is profoundly unequal and inequitable (Oyono et al 2005). Formal participation by Cameroonians in the forestry sector has thus been minimal, with the industry dominated by European and Asian groups (Karsenty 2007). While it seems that the number of national entrepreneurs may have grown since 1992, these are mainly small operators concentrated in the *ventes de coupe*, which are small logging areas of no more than 2,500 hectares. To date, there has been little research into the operations of these small entrepreneurs, particularly with respect to how they contribute to the sector and what share of the local market they represent.

A growing number of Cameroonians are formally involved in and benefiting from the management of forests in two additional contexts. First, participation occurs in the management of community forests, areas established through Cameroon's Forestry Law of 1994 (Cameroon 1994) in the non-permanent forest domain zoned for use by village communities. With technical assistance from the Ministry of Forests' Community Forestry Unit, a village community seeking a forest title can identify a zone not exceeding 5,000 hectares and draft a simple management plan for approval by the ministry. Proceeds from community forest management are used for community development projects. Second, communities can participate in the management of council forests, which are areas zoned within the permanent forest domain and managed according to an approved management plan. The objectives of a council forest, along with its final boundaries, are established during the official classification process. Once allocated, these forests become the private property of a commune. However, the commune, which is the lowest administrative unit in Cameroon, must abide by the management plan in order to retain title to the forest area (Bikié et al 2000).

The Forestry Law of 1994 (Cameroon 1994), seen by many as an innovative and progressive piece of legislation, introduced the concepts of sustainable forest management through forest management plans, of redistributing forest taxes to communities and of council and community forests (Cerutti et al 2010). However, many of the decrees needed to enact the law have not yet been written or voted on, and this has prevented the law from being implemented. It was almost six years before the first community forest licences were finally approved in 2000 (Djeumo 2001). Community forestry in Cameroon has continued for the decade since then, and although empirical research shows benefits for community livelihoods (Beauchamp 2009), a number of problems have arisen, including an increase in local conflicts (De Blas et al 2011) and the appropriation of resources by local elites (Cerutti et al 2010).

An attempt to measure the level of participation in community forestry in Cameroon quantitatively has demonstrated very low participation by local communities, compared with the Ministry of Forests and civil society organizations, in drafting agreements or approving management plans for their forests, and yet the communities are much more active than the other two actors when it comes to implementation and actual work (Ndibi and Kay 1999). This is a familiar scenario for forestry departments, which tend to maintain control of all decision-making roles and delegate the physical work to communities.

In the discussion that follows, we examine two cases of CBNRM with enormous social and economic potential, but limited impact on the communities to date. In exploring the reasons for this, we argue that there is a pressing need to frame community participation in natural resource management within its broader political, cultural and historical contexts, instead of treating it as a romanticized notion of democratic space that remains isolated from, and unaffected by, the real world. Williams (2004, p. 562) adds that uncritically accepting the notion of 'community' as the right and proper site for development may depoliticize development, and thus reinforce the tendency to direct attention away from the wider power relationships that frame local development problems. Below we identify the types of power at play in these two CBNRM models, in order to show how the different dimensions of power operate simultaneously in the invited spaces of community-based projects.

An analysis of power

Visible power

In the forestry sector in Cameroon, and also in the mining sector in Sierra Leone, the state has historically assumed an important role as the protector and facilitator of both natural resources and access to capital. While more recently it has been suggested that global capitalism has eroded the ability of the state to make autonomous decisions concerning the management of its natural resources, the state undeniably remains a visibly powerful actor.

Landownership is the first arena where visible power is exercised. In Cameroon, the state owns the land, and since colonial times indigenous rights to land have not been recognized beyond access and controlled use for self-subsistence. In rural areas in Sierra Leone, the 'native administration' grants communal rights to land to 'sons or daughters of the soil', but the state retains ownership of what the soil might contain. Therefore, for companies to gain access to subsoil mineral resources, some degree of negotiation with the state is required. Though only the state, in theory, has the right to allocate mineral resources for mining, local chiefs are also involved in the negotiations.

In both Cameroon and Sierra Leone, government departments that oversee forestry and mining have become institutions within the state and have traditionally had very powerful links to political figures. Informal links between the private sector, high-level bureaucrats and politicians right up to the presidency are well documented (see Labrousse [2000] for Cameroon and Jackson [2005] for Sierra Leone).

In Cameroon, the corruption networks associated with these links affect community forests in various ways: logging companies that encroach on community-assigned land do so with the implicit protection of forestry staff, and some company officials do not hesitate to strike deals with corrupt village elites, thus depriving the wider community of potential benefits from logging. This lack of respect for the law fuels conflicts at the local level, which in turn create tension within the community, and also between the communities and the bureaucracy.

In the case of Sierra Leone, diamonds have long shaped local and national politics. Patronage politics associated with diamonds have a well-established history in the country, with local elites expecting rewards for party loyalty, since their own supporters depend on their ability to deliver development resources to their local areas. Especially under the leadership of Siaka Stevens, president of Sierra Leone from 1971 to 1985, diamonds were a key strategic resource for the All People's Congress government, and as rent-seeking intensified, the industry became a sphere of corrupt, unregulated private enterprise. Today, many politicians in Sierra Leone continue to have strong links with the diamond sector: the vice-president himself, Samuel Sam-Sumana, is a former diamond businessman.

A second arena of visible power is apparent in the manipulation of the legal framework, predominantly by powerful actors who control access to resources. Though appropriate laws to protect community interests are sometimes passed, their implementation can be slowed down. In the Cameroonian context, the Forestry Law of 1994 (Cameroon 1994) is widely recognized as a progressive framework in relation to community rights and sustainable forest management. However, delays in developing the decrees required by the legislation, as mentioned above, have resulted in most of the progressive gains remaining on paper only.

One of the challenges of the legal context in both countries is the confusion created by the superimposition of different legal frameworks: residual colonial laws, postcolonial laws and customary systems overlap, and all of them may be operating at the same time. This confusion makes it difficult for some people to claim their rights, and easier for others to abuse those rights.

In the context of forests, it is the forestry department that determines the size of community forests, and by law only forests not in the permanent forestry domain (i.e. forests of lesser economic value) can be allocated as community forests. The limited size and profitability of these forests often prevent communities from managing them responsibly, as the costs of management are not offset by the benefits accrued. The power held by the forestry department also permeates the management of community forests, as many management decisions remain the prerogative of the department. This can be defended on the grounds of ensuring sustainability – though of course, in a context of rent-seeking behaviour, it can also encourage more illegal demands on communities by the bureaucracy.

With respect to diamonds, the situation in Sierra Leone is slightly different, in that communities are not allocated portions of mines to be locally managed, but instead given the right to make decisions about how a small share of diamond revenues is to be spent. While most diamond mining in Sierra Leone is small-scale and artisanal, comprising low-tech extraction activities with rudimentary tools, high inputs of unskilled labour and low barriers to entry, in recent years there has been increasing interest in more capital-intensive modes of mechanized extraction by medium-sized companies. Many of these companies have established strong relationships with the government and local chiefs. In many respects, these informal political networks yield much more visible power than that possessed by individuals in community development committees.

Hidden power

A lack of capacity in remote rural communities is often cited as one of the major dimensions of hidden power, as communities have limited financial and human resources to control the management process.

In the context of Cameroon's forestry, communities are responsible for the actual management of forests, which is expensive and requires specialized knowledge and skills. Though indigenous people may have a wealth of environmental knowledge, this does not mean that all rural communities possess the skills required to write management plans and implement them as required by the forestry department. Many observers have commented that the costs of sustainable management (which includes an onerous bureaucratic process) are prohibitive and slow down activities in the field. This also helps explain why community elites, with or without their constituents' support, have been known to sell off stocks of timber to the highest bidder. However, if the money fails to materialize, communities have no recourse to the law, because the sale was illegal in the first place. This means that the management requirements legally imposed on communities with limited capability usually fail to provide those communities with a safe space in which to use their resources.

The lack of information available to communities is also a powerful tool: the state fails to circulate required information, which elites may obtain by capitalizing on their superior literacy skills, and rural people may not be aware of their rights. This also makes it easier for local chiefs to mismanage funds without being challenged by their constituencies.

In Sierra Leone, for example, chiefdoms benefit from the DACDF in accordance with the number of mining licences issued and the value of diamonds recovered from their territory. However, the disbursement of funds and the capacity of communities to monitor development projects have been very contentious matters among critics. In 2002, industry observers produced a series of ad hoc reports (see Temple 2005) which revealed that a number of chiefdoms were not utilizing the DACDF in a competent manner. Jackson (2007, p. 100) has highlighted the continuing misuse of the fund, pointing out that 'there is no accountability mechanism for ensuring that this cash is used for development, and it is extremely common to hear that local people complain of the chief's abuse of the system in pocketing this money'. A report produced by the civil society organization Network Movement for Justice and Development (Network Movement for Justice and Development [NMJD] 2006) argues that because the local-level handling of funds has been carried out independently through community development committees, with no reporting mechanisms in place or any systematic monitoring oversight from the Ministry of Mineral Resources, local power abuses have been allowed to flourish.

Similar outcomes have been documented in Cameroon, where the community share of the forest land tax effectively diminishes as it transits between administrative levels from capital city to village. By the time the revenue eventually reaches the village, much of it has disappeared. This is not only a result of individuals appropriating the money, but also because there is no one competent to maintain a budget ledger at the local level (Morrison et al 2009).

Likewise in Sierra Leone, local governance factors greatly obscure the transparency and accountability of diamond revenues. A recent impact audit of the DACDF carried out in nineteen diamond-producing communities by the Network Movement for Justice and Development (NMJD 2006) notes that a 'consistent trend of poor participation by grassroots stakeholders in project decision-making' is the reason for most DACDF projects being concentrated around chiefdom or district headquarter towns (NMJD 2006, p. 11). With few exceptions, the report continues, the single most important implication of this occurrence is that the majority of local people do not readily identify with DACDF projects.

A combination of the lack of capacity of communities, their limited sense of ownership and pre-existing social norms allows elite capture to occur, while the dynamics of hidden power at local level are often overlooked. This has been one of the central critiques of the CBNRM process: communities are not harmonious entities, but are ridden with power struggles and conflicting interests (Guijt and Shah 1998).

In Sierra Leone, a number of civil society organizations, including the National Advocacy Coalition on Extractives, a well-known alliance of eighteen national and international non-governmental organizations (NGOs) working around issues of mining and the extractive industries, lobbied for the creation of community development committees in order to ensure that project decision-making associated with the DACDF was carried out in a more equitable and accountable manner. In theory, each community development committee was supposed to include a wide cross-section of elected chiefdom residents, representing a broad range of community interests. In practice, though, reports suggest that these committees have more often been composed entirely of rural elites such as section chiefs. This has reportedly stifled the concept of local ownership of the fund and further alienated many stakeholders, such as women and youth (Temple 2005). The fund has thus not been able to address the issue of hidden power differentials.

In the eastern forests of Cameroon, where community forestry is being developed, an ethnic dimension also exists: there is a long history of Bantu people traditionally exploiting Aka Pygmies as labourers on their farms (Lewis et al 2008; Woodburne 1997). Though, as far as can be ascertained, no research has been carried out documenting the differential levels of benefits along ethnic lines, it would not be surprising, given the historical relationships of exploitation, to find indigenous peoples proportionally worse off than more sedentarized agricultural groups.

In Sierra Leone, where unequal power relationships have also always been prevalent in rural communities, a recurrent theme in the history of diamond mining regions, particularly in Kono District, has been confrontation between 'big men' seeking greater control over diamond production and indigenous people seeking to maximize their own returns from the industry and defend their putative rights and entitlements (Fanthorpe and Maconachie 2010). However, since the end of the civil war, as mining has started to shift towards more capital-intensive modes of extraction and has become increasingly attractive to larger companies aiming to pursue mechanized production, community tensions have reportedly worsened.

A final dimension of hidden power that is more directly relevant in the Sierra Leone context can be seen in the burden of recent history. Several commentators have pointed out that in the post-war period in Sierra Leone, the differential between the rural elite and the community at large has fuelled a rift between youths and chiefs in the countryside. Richards (2005), for example, suggests that the war was largely a product of this division, and there continues to be considerable dissatisfaction among youths with the hierarchical political structures in rural Sierra Leone.

However, Brown et al (2006) argue that for all the destruction and disruption caused by Sierra Leone's civil war in the 1990s, its resolution may have opened up new spaces for participation that could potentially challenge and alter existing social rules and institutional practices. On the other hand, they also suggest that many of the patrimonial institutions, practices and networks that were instrumental in shaping natural resource access and control during the pre-war era may have been preserved, and in some cases even strengthened, in the post-conflict period. In such situations, warns Cleaver (2004, p. 271), there is a danger of participation being based on 'over-optimistic notions of agency . . . combined with romantic ideas about groups and institutions'. In the case of Sierra Leone, it remains important to remember that the exercise of agency and citizenship rights is embedded in social relations that are defined by highly unequal power positions located within a complex web of micro-politics.

Invisible power

The third dimension of power, invisible power, is by definition difficult to see, and therefore we need to look closely at the internalized values of social order. Witchcraft, gender and ethnicity are examples of these value systems, which most often remain completely ignored by policy-makers, development professionals and even communities themselves.

Witchcraft is a common practice throughout West Africa, but is little acknowledged or discussed because it often evokes feelings of discomfort (Geschiere 1995; Golooba-Mutebi 2005). Those who do not believe it exists simply do not see or understand that it plays a major role in regulating behaviour; those who do believe it exists are too scared to talk about it. And perhaps there are those who are prepared to acknowledge its power, but because of its secretive nature, do not understand how it works. Yet anthropological work has shown how significant it is: instead of fading away, it has become more prominent in some parts of West and Central Africa and has serious implication for development interventions (Kohnert 1996).

The point relevant to our discussion, however, is that a complex web of internalized fears or beliefs exists which regulates people's behaviour. When practitioners, planners and policy-makers design community-based management models without considering these underlying currents, they ignore a whole facet of reality which may influence outcomes negatively. It is in the inherent nature of witchcraft that outsiders cannot see how it may affect forest or community management; however, the fact that it is practised and has great potential to shape decision-making should at least be noted.

While gender has been mainstreamed in development discourse over the past decade, in practice very few policies or programmes consider *ex ante* how the gender dimensions prevalent locally may affect a new initiative. It is clear from research in Sierra Leone that women are de facto excluded from decision-making at the local level. Similarly, in the forestry context in Cameroon, women's roles and rights are completely absent from the literature and, judging from recent discussions in the field, from people's minds. Gender, of course, is not just a women's issue: both men and women are trapped in gender constructs. Yet in a context of poverty alleviation, it is important to understand how these strictures constrain people and communities working towards a more equitable society. If this is not understood, there is a risk that inequity will merely be reproduced and reinforced, with one group consistently losing out.

Ethnicity is another dimension that is little discussed in relation to development (White 2002), particularly because it is a sensitive issue and is rarely seen as relevant in the context of natural resource management. Yet, as we have argued above, at the local level in Cameroon there is an institutionalized discriminatory relationship between the dominant Bantu groups and the indigenous minority. At the national level, this relationship translates into a Caucasian-dominated logging industry, with direct roots in the recent colonial past, that controls most of the resources owned by Africans. While it would be simplistic and unhelpful to paint all foreign foresters as racists, the appalling working and living conditions that Cameroonian workers are subjected to in the forest concessions cannot be explained solely by a capitalist greed to maximize profits. In the context of community management, these ethnic hierarchies translate, at best, into subconscious attitudes accepting that some groups may be ignored in decision-making or benefit-sharing processes, and, at worst, into conscious beliefs on the part of certain groups that other groups have no rights or claims to a share of the proceeds, based on some notion of their inferiority (Crew and Fernando 2006). Of course this also works the other way around, as subaltern groups internalize power, accept the hierarchy as given, and do not seek to claim their rights.

In the case of Sierra Leone, reports suggest that many local residents in diamondiferous regions have long felt that they are not receiving a fair share of the wealth generated from their land. In the past, concerns have been raised about the ever-increasing influx into diamond mining areas of migrant 'strangers', in particular wealthy members of ethnic trading diasporas (Lebanese, Fula, Mandingo) who support mining but are perceived to be reinvesting the profits elsewhere.

Conclusion

It is apparent from an exploration of recent governance reforms in Sierra Leone's diamond sector that invitations to participate in formally defined decision-making initiatives, such as the DACDF, do not always translate into ideal notions of democratic space. Similar lessons can be drawn from the community forestry sector in Cameroon.

Consequently, one of the most important lessons to be learned is that such arenas of citizen engagement are always mediated by relationships of power between actors, and by the modes of influence that participants exercise. Unequal power relations – whether they be visible, hidden or invisible – are deeply entrenched within communities. This realization yields important insights for scholarship concerned with the role that communities can and should play in natural resource management, and how these relationships are mediated by power. As has been clearly argued in this chapter, an understanding of the power web requires a consideration of a diverse range of issues including the legal framework and management institutions, prevalent social structures and norms, history, capabilities, resources (including access to information), ethnicity and gender.

This, of course, also has broader practical relevance for community-led revenue-sharing programmes initiated for a wide range of natural resources across sub-Saharan Africa and beyond. As Cornwall (2002, p. 51) rightly suggests, community participation may therefore best be treated as a 'situated practice' – one that is shaped and defined by actual political, social, cultural and historical determinants.

As Khadka (2009) demonstrates so effectively in her case study of Nepali community forestry, although there is a high level of awareness concerning the unsatisfactory social outcomes of community forestry, the actors involved in the initiative do not consider it their responsibility to focus on these issues. The donors who have been shaping community forestry believe that the socio-political issues reproducing exclusion are ones for the country to deal with, rather than for donors to interfere in. One of the comments often made by foresters is that their job is about looking after the forests and has nothing to do with the socio-political issues in the communities. Civil society actors locate the shortcomings of community forestry initiatives in the structures and shortcomings of other actors. This tendency to deny responsibility stems not only from a lack of understanding of existing power relationships, but also from a lack of awareness of how actors themselves support and reproduce these inequalities.

In this chapter we have clearly demonstrated how different dimensions of power can operate simultaneously and collide with one another, in the process maintaining or even enhancing situations of power imbalance, social exclusion and inequality. However, the next step, we suggest, is to reflect upon how this understanding of power can be translated into a constructive course of action, in order to contribute to positive social change.

Our intention here has not been to formulate policy prescriptions or to provide a set of tools for change. However, we would argue, based on practical experience and development work and research, that the exercise of power reminds one somewhat of the 'elephant in the room' syndrome. Perhaps the first step towards change, then, is to call the beast by its name while engaging in community-based resource management. Social learning is one approach that has tried to offer people the necessary space to explore issues such as power, gender, ethnicity and exclusion.

The participatory action research and learning process offers another approach to these complex contexts, but the challenge of scaling up and institutionalizing meaningful participation remains as real as ever, and this is in no small part due to the diverse and contested nature of power and its multiple dimensions.

Bibliography

Agrawal, A. and Gibson, C. C. (1999). 'Enchantment and disenchantment: The role of community in natural resource conservation', *World Development*, vol 2, no 4, pp. 629–49

Alexander, A. and J. McGregor (2000) 'Wildlife and politics, CAMPFIRE in Zimbabwe: Case studies of Insiza and Chimanimani districts', *Development and Change*, vol 31, no 3, pp. 605–27

Baker, R., Davies J. and Young, E. (2001) *Working on Country: Contemporary Indigenous Management of Australia's Land and Coastal* Regions, Oxford University Press, Melbourne

Beauchamp, E. (2009) 'Impacts of community forests in exploitation on communities' livelihoods in Cameroon: Lessons learned from two case studies', report in partial fulfilment of MSc/DIC requirements, Imperial College London and Center for International Forestry Research.

Bikié, H., Collomb, J. G., Djomo, L., Minnemeyer, S., Ngoufo, R. and Nguiffo, S. (2000) *An Overview of Logging in Cameroon*, Global Forest Watch and World Resources Institute, Washington, DC

Brown, T., Fanthorpe, R., Gardener, J., Gberie, L. and Sesay, M. G. (2006) *Sierra Leone: Drivers of Change*, IDL Group, Bristol

Buchy, M. and Rai, B. (2008) 'Do women-only approaches to natural resource management help women? The case of community forestry in Nepal', in B. Resurreccion and R. Elmhirst (eds) *Gender, Environment and Natural Resource Management: Livelihoods, Mobility and Interventions*, Earthscan, London, pp. 127–50

Cameroon (1994) Law No 94/01 of 20 January 1994, laying down forest, wildlife and fishery regulations (Forestry Law), and decree No 94/436/PM of 23 August 1994, fixing the modalities of application of forests regime, Yaoundé

Carruthers, E. J. (1985) 'Attitudes to game conservation in the Transvaal, 1900–1910', paper presented to the History Department, University of Cape Town, South Africa, 10 May

Cerutti, P. O., Lescuyer, G., Assembe-Mvondo, S. and Tacconi, L. (2010) 'Les défis de la redistribution des bénéfices monétaires tirés de la foret pour les administrations locales,' *Documents occasionnels no 53*, Center for International Forestry Research

Chambers, R. (1997) *Whose Reality Counts? Putting the First Last*, Intermediate Technology Publications, London

Cleaver, F. (1999) 'Paradoxes of participation; Questioning participatory approaches to development, *Journal of International Development*, vol 11, no 4, pp. 597–612

Cleaver, F. (2004) 'The social embeddedness of agency and decision-making', in S. Hickey and G. Mohan (eds) *Participation: From Tyranny to Transformation? Exploring New Approaches to Participation in Development*, Zed Books, London, pp. 271–7

Cooke, B. and Kothari, U. (eds) (2001) *Participation the New Tyranny?*, Zed, London

Cornwall, A. (2002) 'Making spaces, changing places: Situating participation in development', *IDS Working Paper 170*, Institute of Development Studies, Brighton

Cornwall, A. and Brook, K. (2005) 'What do development buzz words do for development policy?', *Third World Quarterly*, vol 26, no 7, pp. 1,043–60

Crewe, E. and Fernando, P. (2006) 'The elephant in the room: Racism in representations, relationships and rituals', *Progress in Development Studies*, vol 6, no 1, pp. 40–54

De Blas, D. E., Ruiz-Perez, M. and Vermeulen, C. (2011) 'Management conflicts in Cameroonian community forests', *Ecology and Society*, vol 16, no 1, p. 8

Djeumo, A. (2001) 'The development of community forests in Cameroon', *ODI Rural Development Forestry Network Paper 25b*, July, Overseas Development Institute, London

Dzingirai, V. (2003) 'The new scramble for the African countryside', *Development and Change*, vol 34, no 2, pp. 243–64

Eckholm E.P. (1975) *The Other Energy Crisis: Firewood*, World Watch Institute, New York

Fanthorpe, R. and Maconachie, R. (2010) 'Beyond the "crisis of youth"? Mining, farming, and civil society in post-war Sierra Leone', *African Affairs*, vol 109, no 435, pp. 251–72

Gaventa, J. (2006) 'Finding the spaces for change: A power analysis', *IDS Bulletin*, vol 37, no 6, pp. 23–33

Geschiere, P. (1995) *Sorcellerie et politique en Afrique: la viande des autres*, Karthala, Paris

Golooba-Mutebi, F. (2005) 'Witchcraft, social cohesion and participation in a South African village', *Development and Change*, vol 36, pp. 937–58

Guijt, I. and Shah, M. K. (1998) *The Myth of Community: Gender Issues in Participatory Development*, Intermediate Technology Publications, London

Haugaard, M. (2003) 'Reflections on seven ways of creating power', *European Journal of Social Theory*, vol 6, no 1, pp. 87–113

Hobley, M. (2007) *Where in the World is There Pro-Poor Forest Policy and Tenure Reform?*, Rights and Resources Initiative, Washington, DC

Jackson, P. (2005) 'Chiefs, money and politicians: Rebuilding local government in Sierra Leone', *Public Administration and Development*, vol 25, pp. 9–58

Jackson, P. (2007) 'Reshuffling an old deck of cards? The politics of local government reform in Sierra Leone', *African Affairs*, vol 106, no 422, pp. 95–111

Karsenty, A. (2007) *Overview of Industrial Forest Concessions and Concession-Based Industry in Central and West Africa and Considerations of Alternatives*, Centre de coopération internationale en recherche agronomique pour le développement (CIRAD), Rights and Resources Initiative

Khadka, M. (2009) *Why Does Exclusion Continue? Aid, Knowledge and Power in Nepal's Community Forestry Policy Process*, Shaker Publishing, Maastricht, The Netherlands

Kohnert D. (1996) 'Magic and witchcraft: Implications for democratisation and poverty-alleviating aid in Africa', *World Development*, vol 24, no 8, pp. 1,347–55

Labrousse, A. (2000) *Le Silence de la Forêt: Réseaux, Mafias et Filière Bois au Cameroun*, L' Harmattan, Paris

Larson, A. M. (2005) 'Democratic decentralisation in the forestry sector: Lessons learnt from Africa, Asia and Latin America', in C. J. P. Colfer and D. Capistrano (eds) *The Politics of Decentralisation: Forests, Power and People*, Earthscan, London, pp. 32–62

Lewis, J., Freeman, L., and Borreill, S. (2008) 'Free, prior and informed consent and sustainable forest management in the Congo Basin: A feasibility study conducted in the Democratic Republic of Congo, Republic of Congo and Gabon regarding the operationalisation of FSC principles 2 and 3 in the Congo Basin', Anthroscape, London

Logan, B. I. and Moseley, W. G. (2002) 'The political ecology of poverty alleviation in Zimbabwe's Communal Areas Management Programme for Indigenous Resources (CAMPFIRE)', *Geoforum*, vol 33, no 1, pp. 1–14

Maconachie, R. (2010) '"New spaces" for change? Diamond governance reforms and the micro-politics of participation in post-war Sierra Leone', *Public Administration and Development*, vol 30, no 3, pp. 191–202

Maconachie, R. (2012) 'The Diamond Area Community Development Fund: Micropolitics and community-led development in post-war Sierra Leone', in P. Lujala and S. A. Rustad (eds) *High-Value Natural Resources and Post-Conflict Peacebuilding*, Earthscan, London, pp. 261–73

Morrison, K., Cerutti, P. O., Oyono, P. R. and Steil, M. (2009) 'Broken promises: Forest revenue-sharing in Cameroon', *WRI Forest Note*, World Resources Institute, Washington, DC

Ndibi, B. P. and Kay, E. J. (1999) 'Measuring the local community's participation in the management of community forests in Cameroon', *Biodiversity and Conservation*, vol 8, no 2, pp. 255–71

NMJD (2006) *An Impact Audit Study on the Diamond Area Community Development Fund (DACDF)*, Network Movement for Justice and Development, Sierra Leone, October

Oyono, P. R., Kouma, C. and Mala, W. (2005) 'Benefits of forests in Cameroon: Global structure, issues involving access and decision-making hiccoughs', *Forest Policy and Economics*, vol 7, pp. 357–68

Pearce, J. (2006) *From 'Empowerment' to 'Transforming Power': Can a Power Analysis Improve Development Policy, Practice and Impact?*, FRIDE, Madrid www.fride.org/uploads/Empowerment_Jenny.Pearce_EN.pdf, accessed 1 March 2011

Ribot, J. C., Agrawal, A. and Larson, A.M. (2006) 'Recentralizing while decentralizing: How national governments reappropriate forest resources', *World Development*, vol 34, no 11, pp. 1,864–86

Richards, P. (2005) 'To fight or to farm? Agrarian dimensions of the Mano River conflicts (Liberia and Sierra Leone)', *African Affairs*, vol 104, no 417, pp. 571–90

Temple, P. (2005) *Improving the Effective Use of the Diamond Area Community Development Fund (DACDF)*, report by the Integrated Diamond Management Program (IDMP) for submission to the Government of Sierra Leone High Level Diamond Steering Committee (HLDSC), Management Systems International, Washington, DC

White, S. (2002) 'Thinking race, thinking development', *Third World Quarterly*, vol 23, no 3, pp. 407–19

Williams, G. (2004) 'Evaluating participatory development: Tyranny, power and (re)politicisation', *Third World Quarterly*, vol 25, no 3, pp. 557–78

Woodburne, O., (2012) 'Navigating moral dilemmas: Participatory conservation and development amongst the egalitarian BaAka of the Central African Republic', PhD thesis, University of Kent, Canterbury

World Bank (2011) *Cameroon: Country Brief*, http://go.worldbank.org/I7KMKA50S0, accessed 6 June 2012

6 Wildlife paying its way?

A critical analysis of community-based natural resource management in the Chobe Enclave, Botswana

Clare Gupta

Introduction

For several decades, the conservation community has sought ways to transform the burden that wildlife represents to rural livelihoods into a benefit, particularly in areas where humans and wildlife come into conflict. Rural communities across Africa have witnessed the proliferation of community-based models of conservation – models intended to ensure that protected areas and species yield an economic return for local people in particular and the wider economy in general (Adams and Hulme 2001). Unlike those former protectionist strategies that have often been termed 'fortress conservation' (Brockington 2002), community conservation is based, in theory, on local participation and partnership. The model predicts that if communities are given management responsibility over their local natural resources and obtain economic benefits that exceed the costs of management, they will be encouraged to use these resources sustainably, so that both conservation and development goals can be met.

In most of southern Africa, the community conservation approach has fallen under the rubric of community-based natural resource management (CBNRM) (Child 2009a). CBNRM in Botswana represents an attempt to re-establish wildlife-based livelihoods[1] in rural places where most traditional uses of wildlife (hunting for food or sale) were outlawed over a century ago and where agricultural activities are now frequently hindered by wildlife. The key principles of CBNRM in southern Africa draw from theories of wildlife economics and common-pool resource management (Child 2004). Evidence emerged from southern Africa in the 1960s that devolving proprietorship over wildlife to private (mainly white) landholders allowed and encouraged them to maximize their profits from using the resource sustainably (through commercial hunting, ranching and tourism ventures). This led to the slogan 'use it or lose it' as an effective, market-based approach to wildlife conservation in Africa. It also led to the notion that if communities could be given similar rights, through which wildlife could become 'community private property', then a similar model of sustainable use for conservation and development purposes could be achieved on communal[2] as well as on private lands.

In Botswana, the first CBNRM project was implemented in 1993 through the United States Agency for International Development (USAID) as a strategy for achieving both conservation and poverty alleviation in the Chobe Enclave (Figure 6.1). This project established the Chobe Enclave Conservation Trust (CECT), a community organization set up to manage natural resources in the five enclave villages. As in most CBNRM projects in Botswana and much of southern Africa, the CECT's management of natural resources has come to mean engagement in wildlife tourism, either directly or through a joint venture business partner.

Case study research in 2009 and 2010 in the Chobe Enclave[3] revealed that despite 16 years of project efforts, villagers still associate wildlife with the decline of agricultural livelihoods.[4] This perception is driven by conflict with wildlife such as elephants and lions, which eat crops and cattle, respectively. The CECT's members are vaguely cognizant of the fact that safari companies lease land from their community and that their organization is involved in the management of tourism-related funds. However, few people interviewed in 2009 and 2010 indicated that CBNRM had contributed much to their own survival. The promises of the CBNRM project to devolve management and improve wildlife-based livelihood activities as a buffer against wildlife-related damages and dangers remain unfulfilled (Alexander et al 1999; Ecosurv 1996; Jones 2002; Rozemeijer 2003). Villagers have not escaped a colonial legacy in which wildlife management remains primarily in the hands of state wildlife officials and wildlife is widely regarded as the property of the state. While wildlife tourism has developed in Chobe, its economic benefits accrue primarily to experienced non-local safari operations. Indeed, as this chapter highlights, CBNRM in Chobe plays a smaller role in compensating for wildlife-related damage to livelihoods in the Chobe Enclave than government

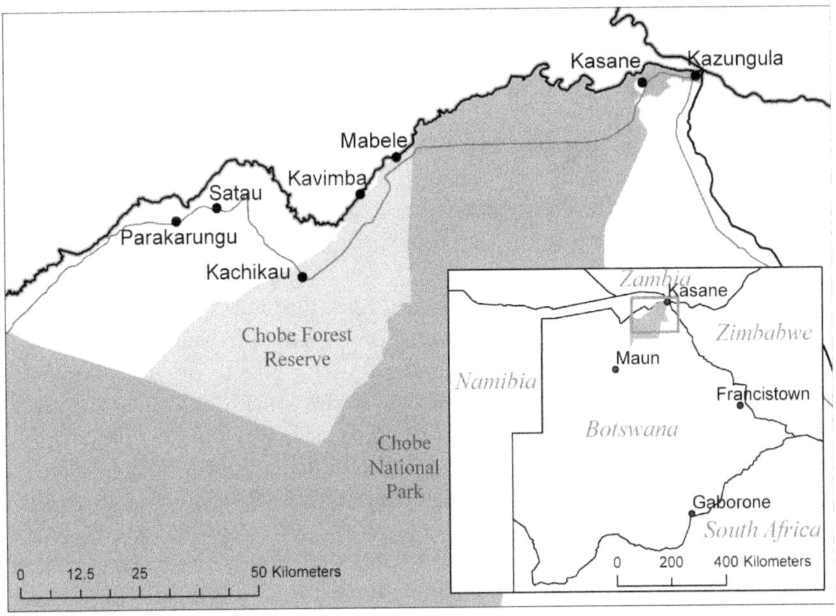

Figure 6.1 Maps of Chobe National Park and Chobe Enclave.

welfare provisioning policies and the remittances that flow to villagers from family members working outside the villages.

Nevertheless, CBNRM remains the dominant paradigm for environmental governance in southern Africa (Blaikie 2006). The CBNRM model is particularly resilient because when CBNRM projects fail, its advocates interpret this as weak implementation of CBNRM principles rather than demonstrating weakness in the principles themselves (Child 2009b). This chapter challenges that narrative by suggesting that the fact that these principles have proven so consistently difficult to implement indicates that the CBNRM model *is* flawed and in need of revision to make it more appropriate for certain local realities.

CBNRM relies on a number of assumptions about resource tenure, market access, cultural plasticity and social organization that are seldom met. Using the case of Botswana's Chobe Enclave, I highlight one set of circumstances in which a CBNRM programme may fail to bring a community together to manage a given resource. I do this by examining the disabling conditions for this kind of collective action on two levels.

First, I analyze livelihoods in Chobe to demonstrate how, in this region, residents are no longer dependent on wildlife: households here have other ways of diversifying their livelihoods that do not require them to re-establish collective natural resource management institutions and engage in modern markets. Specifically, because Botswana is a highly centralized developmental welfare state[5] and has a historical legacy of strong rural-urban linkages, social safety nets and remittances make a large contribution to household livelihood portfolios in the Chobe Enclave. These sources of income can be accessed by individual households and do not require collective action or the ability to develop profitable wildlife-based enterprises.[6]

Second, I use an access-analysis framework (Ribot and Peluso 2003) to explain how the resource property *rights* established under Botswana's CBNRM policy do not give communities the *ability* to use wildlife in ways that would allow them to derive benefits from wildlife that outweigh its costs. As a result, livelihoods based on direct wildlife use have not been re-created, and a vested interest in the sustainable use of wildlife has not materialized in village communities as it did in the private sector.

The findings presented in this chapter suggest that the conceptual underpinnings of CBNRM – which assume that the ecological and economic success that resulted from the devolution of land and resource rights to private landowners in Zimbabwe in the 1960s can be replicated in the context of a village or multi-village communal lands – may need revisiting. Specifically, a market-based model for wildlife conservation may not always be the most appropriate or effective alternative to fortress conservation. The failure of CBNRM forces us to consider whether, in some places, there may be more effective ways to achieve sustainable conservation than the 'use it or lose it' utilitarian approach that characterizes CBNRM in southern Africa and much of international biodiversity conservation today (McCarthy 2005; Büscher and Dressler 2007; Igoe and Brockington 2007; Igoe et al 2010). This is a critical point because it challenges the discursive power surrounding CBNRM, whose proponents posit a compelling causal argument regarding the links between the devolution of ownership over natural resources, the creation of benefits and the successful conservation of wildlife (Blaikie 2006).

The fact that the local CBNRM project has done little to create a more socially just form of conservation makes clear the need to develop alternative practical strategies for environmental management which sustain both people and wildlife. The point of this chapter is not to argue that welfare provisioning and remittances provide an alternative conservation-development strategy which is better than CBNRM, but to highlight conditions under which CBNRM is particularly likely to fall short. It is also to encourage the conservation community to consider how programmes and policies for sustainable and socially just conservation might be designed to build upon existing conditions and institutions: a process that will involve thinking outside of the conventional CBNRM model.

Background

Theoretical roots of community-based natural resource management: Common property theory and neoliberalism

CBNRM initiatives have gained popularity over the past two decades (Adams and Hulme 2001). However, the notion that communities are capable of sustainably managing their own resources according to local custom, knowledge and technologies is not new (Blaikie 2006). There is a long history of scholarship on the commons showing that resource users often create institutional arrangements and collective management regimes that help them allocate use rights and benefits over long periods (Berkes 1989; Ostrom 1990). Common property theorists argue that resources can be used collectively and sustainably provided certain principles are in place, including the recognition of local resource users by external governmental authorities, tenure rights for local institutions and incentives in the form of benefits that exceed the costs of managing resources (Bromley 1992; Ostrom 1990). This research has influenced how policy-makers think about and make policy regarding natural resource management (Agrawal 2001), particularly in the global South, where 'fortress conservation' and state-controlled management frequently have proved ineffective and/or inequitable.

Demands for 'local involvement' in the governance of resource commons have been strengthened by scholarship that has analyzed hundreds of case studies of successful common-pool resource governance (Agrawal and Benson 2010). Academics and practitioners use examples of successful self-governing common-pool resource institutions to suggest that if local people are given rights to own, use and manage natural resources, they are likely to conserve them (Agrawal 2001; Brosius et al 2005; Hulme and Murphree 2001; Mbaiwa 2011).

When CBNRM was introduced, practitioners assumed that 'community-based' natural resource management would entail a process of facilitating and building on local interests and management capabilities (Berkes 1989; Berkes 2004; Dressler et al 2010; Western et al 1994; Wittayapak and Dearden 1999). In theory, policies that devolved environmental governance to local communities would rebuild local natural resource management institutions and, in doing so, create more socially just relationships between communities, the state and access to natural resources. CBNRM advocates set out to rectify a situation in which centralized state control over resources resulted in communities' loss of ownership and thus of motivation to collectively manage resource sustainably.

The CBNRM movement's push to devolve natural resource management to local communities developed out of a rights-based approach towards democratic decentralization, but also intersected with the rise of a neoliberal faith in markets as a means to produce development globally, including in Botswana and sub-Saharan Africa generally (Duffy and Moore 2010). In the context of natural resource management, neoliberal ideology supposes market principles to be the best guide for the efficient management of resources, and a decentralized network of stakeholders to be able to govern resources better than the state (Hulme and Murphree 1999; Ribot 2004). Implicit in the southern African CBNRM model is the neoliberal premise that if institutions can be established that 'get the prices right' for natural resources, then landowners, including communities, that are given property rights over this resource will use it efficiently.

CBNRM's neoliberal roots in southern Africa lie in 1970s legislation that conferred strong proprietary rights over wildlife on owners of private land in the commercial sector. This legislation is credited with promoting successful wildlife conservation on private lands (Child 2009c), as ranchers began to sustainably utilize wildlife on their property for game ranching, hunting and tourism. After independence, leading wildlife officials argued for the extension of these policies beyond the commercial and largely white farming sector to black rural communities (Child 2009b). This effort led to legislative changes that paved the way for CBNRM programmes to emerge.

Specifically, the search for an economic unit equivalent to the private landholder led to the concept of a collective property regime. Informed by common property theory, 'private community property' was understood to mean a common property resource collectively managed and exploited by a community group within an (ideally) self-defined jurisdiction. CBNRM proponents thus sought, and still seek, to replicate conservation-development success in the private sector on communal lands, through the devolution of rights to localized units and through the establishment of localized management regimes adapted to operating under modern market conditions (Jones and Murphree 2004).

While conservation initiatives in southern Africa were designed to be about wise land use overall rather than wildlife conservation specifically (Child 2009b), CBNRM programmes to date have focused on the wildlife sector in its various forms – hunting, game ranching and photographic tourism. CBNRM practitioners have worked to enact policies that enable the wildlife and tourism industry to be an economically competitive form of land use through the establishment of the infrastructure and competitive pricing required for the wildlife and tourism industry (Jones and Murphree 2004). CBNRM is based on the idea that wildlife must 'pay its way' (Eltringham 1994) if it is to survive and presupposes the existence of an enterprising community ready to capitalize upon its newfound ownership of wildlife.

Within this framework, local communities living near wildlife are entitled to benefits from wildlife which are contingent upon their ability to successfully develop market-oriented strategies to profit from this resource. CBNRM, like broader neoliberal policies, thus places faith in market-based principles to achieve devolved community-based conservation (Büscher and Dressler 2010).

Revisiting community-based natural resource management: Lessons learned from common-pool resource institutional analysis

As commons scholar Elinor Ostrom has noted, the probability that users will engage in collective action is high only when the expected benefits of managing a resource exceed the perceived costs of investing in better rules and norms (Ostrom 2009). Similarly, Robert Wade (1988) has shown that whether or not villages organize collectively to manage their common-pool resources is not just about the social structure of the community, but also particularly dependent on ecology – specifically scarcity and risk. When the individual benefits of organizing to collectively manage a resource are big, and when there are high risks to individuals if they fail to come together to manage that resource, stakeholders are more likely to organize. In other words, they will organize when intensely felt needs cannot be met by individual responses (Wade 1988). Therefore collective action around CBNRM is unlikely in circumstances where households have alternative individualized means to meet their immediate needs.

These theories are useful because they help explain why common-pool resource regimes have failed to emerge endogenously in rural parts of Botswana such as the Chobe Enclave. In Chobe, traditional uses of wildlife have been outlawed for over a hundred years, so that people no longer consider wildlife to be an important means of survival. Moreover, wildlife-based livelihood benefits from CBNRM are relatively inaccessible compared with the availability of livelihood support from family members and the government.

Together, these conditions mean that people do not see new forms of collective wildlife management as necessary or beneficial. The availability of alternative institutions that provide financial support to livelihoods means that in Chobe, 'intensely felt needs' regarding livelihoods and household survival *can* be met by individual responses. That is, villagers can access channels of financial and physical capital either through government support programmes or through remittances from family members, many of whom work in nearby safari camps and lodges. As Ostrom (2009) argues, in successful cases of self-organization, users are either dependent on the resource system for a substantial portion of their livelihoods or attach high value to the sustainability of the resource. In Chobe and much of the rest of Botswana, other sources of livelihood support remain easier and less costly for most households to access than market-based capital from wildlife tourism. Thus people are less likely to organize, because the costs of organizing and maintaining a self-governing system are not perceived to be worth the effort (Ostrom 2009).

The decline of wildlife-based livelihoods in Chobe

State conservation laws initiated when Botswana became a British protectorate in 1885 restricted wildlife hunting and use by the local human populations (Spinage 1991). In northern Botswana specifically, the creation of an extensive network of protected areas that culminated in the establishment of Chobe National Park shortly after independence in 1968 had major consequences for the lives and livelihoods of the human communities living in this region.[7] Apart from direct and indirect threats to people and livestock from wild predators that were now protected under conservation law, many farmers had to abandon grazing areas and cattle posts located within the boundaries of the game park and forest reserve. Hunting wild animals, fishing and collecting natural resources such as wood were prohibited within the nearby forest reserves. Outside of the reserves, villagers were required to buy a permit from the district government to conduct these activities (Gumbo 2002).

In Chobe, as in much of Botswana, households no longer rely heavily on wildlife and natural resources for their survival. In 2001, natural resources were estimated to contribute only 5 per cent of household total income (including the value to households of their own production) from all livelihood activities in the Ngamiland and Chobe districts of northern Botswana (Botswana Institute for Development Policy Analysis [BIDPA] 2001). As one villager in Chobe explained, 'The natural resources which are supposed to be important to us are wildlife, but government took wildlife to make money. Now it's not important for us, it's only for tourists and government' (21 October 2009). The Chobe case is characteristic of Botswana and other parts of southern Africa, where past state conservation policies and colonial histories largely eliminated human dependence on wildlife and community wildlife management.

The legacy of the state taking control over this 'national' resource is that today in Chobe, most residents do not have a sense of ownership over wildlife. Norms regarding reciprocity between villagers and collective responsibility for wildlife management have been replaced by an overall sentiment that wildlife is the property of the state and that the state, not the villagers, should be responsible for wildlife management. Interviews reveal that many younger residents do not remember a time when communities had collective responsibility for land and resource management, given that the colonial state took resource use and management rights away from communities a century ago. In the pre-colonial era, villages organized regiments of men to control flooding to prevent crop destruction, to kill lions harassing villagers or to scare elephants away from the fields. Elders in the village reported to me that such groups had gradually dissolved under colonial rule and no longer existed. When asked if village groups existed today to scare off crop-raiding wildlife, one woman from Kachikau explained:

> [T]here is no one. Such things are no longer there [for] my children . . . there are no people [in the village] who would do that. People have left. Even the chiefs have stopped practicing laws from the past. . . . If they [villagers] would

> kill the animals, [they would] go to jail . . . it comes down to money. And Wildlife Department is the one that takes care of animals.
>
> (Personal communication 2010)

In Chobe today, not only have wildlife-based livelihoods been curtailed, but wildlife now poses a threat to agricultural livelihoods. Chobe's abundant wildlife resources and the extensive Chobe-Zambezi river system make it a primary safari destination in southern Africa. This means that today, Chobe Enclave residents face the costs of living near a protected area and in a region zoned for wildlife management and oriented towards tourism.

Villagers complain that the presence of these reserves limits where they can graze their cattle and that their livestock continue to be vulnerable to predation by wild animals. Furthermore, the village's close proximity to wildlife prevents livestock owners from certifying their cattle against foot-and-mouth disease and has thereby halted the sale of their livestock to the subsidized European Union market (a lucrative market for other livestock producers in Botswana). Crop destruction by wild animals is a growing problem. The elephant population in the country has increased sixteenfold over the past 50 years to an estimated 133,829, with 70–80 per cent of the elephants living outside the national parks (World Bank 2009). Correspondingly, the number of elephant conflict reports for the Chobe Enclave has increased steadily from nine reports in 1994 to 144 in 2004 (unpublished data from Dr Kathy Alexander of the Center for African Resource: Animals, Communities and Land Use).

For the past 20 years, the Chobe CBNRM programme has sought to recreate wildlife-based livelihoods and re-establish the collective will for sustainable wildlife management, but today most village residents still do not view wildlife as accessible or important to their livelihoods. Although people recognize that the CECT provides the village with a few community benefits, such as tractors for ploughing and financial support for various civic groups in the village, most villagers reported in interviews that they did not see the CECT as a source of individual income or livelihood support.

In focus group discussions and informal interviews about the national park in my two village field sites (N =100), just 11 community members mentioned the benefits brought about by wildlife-based tourism in the region, and only one directly mentioned the CECT as a stimulus for village development. In my 2010 survey of two Chobe Enclave villages, neither wildlife nor CECT/CBNRM were cited as one of the top three sources of income for a household. Village residents accept state-centralized control over natural resources, including land and wildlife. As a Kachikau man who had moved to the nearby town of Kazangula told me, 'This entire land belongs to the government. This government controls everything. They tell you where to stay and where to plough . . . the land is under government. It is not controlled by us' (7 May 2010).

The failure of the CBNRM project in Chobe to galvanize residents to develop self-organized institutions for the collective governance of wildlife is typical of CBNRM programmes in Botswana generally. In a recent assessment of CBNRM in Botswana, Rozemeijer reported that for all CBNRM projects in the country,

Figure 6.2 Author soliciting community feedback on the accuracy of previous studies of livelihood sources in the Chobe Enclave (photo: Prince Mbeha).

> it is very unlikely that in these projects benefits to the average community member will exceed the costs to the same individual. It is therefore equally unlikely that this will prompt the conservation of natural resources, especially by those who have reduced access to them for subsistence purposes.
>
> (Rozemeijer 2009, p. 253)

In other words, residents have little incentive to self-organize to manage and utilize wildlife sustainably.[8] Not surprisingly, then, there is little evidence that land use patterns in resource-rich areas have changed to the benefit of wildlife (e.g. through the adoption of less damaging arable agriculture and livestock grazing models) (Rozemeijer 2009).

Understanding livelihoods: The role of state transfers and remittances in household decision-making

The Chobe Enclave is a place where wildlife is now more of a hindrance than an asset to livelihoods. Households in Chobe are poor, but have some mechanisms for buffering the costs of living near wildlife. Botswana's political economy is characterized by strong rural-urban linkages and robust social services provisioning.

This section highlights the broader political economy within which the enclave is situated in order to explain in part why the Chobe Enclave CBNRM project has not been more successful.

In Chobe, household decision-making takes place in the context of a relatively well-functioning welfare state and in a rural locale tightly linked to the resources of Botswana's urban sphere. Rural household dependence on remittances and government support in Chobe is not unique to the region, but reflects rural livelihood strategies in Botswana generally. Chobe Enclave residents are able to draw upon these entitlement programmes and kinship networks in order to make ends meet. The relationships between the state and its citizens, and between urban and rural kin, mean that Chobe residents are able to sustain their livelihoods in ways that do not necessarily demand the time, effort and resources that would be required to make a common-pool resource regime function effectively.

It is difficult to determine whether the dependence of Chobe Enclave households on the government, and on family members working outside the village, simply reflects national-level relations between the state and its rural citizens and between rural and urban kin, or whether it has increased beyond the national norm because the community lives near land designated for wildlife conservation. However, the larger point is that the choices of Chobe Enclave residents regarding participation in CBNRM projects are very much mediated by the availability of and access to resources from the state and remittances from family. Below, I elucidate how the political economy of Botswana makes it easier for Chobe residents, and thus gives them more incentive, to access capital from the state and urban kin than to do so from participation in community wildlife management institutions established under CBNRM.

Social safety nets in Botswana

Botswana is often described as a developmental state that provides for its citizens in a relatively efficient manner (Acemoglu et al 2001). Botswana's lucrative diamond mining industry, good governance and relatively small population mean that it can provide a level of social services and financial support to its citizens that few other African countries can deliver. Since the discovery and exploitation of diamonds shortly after independence, and the subsequent surge in the economy (Colclough and McCarthy 1980), Botswana has strategically used its economic growth and access to donor funds to achieve impressive levels of social service delivery. Since 1966 it has invested in the health and education of its citizens and guaranteed them access to a number of institutionalized social safety nets.

Today, Botswana's National Strategy for Poverty Reduction, adopted in 2003, comprises three clusters of targeted interventions: income generation strategies such as microcredit and agricultural support programmes (e.g. the Integrated Support Programme for Arable Agricultural Development, ISPAAD); safety nets such as drought-relief food aid and labour-based public works (the Ipelegeng programme) and entitlements such as old-age pensions and destitute allowances (BIDPA 2001). Other safety nets include school feeding, orphan rations and community home-based care for AIDS patients, all of which has become increasingly necessary since the HIV/AIDS crisis emerged in the late 1980s.

These social safety net programmes are particularly prevalent in Chobe: 71.1 per cent of surveyed households in northern Botswana (Ngamiland and Chobe districts) said that government support programmes provided them with direct access to food, in comparison with the national average of 19.6 per cent (BIDPA 2001). In my survey of the Chobe Enclave, approximately 17.8 per cent of households stated that their main income source fell into the category of government assistance. A 2001 survey conducted by the Botswana Institute for Development and Policy Analysis revealed that government transfers contributed 55 per cent of the total income from all livelihood activities in northern Botswana (BIDPA 2001).

Chobe residents access government transfers through a number of different support programmes. For example, farming is heavily subsidized for Chobe Enclave households that decide to farm. Through the agricultural support programme Integrated Support Programme for Arable Agricultural Development [ISPAAD], the government spends 700 pula (equal to roughly US$100 at the time of writing) per hectare during the ploughing season to prepare the soil, plough the fields and plant seeds for village farmers in the Chobe Enclave. Agricultural extension agents working in the Chobe Enclave report that all farmers in the enclave who have cultivated their fields in the past few years have received support from ISPAAD. Also, the Botswana government's labour-based public works programme, Ipelegeng, provides a source of (at least temporary) employment and income to many Chobe Enclave residents. Recent records from the Chobe District Council show that in the year 2008/09, there were 572 beneficiaries of Ipelegeng in the two surveyed villages, or more than one person per household working for Ipelegeng. In interviews and focus groups, villagers emphasized the importance of Ipelegeng in providing temporary work, especially to young people facing a lack of opportunities in farming and formal employment.

Rural-urban linkages

Livelihood strategies in Chobe are influenced by the flow of resources, not only between the state and rural households, but also between urban and rural kin. Botswana has a long history of human mobility and rural-urban social and economic linkages dating back to the late nineteenth century (Kerven 1980). At the same time that colonial authorities were curtailing villagers' use of wildlife, urban migration was increasing, as men migrated to South Africa to work in the mines in order to pay the imposed colonial 'hut tax' (Schapera 1948). Ultimately migration and movement between the rural and urban or semi-urban areas determined the structure of family social relations and household livelihood portfolios in Botswana and southern Africa. Since the colonial era in the early twentieth century century, residents of Chobe Enclave, like most rural dwellers in Botswana, have relied on remittances from wage-earning family members in urban areas for their survival (Parson 1984).

The majority of surveyed households (54.4 per cent) in the Chobe Enclave in 2010 received remittances from family members, and these remittances made up, on average, 20 per cent of household income. Eighty-three per cent of households reported that they had at least one family member living and working outside the village. The actual percentage of households receiving remittances is likely to be higher than the recorded level, given the tendency for the under-reporting of income in rural household surveys (Devereux and Hoddinott 1993). Past studies show that most migrants in southern Africa are expected to send remittances back to their family in their home villages (Kerven 1980; Lucas 1978; Pendleton et al 2006). Two-thirds of migrants interviewed (including students, who are not expected to remit) stated that they regularly sent home remittances to their village families (66.7 per cent; N = 147).

The significant flow of funds from the state and from urban kin to the Chobe Enclave stands in stark contrast to the minimal contribution of CBNRM-related benefits to household survival. Chobe households are accustomed to accessing resources through subsidies and safety net programmes provided by the state, and through remittances from family members. In comparison with CECT benefits, the delivery of social services to villagers in the enclave is regular, reliable and transparent (Seleka et al 2007). Botswana is efficient in its social service provisioning (Acemoglu et al 2002), and qualifying individuals and households can register relatively easily to receive entitlements and various forms of assistance. Norms regarding kinship obligations mean that family members who move away to work are expected to send money and/or goods home regularly. While social safety nets and remittances may not necessarily provide large sums of money, they both represent established institutions that villagers know how to access.

As noted above, community members must see great benefits in collective organization and great risks in failing to organize for such a community to come together to manage a resource (Wade 1988). In Chobe, it is not imperative for communities to manage wildlife enterprises collectively in order for households to survive. Even without collective action, they can access some indirect livelihood benefits from wildlife, including remittances from family members working in the safari industry. Some villagers even view government social safety nets as an indirect compensation for living near wildlife. Citizens thus have alternative sources of capital with higher (or at least less risky) perceived individual benefits (e.g. cash or goods flow to households) and relatively low transaction costs or barriers to entry compared with CBNRM engagement. This scenario results in individualized household livelihood strategies that make it especially difficult to garner widespread community interest and involvement in CBNRM.

Evaluating access to capital from wildlife under community-based natural resource management

Household decision-making is structured by the perceived availability of alternative sources of livelihood support *relative* to that provided through CBNRM. Collective resource management regimes have not emerged in Chobe and the rest of

Botswana, not only because households are no longer dependent on wildlife and have alternative sources of income, but also because households do not see the benefits of new forms of collective resource management.

CBNRM's failure to re-establish collective wildlife management and livelihood benefits from wildlife is generally agreed to be due to several key deficiencies: the lack of complete devolution of management authority, the lack of management expertise, and the lack of accountability and democratic participation (Agrawal and Gibson 2001; Blaikie 2006; Brosius et al 2005; Songorwa et al 2000; Twyman 1998). I will now review these key challenges in the context of Chobe through the lens of an access analysis (Ribot and Peluso 2003).

Access analysis is a method of studying people and resources that differentiates access from property. 'Access' is defined as the '*ability* to benefit from things' (i.e. resources), which broadens the definition of 'property' as 'the *right* to benefit from things' (Ribot and Peluso 2003). This theory of access gains traction when applied to settings where people may have rights to resources, but are part of broader social relationships that differentially constrain or enable people setting out to benefit from those resources. In these situations, an analysis of the set of factors that constitute and configure access provides a deeper understanding of social and environmental outcomes than a focus on property relations alone. An access analysis of the CECT is useful here because it highlights a shortcoming of the CBNRM model: that the devolvement of *rights* to a resource, especially highly circumscribed rights, by no means guarantees *access* to that resource. Access to livelihood benefits from wildlife is constrained under CBNRM in three primary ways. First, community access to wildlife is circumscribed because community rights to wildlife remain incomplete, despite policy shifts towards devolution. Second, mismanagement has deprived the majority of villagers of access to the financial benefits from wildlife that accrue to community trusts. Third, most villagers do not have access to the social and business networks that facilitate successful entry into wildlife-oriented enterprises (e.g. tourism, game ranching and commercial hunting).

Barriers to access I: Circumscribed rights to wildlife

As studies of common-pool resource regimes have shown, when users have full autonomy at the collective-choice level to craft and enforce their own rules, they are more likely to self-organize to manage the resource in question (Ostrom 2009). Successful commons management depends upon the rights of resource users to devise their own institutions without being challenged by government authorities and obstructed by minimal recognition of their rights to organize (Ostrom 1990).

While CBNRM policies in theory decentralize resource governance to local communities, village residents' management rights in Botswana are, in fact, still highly circumscribed. Villagers remain subject to strictly enforced laws prohibiting the hunting of globally and nationally protected wildlife species (including problem animals such as crop-raiding elephants) and the gathering of various forest products.

Communities in a controlled hunting area of Botswana like the Chobe Enclave may be granted resource leases over wildlife and tourism on their land for a period of up to 15 years by a tribal land board if they have formed a community trust. Holders of these leases can be awarded a wildlife quota for hunting purposes by the Department of Wildlife and National Parks (DWNP). Legally, they can decide how to use this quota: whether to put the quota and the tourism concession lease out to tender in the private sector or to manage tourism operations themselves. However, the lack of local capacity to run the high-end hunting and tourism safaris that characterize the Botswana tourism industry means that there is essentially no option besides the former if the trust wants to maximize its income from wildlife and tourism. Also, the trust has no control over the quota-setting process itself. Instead, the centralized DWNP decides, based on its animal census data.

Chobe villagers cannot come together to decide, for example, to cull the local elephant population. Restrictions upon their ownership rights mean that the rights villagers hold to 'manage' wildlife are essentially limited to creating and managing revenue from the wildlife safari industry and related tourism enterprises. As a result, it has been difficult for CBNRM projects to engender motivation within the villages to self-organize to manage their community trust when the authority over wildlife use still remains primarily in the hands of centralized state institutions.

Barriers to access II: Lack of financial management expertise and leadership

Not only are rights to wildlife broadly circumscribed, but the few rights that are devolved to the community under CBNRM do not translate into *access* to the financial benefits from permitted wildlife use (i.e. tourist photographic and hunting ventures) for the average villager. Under CBNRM, the community trust is expected to manage the revenue generated from the rent paid by the wildlife safari industry as a cooperative business might. However, most village residents, including traditional village leaders, do not have the financial experience or skills needed to manage and invest large sums of money. To say that these tasks require a capacity to exercise management authority not present in the village is not to say that village communities cannot organize effectively: the thriving burial societies and lending circles are sufficient proof that they can. The real point is that the terms of rights granted under CBNRM came from above in the form of a complicated recipe (Rozemeijer 2009) not readily compatible with traditional or locally derived institutions already in place.

The resulting lack of sound financial management means that funds from the trust are frequently misused or inefficiently allocated. As Ostrom (2009) has explained, self-organization is more likely when some of the users of a resource system have entrepreneurial skills and have gained respected as local leaders through prior organizational activities. In Botswana, the village chief and headmen are still considered the 'leaders' of the village, but in actual fact have little power over resource management and generally no experience managing revenue from wildlife tourism. Those village residents who do possess these skills – those with higher education – generally leave the village for better opportunities to use those skills elsewhere. There is a dearth of individuals in the village with the expertise to help establish or lead a successful CBNRM trust, and those who are in charge often waste or misuse revenue.

Like a number of CBNRM trusts in Botswana, the CECT is prone to elite capture by the elected management committees (Suich et al 2009). Involvement in CBNRM is complicated, and the CECT is made up of a board of trustees and representatives from each village that is not downwardly accountable to its constituents. Community trust meetings are transparent in theory (minutes are taken and made public), but most villagers have very little idea of what goes on during these meetings, especially since they are always held in the same village, making it difficult for most residents of the other four enclave villages to attend.

As a result, although the CECT can decide how to use its income from the rent paid by wildlife tourism partners, which can be quite substantial, benefit distribution within the five enclave villages has been minimal. Under the CECT constitution, 85 per cent of revenue is supposed to be divided evenly among the five village trust committees, while 15 per cent is supposed to remain with the CECT for operating costs. Each village trust committee decides how to spend its share of the revenues, theoretically focusing on investments that provide community services or employment. However, in the 2009/10 fiscal year, despite generating over half a million US dollars in revenue, the CECT had a deficit of almost US $40,000, due to operating and administrative costs, many of which the trust could not account for (CECT 2010). These figures represent extreme financial inefficiency, and demonstrate that devolved (and admittedly limited) management rights do not guarantee access to benefits from wildlife tourism for the average villager, because skilled and accountable leadership within the trust is lacking.

Barriers to access III: Lack of market-based networks and knowledge

Not only is the CECT's revenue from leasing its land and wildlife quotas difficult for most Chobe villagers to access, but CBNRM also offers few viable opportunities for Chobe households to develop wildlife-based livelihoods. The CBNRM model emphasizes the importance of devolved ownership over resources as a precondition for the local entrepreneurial development of products and markets for wild resources (International Resource Group [IRG] 2009). But property rights over wildlife have little practical value if people do not have access to the skills or knowledge to develop that resource in the marketable, revenue-generating way that is permitted to and expected of them.

Botswana maintains a high-end, low-volume tourism policy that promotes the dominance of the tourism industry by foreign companies with tourism expertise (Rozemeijer 2009). These companies typically have connections to international markets and start-up capital that local villagers do not possess. This makes it difficult for villagers to start small-scale tourism businesses, as there is virtually no market for budget travel in Botswana. There are also few economic linkages between tourism operations and local businesses of any kind, because of the high quality of goods demanded by the tourist clientele. Handicrafts are often imported from wholesalers in neighbouring countries such as Zimbabwe, and lodge employees are brought in from other districts of Botswana where training is superior (Chobe Game Lodge manager, personal communication, 23 April 2010). There is a widespread sentiment among villagers that they are deliberately excluded from participating in the tourism industry and that hiring practices are characterized by nepotism. The District Development Plan for Chobe states that the tourism sector is still monopolized by foreign-owned safari tourism operators who are believed to be racist and intent on keeping the indigenous population away from their operations (Chobe District Development Plan 2003). The tourism companies that operate on land leased to them through the CECT have not facilitated the development of spin-off small enterprises supporting their larger operations, as CBNRM advocates predicted would happen. Wildlife tourism therefore remains primarily the domain of outsiders with business expertise.

The Chobe Enclave community thus lacks not only the leadership to manage trust funds effectively, but also the knowledge to 'use' wildlife successfully through the type of resource use that CBNRM facilitates: wildlife-based tourism. CBNRM fosters the creation of wildlife-derived benefits contingent upon greater market integration and intensification, an approach that simply makes poverty more complicated (Büscher and Dressler 2010). As a result of these barriers to access, the perception of benefits from CBNRM remains low among community members. Unlike most government programmes or remittance networks, CBNRM is a complicated, risky and newly formed institution that to date has provided few household-level pay-offs.

Conclusion

Findings from Chobe, Botswana, challenge the conceptual underpinnings of CBNRM: that local people must find a way to benefit directly from a resource like wildlife if it is to be conserved. As Adams and Hulme (2001) have argued, there may be a number of circumstances in which CBNRM is in fact *not* the answer for conservation policy. The fact that CBNRM programmes have so rarely succeeded in creating collective units of villagers that behave in similar ways to the private landowners who profit from wildlife in South Africa (and previously Zimbabwe) suggests that CBNRM's theoretical mix of economic instrumentalism and common property theory may contain problematic assumptions about collective engagement with markets. CBNRM assumes that the instrument of the market can and should be introduced to produce value from common property natural resources for local people (Büscher and Dressler 2010). However, most communities in CBNRM project areas do not have the knowledge, power or social connections – what Ribot and Peluso (2003) call 'structural or relational mechanisms of access' – that would give them the ability to benefit from (i.e. to access) a natural resource such as wildlife. In Botswana, the community trust is an institution in which village-level control over natural resources is still highly circumscribed, and in which the benefit stream that might come from collective organization is not guaranteed or obvious.

Instead, the case of Chobe shows that the state can provide a buffer against livelihood costs incurred by the presence of a protected area and its wildlife. It may be true that these state transfers merely alleviate poverty rather than develop productive livelihoods, but the fact that in Chobe state transfers contribute more to an average household's livelihood portfolio than does CBNRM calls into question, at the very least, the southern African CBNRM paradigm. It suggests that there may be other models for sustainable conservation, especially in places with strong central state provisioning and a mobile population that maintains social and economic links with its rural kin. While Botswana may be unique in its extensive provisioning of social services to its rural population, it still exemplifies how state-led rather than market-led mechanisms can play a role in the amelioration of conflicting wildlife and human interests.

Furthermore, remittances from family working in the safari tourism industry also provide the enclave with an indirect source of income from wildlife that is not contingent upon community-level organization and collective markets. To make this point is not to argue that welfare programmes and remittances represent alternative pathways to sustainable conservation; what it does is identify the conditions under which CBNRM does not function effectively and emphasize that alternatives suited to local political-economic and social conditions must be considered.

State transfers and remittances are not, by any means, ideal vehicles for achieving a just form of conservation. While critical to household basic survival, they do not spur the creation of institutions for devolved environmental governance. However, governmental social support programmes do compensate to a degree for living near wildlife, and remittances ostensibly represent an indirect benefit from wildlife (since a number of remitting migrants work in the tourism industry in the nearby town) – but these economic relationships do not give local villagers a sense of ownership or authority over wildlife. Within such relationships, residents of Chobe remain passive recipients of indirect wildlife benefits, and have little control over the nature or longevity of the benefit stream or over the lucrative resource (wildlife) itself.

Villagers may receive a relatively large amount of state support, but this support is often viewed locally as a way to placate villagers living near conservation zones, as the conservation of wildlife is given priority over other competing demands for land use in Chobe. While important, state transfers and remittances are undergirded by institutions operating within a centralized form of governance in which rural dwellers have little control over the economic health of the state, the tourism industry or the management of the natural resource that sustains it.

A socially just form of conservation still remains better formulated in theory than in practice. In theory, CBNRM provides a pathway for achieving environmental conservation in a socially equitable way, by designating local people living near wildlife to be the managers and beneficiaries of the resource and its market value. In reality, these empowered roles for local communities have often failed to materialize. Ultimately, to push back against the CBNRM model is to question the type of relationship that communities should have with their environment and to interrogate the normative view espoused by CBNRM proponents: namely, that for wildlife conservation to be effective and sustainable, local people living near wildlife must engage in market relations that make wildlife an economic asset. The case of Chobe reveals to us that in some places, it is the overlooked relationships – those between state and society or between urban and rural kin, for example – rather than market-based relations between communities and wildlife, that provide the safety net to bolster the livelihoods of those who suffer the costs of environmental and, in particular, wildlife conservation.

Notes

1 In this chapter, a livelihood is understood to comprise 'the assets (natural, physical, financial, human and social capital), the activities, and the access to these (mediated by institutions and social relations) that together determine the living gained by the individual or household' (Ellis 2000).
2 Communal land in Botswana is referred to as tribal land and constitutes about 71 per cent of the country's total land area. Those who have been allocated land (through one of the land boards) do not own it but have usufruct rights (Jones 1999).
3 The data presented in this chapter were collected during nine months of 2009 and 2010 in two villages in the Chobe Enclave, where I conducted participant observation, focus-group workshops, semi-structured interviews with both village residents and local government officials, and a household survey. The survey was administered to a random 30 per cent sample of households (sample size: 90 households) selected from a list of village plot holders (all residential plots have to be registered, so the list is relatively comprehensive) from the district land board office using a random number generator. The survey was administered to the head of each household (including female-headed households) and the spouse of the head of the household if there was one (for male-headed households). The survey data used in this article were collected only from the head of the household, as prior comparison of spousal data indicated enough similarity to rely on the household head data as accurately portraying the household. Data were also collected from the Botswana National Archives.
4 Thirty-nine per cent of migrants originating from the two villages cited elephants as the cause of agricultural decline in their home villages (N = 147 unstructured interviews; no prompt regarding elephants or agriculture), and more than half of residents interviewed cited wildlife as either a disturbance to livelihoods (11 out of 40 respondents) or an inaccessible livelihood source, generally describing current restrictions against the hunting of wildlife (10 out of 40 respondents).
5 I am using Bertram's (2011) definition of 'welfare-state regime' as government policies and expenditures aimed at securing, to individuals or groups, measurable benefits which those individuals or groups could not secure directly through participation in the market economy, including payment in cash of income transfers and the provision in kind of key basic services, such as health, education and housing.
6 While wildlife-derived remittances may financially link wildlife to rural livelihoods, the employment is mainly with safari companies based in a nearby town or on land disconnected from the Chobe Enclave CBNRM programme. As such, they provide individualized indirect benefits from wildlife, and do not represent a livelihood derived from community-based wildlife management.
7 See Gumbo (2002) for extensive historical and contemporary documentation of the effects of conservation policy on local economic activities in the Chobe region.
8 The Chobe Enclave falls short not in its general capacity for community organization, but in its community organization around wildlife specifically. A number of community organizations (church groups, lending circles and burial societies) are active in the enclave, demonstrating that Chobe village residents clearly have the capacity to organize.

Bibliography

Acemoglu, D., Johnson, S. and Robinson, J. (2001) 'How Botswana did it: Comparative development in sub-Saharan Africa', unpublished paper

Acemoglu, D., Johnson, S. and Robinson, J. (2002) 'An African success story: Botswana', CEPR Discussion Paper 3219, Centre for Economic Policy Research

Adams, W. M. and Hulme, D. (2001) 'If community conservation is the answer in Africa, what is the question?', *Oryx*, vol 35, no 3, pp. 193–200

Agrawal, A. (2001) 'Common property institutions and sustainable governance of resources', *World Development*, vol 29, no 10, pp. 1,649–72

Agrawal, A. and Benson, C. S. (2010) 'Common property theory and resource governance institutions: Strengthening explanations of multiple outcomes', *Environmental Conservation*, vol 38, no 2, pp. 199–210

Agrawal, A. and Gibson, C. (2001) *Communities and the Environment: Ethnicity, Gender, and the State in Community-Based Conservation*, Rutgers University Press, New Brunswick, NJ

Alexander, E. M, Molokomme, I. M. and Murray, M. (1999) 'Chobe Enclave end of project evaluation', final report to Botswana Natural Resources Management Project, Gaborone

Berkes, F. (1989) *Common Property Resources: Ecological and Community-Based Sustainable Development*, Belhaven Press, London

Berkes, F. (2004) 'Rethinking community-based conservation', *Conservation Biology*, vol 18, no 3, 621–30

Bertram, G. (2011) *Assessing the Structure of Small Welfare States*, Social Policies in Small States Series No. 4, United Nations Research Institute for Social Development

BIDPA (2001) *Consultancy on the Review of the Rural Development Policy: Final Report*, Botswana Institute for Development Policy Analysis, Gaborone

Blaikie, P. (2006) 'Is small really beautiful? Community-based natural resource management in Botswana and Malawi', *World Development*, vol 34, no 11, pp. 1,942–57

Brockington, D. (2002) *Fortress Conservation: The Preservation of the Mkomazi Game Reserve, Tanzania*, University of Indiana Press, Bloomington, IN

Bromley, D. (1992) *Making the Commons Work*, Institute for Contemporary Studies, San Francisco, CA

Brosius, P., Tsing, A. and Zerner, C. (2005) *Communities and Conservation: Histories and Politics of Community-Based Natural Resource Management*, AltaMira Press, Walnut Creek, CA

Büscher, B. and Dressler, W. (2007) 'Linking neoprotectionism and environmental governance: On the rapidly increasing tensions between actors in the environment-development nexus', *Conservation and Society*, vol 5, no 4, pp. 586–611

Büscher, B. and Dressler, W. (2010) 'Commodity conservation: The restructuring of community conservation in South Africa and the Philippines', *Geoforum*, vol 43, no 3, pp. 367–76

CECT (2010) Annual General Meeting, Chobe Enclave Conservation Trust, Kavimba, Botswana, 27 March

Child, B. (ed.) (2004) *Parks in Transition: Biodiversity, Rural Development and the Bottom Line*, Earthscan, London

Child, B. (2009a) 'Community conservation in southern Africa: Rights-based natural resource management', in H. Suich, B. Child and A. Spenceley (eds) *Evolution and Innovation in Wildlife Conservation: Parks and Game Ranches to Transfrontier Conservation Areas*, Earthscan, London

Child, B. (2009b) 'Conservation in transition', in H. Suich, B. Child and A. Spenceley (eds)

Evolution and Innovation in Wildlife Conservation: Parks and Game Ranches to Transfrontier Conservation Areas, Earthscan, London

Child, B. (2009c) 'Game ranching in Zimbabwe', in H. Suich, B. Child and A. Spenceley (eds) *Evolution and Innovation in Wildlife Conservation: Parks and Game Ranches to Transfrontier Conservation Areas*, Earthscan, London

Chobe District Development Plan 2003–2009 (2003), Government Printer, Gaborone

Colclough, C. and McCarthy, S. (1980) *The Political Economy of Botswana: A Study of Growth and Distribution*, Oxford University Press, New York

Devereux, S. and Hoddinott, J. (1993) *Fieldwork in Developing Countries*, Lynne Rienner Publishing, Boulder, CO

Dressler, W., Büscher, B., Schoon, M., Brockington, D., Hayes, T., Kull, C., McCarthy, J., and Streshta, K. (2010) 'From hope to crisis and back? A critical history of the global CBNRM narrative', *Environmental Conservation*, vol 37, no 1, pp. 5–15

Duffy, R. and Moore, L. (2010) 'Neoliberalising nature? Elephant-back tourism in Thailand and Botswana', *Antipode*, vol 42, no 3, pp. 742–66

Ecosurv (1996) 'Chobe Enclave socio-economic study', report prepared for the Natural Resources Management Project and USAID, Gaborone

Ellis, F. (2000) *Rural Livelihoods and Diversity in Developing Countries*, Oxford University Press, Oxford

Eltringham, S. K. (1994) 'Can wildlife pay its way?', *Oryx*, 28, 163–8

Gumbo, G. B. (2002) 'The political economy of development in the Chobe: Peasants, fishermen and tourists, 1960–1995', MA thesis, University of Botswana, Gaborone

Hulme, D. and Murphree, M. (1999) 'Communities, wildlife and the "new conservation" in Africa,' *Journal of International Development*, vol 11, no 2, pp. 277–85

Hulme, D. and Murphree, M. (2001) *African Wildlife and Livelihoods: The Promise and Performance of Community Conservation*, James Currey, Oxford

Igoe, J. and Brockington, D. (2007) 'Neoliberal conservation: A brief introduction', *Conservation and Society*, vol 5, no 4, pp. 432–49

Igoe, J., Neves, K., and Brockington, D. (2010) 'A spectacular ecotour around the historic bloc: Theorizing the current convergence of biodiversity conservation and capitalist expansion', *Antipode*, vol 42, no 3, pp. 486–512

IRG (2009) 'Part II, chapter 2: Community-based natural resource management', in *Environmental Guidelines for Small-Scale Activities in Africa*, SD Publication Series, Bureau for Africa, Office of Sustainable Development, USAID

Jones, B. (1999) 'Community-based natural resource management in Botswana and Namibia: An inventory and preliminary analysis of progress', Evaluating Eden Series Discussion Paper No. 6, International Institute for Environment and Development

Jones, B. (2002) 'The Chobe Enclave, Botswana: Lessons learnt from a CBNRM project 1993–2002', IUCN World Conservation Union

Jones, B. and Murphree, M. (2004) 'Community-based natural resource management as a conservation mechanism: Lessons and directions', in B. Child (ed.) *Parks in Transition: Biodiversity, Rural Development and the Bottom Line*, Earthscan, London

Kerven, C. (1980) 'Rural-urban interdependence and agricultural development', in R. Hitchcock and M. Smith (eds) *Settlement in Botswana: The Historical Development of a Human Landscape*, Botswana Society, Gaborone

Lucas, R. (1978) 'Outmigration, remittances and investment in rural areas', in *National Migration Study*, vol 13, Central Statistics Office, Gaborone

Mbaiwa, Joseph. (2011) 'From collaboration to conservation: Insights from the Okavango Delta, Botswana', *Society and Natural Resources*, vol 24, no 4, pp. 400–11

McCarthy, J. (2005) 'Devolution in the woods: Community forestry as hybrid neoliberalism', *Environment and Planning*, vol 37, no 6, pp. 995–1,014

Ostrom, E. (1990) *Governing the Commons: The Evolution of Institutions for Collective Action*, Cambridge University Press, Cambridge

Ostrom, E. (2009) 'A general framework for analyzing sustainability of social-ecological systems', *Science*, vol 325, no 5,939, pp. 419–22

Parson, J. (1984) *Botswana: Liberal Democracy and the Labor Reserve in Southern Africa*, Westview Press, Boulder, CO

Pendleton, W., Crush, J., Campbell, E., Green, T., Simelane, H., Tevera, D. and de Fletter, F. (2006) *Migration, Remittances and Development in Southern Africa*, Migration Policy Series No. 44, Southern African Migration Project, Idasa, Cape Town

Ribot, J. C. (2004) *Waiting for Democracy: The Politics of Choice in Natural Resource Decentralizations*, World Resources Institute, Washington, DC

Ribot, J. and Peluso, N. (2003) 'A theory of access', *Rural Sociology*, vol 68, no 2, pp. 153–81

Rozemeijer, N. (2003) 'CBNRM in Botswana: Revisiting the assumptions after 10 years of implementation', IUCN/SNV CBNRM Support Programme in Botswana

Rozemeijer, N. (2009) 'CBNRM in Botswana', in H. Suich, B. Child and A. Spenceley (eds) *Evolution and Innovation in Wildlife Conservation: Parks and Game Ranches to Transfrontier Conservation Areas*, Earthscan, London

Schapera, I. (1948) *Migrant Labor and Tribal Life*, AMS Press, New York

Seleka, T., Siphambe, H., Ntseane, D., Mbere, N., Kerapeletswe, C., and Sharp, C. (2007) *Social Safety Nets in Botswana: Administration, Targeting and Sustainability*, Botswana Institute for Development Policy Analysis, Gaborone

Songorwa, A., Buhrs, T. and Hughey, K. (2000) 'Community-based wildlife management in Africa: A critical assessment of the literature', *Natural Resources Journal*, vol 40, no 3, pp. 603–43

Spinage, C. (1991) *History and Evolution of the Conservation Laws of Botswana*, Botswana Society, Gaborone

Suich, H., B. Child and A. Spenceley (eds). (2009) *Evolution and Innovation in Wildlife Conservation: Parks and Game Ranches to Transfrontier Conservation Areas*, Earthscan, London

Twyman, C. (1998) 'Rethinking community resource management: Managing resources or managing people in western Botswana?', *Third World Quarterly*, vol 19, no 4, pp. 745–70

Wade, R. (1988) *Village Republics: Economic Conditions for Collective Action in South India*, Cambridge University Press, Cambridge

Western, D., Wright, R. M. and Strum, S. C. (1994) *Natural Connections: Perspectives in Community-Based Conservation*, Island Press, Washington, DC

Wittayapak, C. and Dearden, P. (1999) 'Decision-making arrangements in community-based watershed management in northern Thailand', *Society & Natural Resources: An International Journal*, vol 12, no 7, pp. 673–91

World Bank (2009) 'Project appraisal document on a proposed grant to the Republic of Botswana for a northern Botswana human wildlife coexistence project', Environment and Natural Resources Management, Sustainable Development Department, Africa Region

7 Community-based conservation and protected areas

Commons perspectives for promoting biodiversity and social justice in southern Africa

Arthur Hoole

Introduction

Global biodiversity is seriously threatened by widespread habitat loss, overexploitation, invasive species, pollution and climate change. The loss of biodiversity is especially acute in the equatorial belt, where the greatest part of the world's biodiversity is concentrated (Western and Pearl 1989). Protected areas, notably national parks with strict protection regimes, are widely considered a principal means for conserving global biodiversity. However, there is growing recognition that protected areas cannot effectively achieve the conservation of biodiversity needed at the wider landscape levels beyond them. Also, new paradigms in protected areas governance are emerging that embrace a 'humans in nature' view, recognizing the critical roles played by local and indigenous communities in biodiversity conservation (Phillips 2003; Berkes 2007; Dudley 2008; Borrini-Feyerabend et al 2010). There is also an acknowledged need to redress injustices suffered by indigenous people in natural resource access and use (Brechin et al 2003).

Many protected areas have been established to advance Western-conceived notions of conservation at the expense of social justice and livelihoods for local and indigenous communities (Timberlake 1991; Neumann 2002; Dowie 2009). This is true in southern Africa, where protected areas, especially national parks, contributed historically to a racial divide in rights to wildlife resources during European colonization and subsequent white minority rule.

Protected areas have been established and managed in Africa with little or no regard for local community resource access and use. In fact, local and indigenous communities have been disenfranchised and displaced from traditional areas of occupancy and resource use, with serious consequences for community livelihoods and sociocultural survival (Owen-Smith 1987; Timberlake 1991; Western 2002). A fortress approach to conservation in national parks has excluded local and indigenous use and management of water, wildlife, forests and grasslands (Adams and Hulme 2001). 'Fences and fines' measures have contributed to adversarial relationships between local indigenous communities, wildlife populations and protected areas, and such polarization can aggravate the loss of biodiversity (Western 2002).

Since achieving independence, many African states have extended rights in resources to indigenous Africans. The purpose of this chapter is to consider community-based natural resource management (CBNRM) for wildlife in this context, with particular attention to prospects for developing positive institutional relationships between CBNRM and the state's management of protected areas. It is posited that such prospective institutional linkages could more fully achieve biodiversity conservation in southern Africa while restoring or promoting social justice in terms of cultural renewal, governance and livelihoods.

The findings presented here are derived from doctoral research carried out in Namibia in 2006 and 2007 (Hoole 2008). A qualitative research approach was employed (Berg 2004), including detailed semi-structured interviews with a cross section of key informants drawn from non-governmental organizations (NGOs) involved in conservation, government ministries, private enterprises and the national university, supplemented with extensive archival research in Namibia. A community-level case study was conducted in the Ehi-rovipuka communal conservancy, paying particular attention to how the conservancy had been formed, its essential governance arrangements and the relationships between the Herero communities constituting the conservancy and the contiguous Etosha National Park (Figure 7.1). The Herero formerly lived in parts of Etosha and were ousted early in the twentieth century (Hoole and Berkes 2010). The Ehi-rovipuka Conservancy thus provides a specific opportunity to learn about how the designation and management of a national park were implicated in displacing local indigenous people from their traditional resource base.

The case illustrates how CBNRM and protected areas management might be more strongly linked to help restore the local social-ecological system, promoting greater biodiversity conservation and social justice. The term 'social-ecological system' refers to integrated people-environment systems, emphasizing the interdependence and co-evolutionary nature of these interactions (Berkes and Folke 1998). The concept of 'coupled systems', which is invoked later, refers to the feedbacks between the social and ecological subsystems. The terms 'decoupling' and 'recoupling', also introduced later, describe the loss of feedbacks between the subsystems on one hand and the restoration of these feedbacks on the other (Hoole and Berkes 2010).

The case study research involved participant observation, semi-structured interviews with key conservancy informants, structured interviews with 40 conservancy villagers and memory mapping exercises with community elders (Chambers 1997; Tobias 2000) to elicit indigenous occupancy and use relationships within Etosha. An Otjiherero-speaking translator was recruited to assist in these processes, and an open community meeting was conducted to verify preliminary findings in the field. Further methodology details are available in Hoole (2008).

This chapter first sets out the background to Namibia's protected area system and the evolution of its CBNRM programme and then presents the community case study findings. Potential linkages between CBNRM and protected areas management are then elucidated, followed by a summary and concluding remarks.

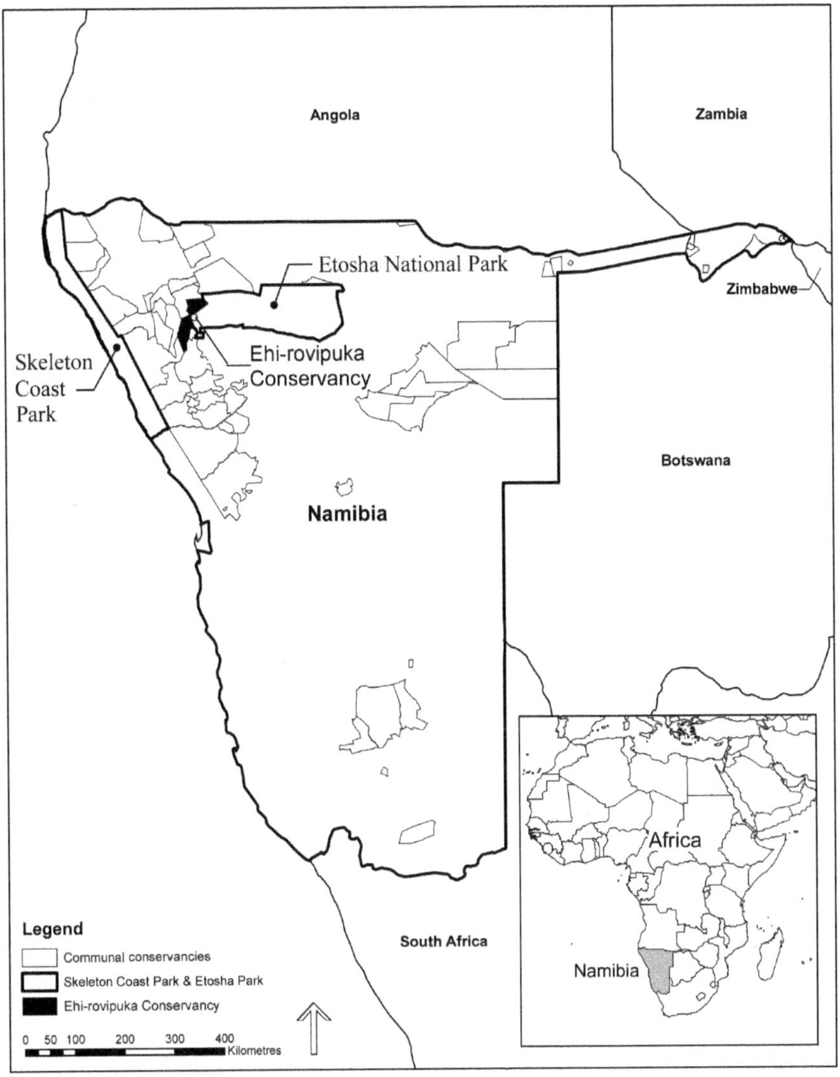

Figure 7.1 Case study area in Namibia.

Namibia's protected areas and Etosha National Park

Namibia's protected areas are established and managed under the enabling authority and provisions of the Nature Conservation Ordinance, No. 4 of 1975. This legislation was promulgated for the Territory of South West Africa, before Namibia's independence, by the South West Africa Administration. It repealed and replaced earlier legislation on protected areas and wildlife management introduced by the South African colonial administration as far back as 1927. The 1975 ordinance continues in force today, notwithstanding Namibia's independence in 1990. New protected areas legislation has been under active consideration for several years, but has yet to be enacted (Michael Sibalatani, SPAN Project Coordinator, personal communication, 29 March 2011). National parks therefore continue to be managed and administered under command-and-control legislation that has treated the indigenous peoples of Namibia as illegal users of parks and wildlife. A brief examination of the legislation illustrates these points.

The 1975 legislation provides for two types of state-protected areas which are not differentiated in terms of protection purpose or management objectives: 'game parks' and 'nature reserves' (South West Africa Legislative Assembly 1975). The legislation uses these two categories in conjunction and interchangeably. The minister may declare any area a game park or a nature reserve

> for the propagation, protection, study and preservation therein of the wild animal life, fisheries, wild plant life and objects of geological, ethnological, archaeological, historical and other scientific interest and for the benefit and enjoyment of the inhabitants of the Territory and other persons.
>
> (Nature Conservation Ordinance No. 4 1975)

Purposes for which permission may be granted to enter the game parks and nature reserves include 'health, study and recreation, travel or transport along prescribed routes and transacting lawful business'. Prohibitions include the hunting of animals, the wilful or negligent injury, capture or disturbance of animals, eggs or nests, the introduction of livestock or any domestic animal, the picking of any indigenous plant and the chopping, cutting or destroying of any tree.

A pattern of protected areas' establishment and management common in southern Africa (Child 2004) is reflected in Namibia's state-protected areas. Local and indigenous peoples have been and remain essentially excluded from the parks, with only one or two exceptions arising either from outright defiance by the indigenous people (Hinz 2003, p. 48) or from recently evolved, limited co-management relationships under CBNRM (Namibian Association of CBNRM Support Organisations [NACSO] 2010, p. 95). Traditional uses of plants, animals, vegetation and water practised prior to protected areas designation are treated as illegal uses under the law.

Beyond its provisions for protected areas, the Nature Conservation Ordinance is largely devoted to prohibitions or prescriptions for permitted uses of wildlife. It sets out a series of rights for the use and enjoyment – by predominantly white landowners – of certain categories of wildlife, and these rights are not extended to communal area residents. This social and economic injustice was not redressed until the emergence of CBNRM in Namibia.

Etosha National Park: A brief retrospective

Etosha National Park is considered the flagship of the Namibian protected areas system and is the most important international tourism destination in the country (Turpie et al 2004; Ministry of Environment and Tourism 2007). Etosha, at 22,270 square kilometres, is one of the largest national parks in southern Africa. It is classified under the Nature Conservation Ordinance as a game park. The International Union for Conservation of Nature (IUCN) recognizes Etosha as a category II protected area or national park (Protectedplanet.net n.d.).

On 1 April 1907, the precursor to Etosha, Game Reserve No. 2, was proclaimed by the German colonial administration (South West Africa Administration n.d.). According to the proclamation, hunting was prohibited, traffic of any vehicles was only permissible with written permission and a fine of up to 5,000 marks could be imposed for a contravention of the proclamation. Game Reserve No. 2 became the largest nature reserve in the world at the time (Ministry of Environment and Tourism 2007).

Following the ousting of the German administration in South West Africa in 1915, the South African administration confirmed the borders of Game Reserve No. 2 through the Prohibited Areas Proclamation, No. 26 of 1928. This created the Police Zone or so-called 'Red Line', which could not be crossed without a permit (Dieckmann 2007). This was a direct consequence of the rinderpest epidemic of 1896 and 1897. The 'Red Line' was reinforced in the early 1960s with the installation of a veterinary fence running all the way from the Skeleton Coast, along the southern boundary of the present-day Etosha National Park and on to Botswana. The intent was to prevent the movement of African free-ranging livestock from the north and hence the spreading of disease into the commercial farm stock of colonial settlers. Indeed, Etosha National Park was completely fenced by 1973 (Dieckmann 2003; Berry 1997), and the fences have served to reinforce the alienation of indigenous people from the park (Dieckmann 2003; Hoole and Berkes 2010).

In 1962, South Africa appointed the Commission of Inquiry into South West Africa Affairs, commonly known as the Odendaal Commission (Dieckmann, 2007, p. 176), which demarcated new apartheid homelands in South West Africa and changed the boundaries of the game reserves (Figure 7.2). Game reserves were reduced significantly in size to make way for the new homelands (De la Bat 1982).

This history of Etosha National Park is very much a thumbnail sketch. What emerges from various archival accounts is that the park area was occupied, used or laid claim to by at least three indigenous peoples: the Owambos residing to the north, the Hai//om to the south of the Etosha Pan and the Herero to the west.

Community-based conservation in southern Africa

Community-based conservation assumes that if conservation and development are simultaneously achieved, the interests of both can be served (Berkes 2004). In the African context, 'community conservation' has been defined as principles and practices stressing conservation goals that emphasize natural resource decision-making by local residents (Adams and Hulme 2001). Community-based conservation has been practiced in many forms, but in the broadest sense includes conservation by,

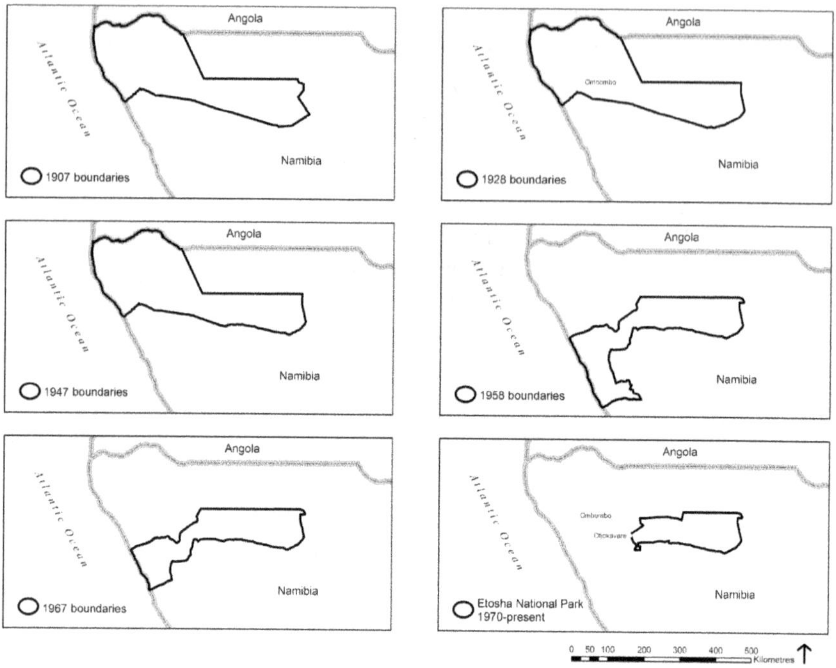

Figure 7.2 Boundary alterations to Game Reserve No. 2 and Etosha National Park.

for and with the local community. The coexistence of people and nature, as distinct from protectionism and the segregation of people and nature, is its central characteristic (Murphree 1994; Western and Wright 1994). Community-based conservation is employed here as an overarching concept that includes CBNRM.

Centrally and internationally conceived approaches in the community-based conservation of wildlife emerged in the 1980s in southern Africa (Hulme and Murphree 2001; Fabricius et al 2004). CBNRM programmes were developed in Zimbabwe, Zambia, Botswana, Malawi and Namibia with international donor funding. These programmes typically featured the devolution of bundles of certain rights in the use of wildlife to local communities, premised on making wildlife pay, the intention being to obtain local benefits that exceeded the costs of living with wildlife. The central theory was that economic incentives would promote conservation by local and indigenous peoples. These approaches, while achieving some conservation, often entailed more co-optation than empowerment. There was limited recognition of and support for traditional and indigenous resource management institutions or an indigenous conservation ethic (Callicott 1994; Infield 2001).

Zimbabwe's Communal Areas Management Programme for Indigenous Resources (CAMPFIRE) and Zambia's Administrative Management Design for Game Management Areas programme (ADMADE) were especially influential in the design of Namibia's CBNRM programme (Jones and Murphree 2004; Garth-Owen Smith, Founding Co-Director of IRDNC, personal communication, 14 July 2006). A brief overview of CAMPFIRE and ADMADE will therefore provide a useful introduction to the more specific consideration of Namibia's CBNRM programme and communal conservancies.

Communal areas management programme for indigenous resources in Zimbabwe

Zimbabwe's Parks and Wild Life Act of 1975 was amended in 1982 to give 'appropriate authority' over wildlife to rural district councils for communal areas (Murombedzi 2001). CAMPFIRE was ultimately diffused to many rural district councils, but strongly tenurial communal property regimes were not acceptable to the councils. They did not want communal lands removed from their authority, along with the wildlife revenue potential from those areas.

Rural district ward boundaries in Zimbabwe were used to define areas for collective action, when in fact there were differing and competing community groups and interests in such bounded areas (Jones and Murphree 2004). The 'hard' boundaries created by formal ward designations, land use and zoning plans were at odds with the 'soft' boundaries that communities traditionally used to enable overlapping and negotiated rights of resource access. Furthermore, the institutional forms adopted in CAMPFIRE tended to be outgrowths of higher-level government agencies and did not originate within or reflect traditional, customary and less formal institutions at the community level.

Bond (2001) concluded that CAMPFIRE needed to achieve a much higher level of proprietorship at the community level. Other research echoed this, noting that communities were not given the right to use wildlife, but only allowed to share some of the benefits from its use by others (Murombedzi 2001). There was little use of local and traditional conservation institutions for land and resource management. CAMPFIRE was quite top-down and did not effectively devolve authority to manage wildlife below the district level. Consequently, it did not comply with the subsidiarity principle: as much local solution as possible and only so much government regulation as necessary (Berkes 2004).

Administrative management design for game management areas in Zambia

ADMADE was initiated by Zambia's National Parks and Wildlife Service in 1987 in the Luangwa Valley with donor assistance from the World Wildlife Fund (US) and United States Agency for International Development (USAID) (Gibson 1999). ADMADE explicitly tried to create a shift from the command-and-control style of colonial administration to a more community-based approach to wildlife management. Revenue from safari concession fees, hunting licences, donor contributions and the profits of activities like wildlife culls were to be shared at community level, to promote wildlife conservation and curtail poaching.

ADMADE ended up creating another layer of bureaucracy for local communities, alienating them with increased enforcement (Gibson and Marks 1995). The programme attempted to change individual behavior by offering incentives that mimicked public goods, such as schools and clinics. However, the programme did not fully appreciate the social significance of hunting, and hunters continued to poach (Marks 2005). Increased enforcement simply altered their tactics and prey selection. For game scouts, the pay and jobs were positive incentives to enforce, but for hunters, the public goods nature of incentives led to free-riding (Gibson 1999). Game scouts were also under considerable social pressure from neighbours, who were often their friends and relatives.

Chiefs oversaw the community projects funded by their communities' share of wildlife revenue, and they selected the individuals to be trained and employed as village game scouts. These practices produced predictable problems of questionable benefits distribution, nepotism and alienation of the game scouts from their communities (Gibson and Marks 1995). The boundaries of the ADMADE programme were drawn according to nationally defined game management areas, not locally negotiated patterns of access and use reflecting local traditions and cultural practices. Thus ADMADE, like CAMPFIRE, did not meet the subsidiarity principle.

As noted above, the experiences with CBNRM in Zimbabwe and Zambia were specifically considered by those implementing CBNRM in Namibia. Steps begun in the early 1980s led to Namibia's CBNRM programme receiving statutory authority over a decade later, in 1996 (Nott and Jacobsohn 2004).

Namibia's community-based natural resource management programme and communal conservancies

The Namibia Wildlife Trust, a national conservation NGO, concerned about severely depleted wildlife in north-west Namibia, appointed Garth Owen-Smith as a conservator for the region in 1982. He possessed long experience in the region and engaged local headmen, who shared the concern about the loss of wildlife (Owen-Smith, personal communication, 14 July 2006). The headmen appointed their own auxiliary game guards, later to be known as community game guards, who were all respected hunters from local communities. The aim was to stop poaching (Margaret Jacobsohn, Founding Co-Director of IRDNC, personal communication, 14 July 2006), so the game guards monitored wildlife, reporting suspicious activities and poaching incidents to the headmen, who in turn informed government wildlife enforcement personnel.

By the late 1980s, regional wildlife populations had noticeably recovered. The ending of armed conflict and improved rainfalls made important contributions to this, but the community game guards were a major factor in stopping poaching, which also helped wildlife to recover. These early successes led to the formation of a new Namibian NGO, Integrated Rural Development and Nature Conservation (IRDNC), which continues to facilitate and support the further development of CBNRM in the Kunene and Caprivi regions of northern Namibia.

With independence in 1990, Namibia's black majority government extended rights in wildlife to communal area residents that had previously only been granted to white farmers on private lands by the South African administration. CBNRM in Namibia received statutory authority under the Nature Conservation Amendment Act, No. 5 of 1996. This legislation did not repeal or replace the Nature Conservation Ordinance as the statutory authority for protected areas designation and management. Rather, it had the more circumscribed purpose of providing for the devolution of certain rights and uses of wildlife to communal area residents. These included rights to hunt, capture, cull and sell 'huntable game' such as springbok, oryx and kudu under quotas established by the Ministry of Environment and Tourism, as well as the right to use quotas of protected game such as elephant for trophy hunting.

Communal area residents are required to form a common property resource institution called a conservancy if they wish to participate in CBNRM and enjoy the rights in wildlife and related tourism development devolved under the legislation. Conservancies must be approved by and registered with the Ministry of Environment and Tourism. Conservancy registration requires a defined conservancy boundary, an identified membership, a representative conservancy committee, a constitution recognized by the government and a commitment to producing a benefits distribution plan (Long 2004; World Resources Institute et al 2005).

Key linkages and partnerships have evolved in Namibian CBNRM, from a few simple ones between local communities, the national conservation NGO and the national government wildlife agency during the initial community game guard programme, to multiple linkages and networks involving several international donors, multiple national NGOs, the University of Namibia, private enterprise, and the Ministry of Environment and Tourism (Hoole 2010). There has been a rapid scaling up of communal conservancies in Namibia, from an initial four in 1998 to sixty-six in 2011 (NACSO 2012).

Design principles for long-enduring common property institutions at local levels (Ostrom 1990; Agrawal 2002) provide an organizing framework for summarizing the comparable and contrasting characteristics of Namibia's CBNRM, Zimbabwe's CAMPFIRE and Zambia's ADMADE programmes (Table 7.1).

Namibia's CBNRM programme and conservancies explicitly consider and apply many of these recognized design features and principles, including defined conservancy boundaries, a defined membership, external legal recognition and government-approved rights to organize. There has been a deliberate effort to avoid predetermined delimitation, such as CAMPFIRE's use of rural district ward boundaries and ADMADE's use of nationally defined game management area boundaries. Rather, communities are required to self-organize and negotiate their own boundaries, thus helping ensure the genuine devolution of wildlife use rights and benefits to the community level.

Another important feature is the formal registration and gazetting of conservancy members, which reinforces the principle of external recognition for community level institutions – again a significant departure from both CAMPFIRE and ADMADE. In addition, the revenues and other benefits under conservancies

Table 7.1 Institutional comparisons between southern African CBNRM programmes

Design principles for enduring common property resource institutions (Ostrom 1990)	*Zimbabwe's CAMPFIRE programme*	*Zambia's ADMADE programme*	*Namibia's CBNRM programme and conservancies*
Clearly defined boundaries for resources used or managed	Wildlife are migratory and distributions do not conform to boundaries of rural district wards and protected areas.	Wildlife are migratory and distributions do not conform to game management areas and protected areas.	Wildlife are migratory and distributions do not conform to conservancy boundaries and protected areas.
Clearly defined boundaries for social groups involved	Established rural district ward boundaries applied; no local community definition.	Established game management area boundaries applied; no definition by local community.	Local communities negotiate and self-define conservancy boundaries.
Agreed rules for resource access and use	Wildlife laws and quotas set by central government; certain wildlife management and benefits devolved to rural district councils; revenues shared between rural district councils and ward/village levels.	Wildlife laws and quotas set by central government; benefits and revenues shared between central government and community chiefs.	Conservancies monitor wildlife and propose quotas for government approval; all revenues and benefits accrue to conservancies.
Collective choice arrangements	Rural district management is limited or there is no community recognition in governance.	Chiefs and headmen make decisions for community.	Conservancy management committees elected to represent community members.
Provisions for monitoring resource and use	Central government monitors the state of wildlife and use.	Central government monitors state of wildlife and use.	Community game guards recruited by conservancies; report state of wildlife and violations to central government; do not enforce.
Graduated sanctions for violations	Central government wildlife laws and central government penalties are applied, and there is enforcement and prosecution of violations.	Village game scouts appointed by chiefs enforce wildlife laws on behalf of central government, which sets penalties and prosecutes violations.	Central government applies wildlife laws, provides enforcement and prosecutes violations.
Provisions for conflict resolution	Central government and rural district councils.	Central government; chiefs and headmen.	Conservancy community committees prepare management and benefits distribution plans; annual general meetings of members.

External recognition of local institutions	Programme defined top-down, with recognition of rural district council level.	Top-down programme; community chiefs recognized by central government.	Conservancies legally recognized by central government; boundaries and members legally registered.
Nested enterprises for appropriation and governance	Vertical linkages dominant; NGO support, central government and rural district councils.	Central government, NGO support, chiefs recognized; vertical linkages dominate.	Central government, multiple NGOs, national CBNRM organization, multiple conservancies, networks.

accrue to conservancy committees and are not shared with the central government or regional authorities, as they were under CAMPFIRE and ADMADE. The advantages of devolving the rights and benefits of managing wildlife to the community level were learned from CAMPFIRE, but so too was the lesson that all revenue from wildlife should be retained at the community level (Jones 2001; Owen-Smith, personal communication, 14 July 2006).

The recognition of local rights to organize by institutions and authorities beyond the local level implies that relationships are needed with other institutions at different levels, beyond local institutions. The term 'nested enterprises' describes different levels of collective action that are organized in mutually reinforcing layers (Ostrom 1990). Clearly, the external recognition of conservancies as provided for in Namibia's legislation, the omnipresence of international donor assistance, the evolution of multiple national NGOs facilitating and supporting community-based conservation, and conservancy partnerships with private enterprises are all evidence of this principle. Multiple linkages and increasingly dense networks among international, national and local agents are all evident in Namibia (Hoole 2010).

The Ehi-rovipuka Conservancy case

Ehi-rovipuka – the name means 'place of wildlife' in Otjiherero – was officially registered as a communal conservancy in January 2001. It has an overall area of 1,980 kilometres squared (NACSO, 2010) and is irregular in shape, extending more than 100 kilometres from north to south, with a variable width ranging up to about 20 kilometres (Figure 7.3). The Herero population is approximately 2,500, with densities ranging from less than one person per square kilometre to about ten in the larger village areas such as Otjokavare. There are approximately 30 villages, most of them in the northern half of the conservancy. Generally the villages are very small, with between 50 and 100 persons comprising only a few extended family groups.

People move from village to village, depending on marriage and family relationships, as well as localized drought conditions and available grazing for their cattle. The Ehi-rovipuka Conservancy has about 700 registered members (Asser Ujaha, Field Officer for Ehi-rovipuka Conservancy, personal communication, 17 May 2007). Membership, which is voluntary, is open to all adults (18 or older) who have lived in the conservancy for at least three years (Ehi-rovipuka Conservancy 2000). Members must also hold Namibian citizenship, be permanent residents in the conservancy area and not be members of another conservancy. Membership is gained by signing a registration form signifying that the member accepts the conservancy's constitution and its land use and wildlife management plans, and is willing to uphold them.

The Ehi-rovipuka Conservancy is situated on an upland plateau lying east of the north-western escarpment and along the western boundary of Etosha National Park. Elevations range from about 1,400 metres above sea level to 1,200 metres above sea level along its shared boundary with the Etosha National Park. The conservancy lies within the upper reaches of the Hoanib River watershed. Several

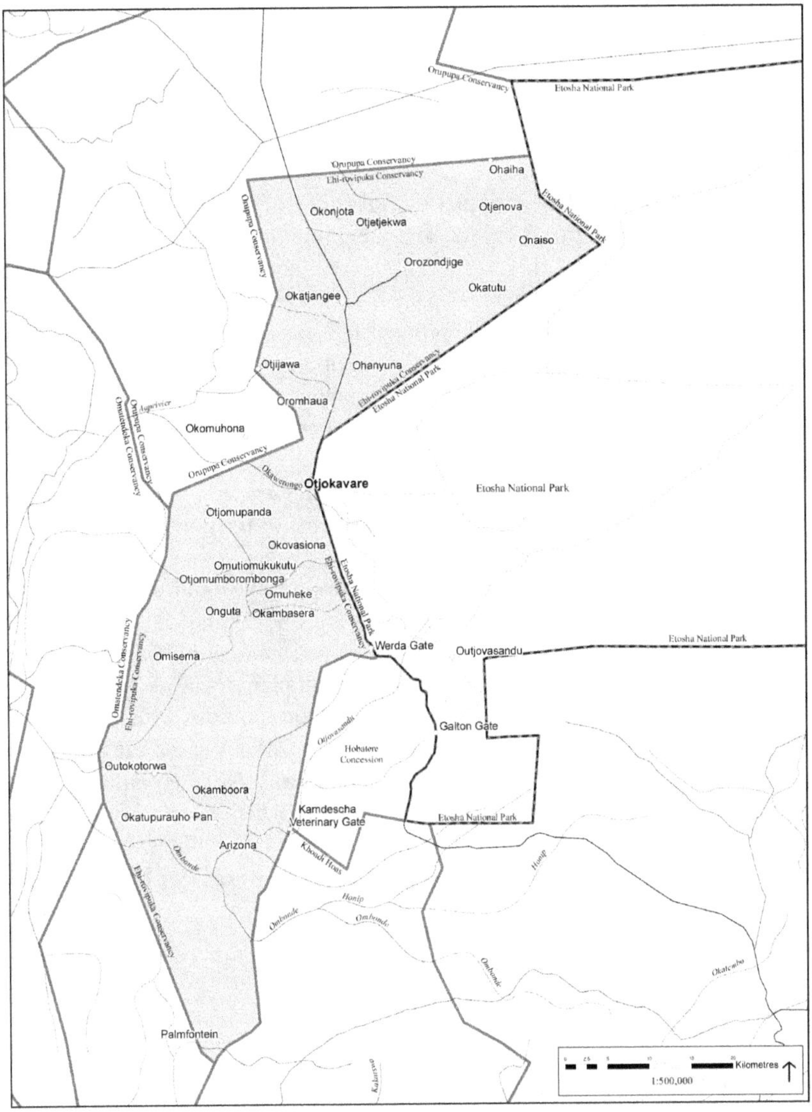

Figure 7.3 Ehi-rovipuka Conservancy and west portion of Etosha National Park.

tributaries of the Hoanib River drain from east to west, notably the Ombonde River and its tributary the Otjovasandu River, as well as the Okawerongo River further north. These are ephemeral rivers that flow only during the rains and provide important connectivity corridors for wildlife, livestock and people. Groundwater springs and riparian vegetation are also concentrated along these rivers, both within the conservancy and throughout the wider Kunene region. Annual precipitation ranges from about 250 millimetres to 350 millimetres (Jacobson et al 1995) with higher rainfalls on the eastern margins, declining to the west. Drought is a regular occurrence, and the entire area of the conservancy is semi-arid.

Conservancy governance, wildlife conservation and community benefits

The constitution of the Ehi-rovipuka Conservancy at the time of this research (2006 and 2007) made provision for a conservancy management committee of 12 members elected every three years by the conservancy membership at large (Ujaha, personal communication, 17 May 2007). Ten of the management committee members represent five different village blocks, two representatives for each block, and two additional members are appointed by the two traditional authorities that share territory with the Ehi-rovipuka Conservancy area (Figure 7.4). Recent accounts suggest that these governance provisions remain essentially the same (NACSO 2010).

The village blocks are each named after a wildlife species, reflecting both the prevalence of different species and their importance to the conservancy purpose (Figure 7.5).

The conservancy management committee is responsible for approving the policies, programmes and projects of the conservancy, and is financially accountable to community members. The committee must develop and uphold the conservancy's constitution, prepare land use and management plans, compile a benefits distribution plan and conduct regular meetings and consultations with conservancy members, including an annual general meeting (AGM). A small administrative staff reports to the committee, carrying out activities such as community game guard patrols and annual wildlife monitoring (Figure 7.6). The conservancy game guard programme, the annual monitoring of wildlife populations and the community use of wildlife within quotas set by the central government have all helped sustain and increase wildlife populations in the conservancy and wider Kunene region (Stuart-Hill et al 2005; Hoole 2008; NACSO 2010).

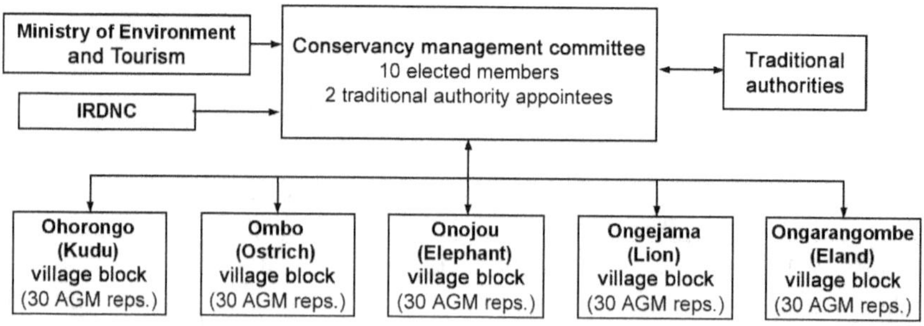

Figure 7.4 Ehi-rovipuka Conservancy governance structure.

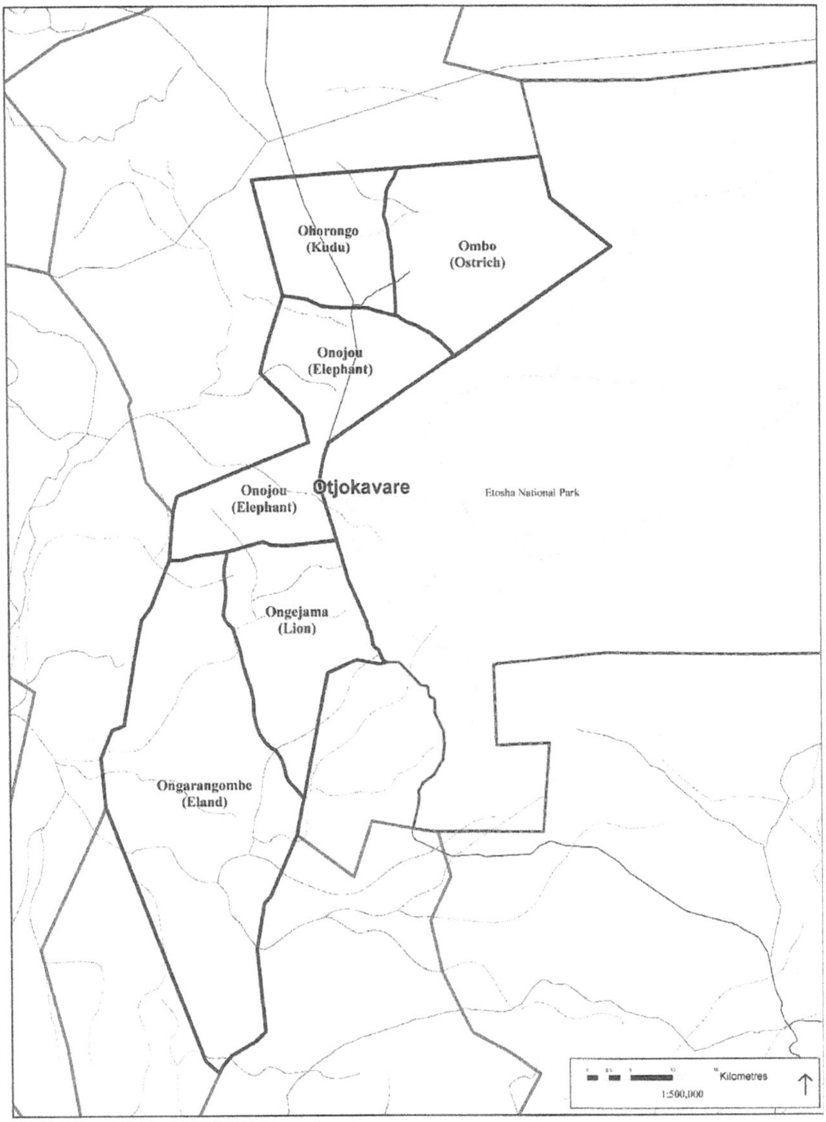

Figure 7.5 Village blocks in Ehi-rovipuka Conservancy.

The Ehi-rovipuka Conservancy receives most of its operational funding from international donors, channelled through IRDNC as lead regional conservation NGO. Villager and key informant interviews confirmed that the most important community benefits from the conservancy were wild meat distributed from a trophy hunting operation, some direct revenue from the shooting and sale of wild

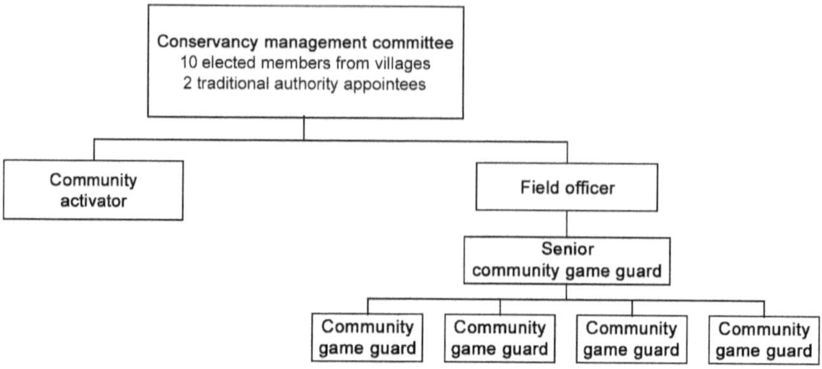

Figure 7.6 Ehi-rovipuka Conservancy organizational structure.

game, and a few full-time jobs in the conservancy management. More recent data on the status of communal conservancies in Namibia shows that these field findings remain generally accurate, although there has been an increase in own-use hunting for wild meat (NACSO 2010, p. 28).

Joint venture tourism premised on wildlife has produced the greatest overall benefits for conservancies (NACSO 2006; NACSO 2010). Interviews with key informants in the Ehi-rovipuka Conservancy, including conservancy staff and management committee members, revealed the strong shared goal of developing a joint venture tourism lodge, as other conservancies had done successfully. Conservancy members particularly wanted to establish such a venture within the Etosha National Park. This goal was not met during the period of field research and evidently it remains unfulfilled (NACSO 2010). Reportedly (Brian Jones, Environment and Development Consultant, personal communication, 29 December 2011) the conservancy recently concluded an agreement for a tourism enterprise within the Hobatere state tourism and hunting concession area, which is contiguous with Ehi-rovipuka and Etosha National Park (see Figure 7.3).

Institutional effectiveness of Ehi-rovipuka Conservancy

A specific research objective was to learn what level of community self-organization and participation had contributed to the establishment of the conservancy. External interventions proved to have been of central importance: the early collaboration of Garth Owen-Smith with headman Kephas Muzuma in the early 1980s was well known to all villagers interviewed, reinforcing the importance of leadership and cross-cultural communication in initiating community-based conservation (Stern et al 2002). In fact, Kephas Muzuma was one of four headmen that Smith had worked with during the 1980s in the precursor community game guard programme.

The roles played by the central government and NGOs were also reinforced in villager responses. Most noted that a community task force of villagers had been created by the traditional authority and received training from IRDNC. This task force, which included both men and women, took the conservancy idea out into the conservancy villages, promoting understanding of and support for the concept and helping negotiate the boundaries with neighbouring communities. An important point emphasized by some villagers was that conservancy boundaries defined rights of access to wildlife only, and the conservancy included only those communities that had agreed to share wildlife: access to grazing, water and other resources was not subject to the exclusivity defined by conservancy boundaries. Villager interviews confirmed that the boundaries were well known at the community level: 80 per cent of those interviewed indicated that they knew the boundaries, or at least which villages made up the conservancy.

In sum, conservancy start-up had both top-down and bottom-up dimensions. The idea originated and was legally enabled from outside and at higher levels of organization than the local community level. Yet there was a high degree of self-organization and collective action at community level in the implementation of conservancy institutional arrangements, notably in negotiating the boundaries.

The participation of conservancy members in decision-making was also explored. Villager responses emphasized AGM attendance, participation in other meetings with the conservancy management committee, and direct representation on the conservancy management committee as the principal means of participation. Actual registered membership of the conservancy was not mentioned by many, and some expressed frustration at difficult access to meetings, including the AGM.

Villagers were asked if they thought the conservancy would be working well in ten years' time. Most felt that this depended upon whether or not there was proper management, capacity and transparency. Others thought the conservancy would be working well and would increase revenues to the conservancy from more tourism enterprise projects. These and other responses revealed that villagers recognized some early benefits from the conservancy, including increased wildlife numbers, an improved meat supply, more positive attitudes towards wildlife and a few revenue-producing projects. However, some villagers interviewed reported that as conservancy revenues and infrastructure started to build (e.g. a local conservancy office and vehicles for community game guard patrols), they had experienced frustrations concerning poor financial management, accountability and transparency in decision-making by the management committee. Nevertheless, the villagers remained hopeful about future prospects for the conservancy and for community livelihoods based on wildlife conservation at the time of the field research.

Overall, the Ehi-rovipuka Conservancy has developed positively as a common property institution for wildlife conservation since its inception in 2001, and continues to do so (NACSO 2006; NACSO 2010). The conservancy has offered communities a voice in wildlife conservation and has begun to realize benefits in terms of direct revenues from wildlife-based enterprises, as well as food security contributed by wild meat. It has fostered positive attitudes towards wildlife as well. The conservancy has developed fruitful and reciprocal working relationships with the government, donors and conservation NGOs in Namibia, most notably IRDNC.

Capacities in financial and programme management were relatively weak at the time of this research, and community members were seeking greater transparency in decision-making from their elected representatives. There was also an inherent limitation evident in the negotiated boundaries between neighbouring conservancies. The communal conservancies all depend upon wildlife species that range widely across the semi-arid Kunene region. Just as there has been a rapid scaling up in the number of communal conservancies successfully negotiated and gazetted to date, there remains an important challenge in further developing effective regional collaborations among conservancies and central government based upon sustaining and improving ecological connectivity for wildlife and biodiversity in the wider landscape. In other words, the small size of individual conservancies such as Ehi-rovipuka creates a scale mismatch with the fugitive nature of the wildlife upon which they are premised (Cash and Moser 2000; Stern et al 2002). This point is also highly pertinent to prospective conservancy relationships with Etosha National Park.

The Hereros and Etosha National Park

Through stories and memory mapping, conservancy elders attest (Hoole 2008; Hoole and Berkes 2010) that the Hereros occupied the western parts of Etosha prior to its original designation as a game reserve in 1907. Indeed, elders' stories relate how Herero people occupied and used the western part of present-day Etosha National Park prior to game reserve designation up to about 1928 or 1929. At the time of the field research there were surviving Herero elders in the conservancy villages from this earlier period, some of whom were interviewed and shared memory maps. Younger residents testified that their parents or grandparents had been born in the park area, suggesting that Herero people lived there before moving into central Namibia, and their descendants returned during the German-Herero war. Elders reported that their families moved back into the western part of Etosha in 1907 and 1908, doubtless as part of the Herero diaspora that resulted from the genocide by the German colonial forces around 1905 (Pakenham 1991).

An especially significant revelation in elders' stories was the reason given for the initial displacement of the Hereros from present-day Etosha. Specifically, the people were 'chased out of the park' when 'whites' came from Angola in 1928 and 1929. In other words, the potential competition for place and space with colonial settlers was the pretext for relocating the Hereros from present-day Etosha; they were not initially displaced by a Western conservation agenda. These stories revealed that the Hereros were relocated to the Ombombo area, which at the time was still within Game Reserve No. 2 (Figure 7.2). The Hereros were not removed from the original game reserve but relocated to a more remote part of it, away from the area intended for colonial farm settlements and cattle production.

It was only with the evolution of a park conservation agenda following the Second World War and the fencing off of Etosha in the early 1970s that the Hereros were denied access to the park area on the grounds of wildlife conservation (Hoole 2008). Where once they had grazed and watered their cattle as semi-nomadic pastoralists, hunted wild game, gathered 'field foods' such as mopane worms and honey and tended their ancestral graves, the Hereros were denied access to the places and practices of their culture and livelihood within the park. In fact, it became an unfulfilled vision of successive headmen to return the Hereros to their ancestral birthplaces, burial sites and traditional use areas in Etosha (Hoole 2008; Hoole and Berkes 2010).

One villager's comment captures the message from most respondents when the topic of living next to Etosha National Park was introduced in interviews: 'There you come to the wound. People get much pain when they hear of the park'.

Table 7.2 summarizes 'decoupling' and 'recoupling' mechanisms for the coupled social-ecological system that existed between the Herero and Etosha National Park. Decoupling mechanisms reflect the felt consequences of the Herero being excluded from Etosha. Recoupling mechanisms are based upon the prospective benefits from Etosha that Ehi-rovipuka villagers strongly seek, as identified in interviews.

The findings underscore a divide between the recently formed common property conservation institution, the Ehi-rovipuka Conservancy, and the state conservation institution of Etosha National Park. In the circumstances this divide is ironic – even more so when one considers that the national CBNRM programme and Namibia's protected areas are enabled under statutes administered by the same Ministry of Environment and Tourism. Yet mechanisms such as those summarized in Table 7.2 could prospectively help recouple the local social-ecological system (Berkes and Folke 1998; Hoole and Berkes 2010). Such mechanisms could link CBNRM and protected areas more strongly, promoting biodiversity conservation and social justice.

Most of the recoupling mechanisms identified in Table 7.2 demand more collaborative and adaptive management on the part of the conservancy and the national park authorities, at multiple levels (Cash et al 2006; Armitage et al 2009). Indeed, such an approach should extend beyond the individual case to region-wide collaborative and adaptive management approaches among multiple communal conservancies, conservation NGOs and central government ministries.

Summary and concluding remarks

Figure 7.7 illustrates trajectories for institutional relationships between state-protected areas and community-based conservation in southern Africa, and serves to summarize and generalize findings and ideas in this chapter.

What follows is an explication of the trajectories set out in Figure 7.7. Local and indigenous resource-use systems for collective action long preceded European colonization and the designation of protected areas in the nineteenth and twentieth centuries (historical traditional period). Protected area designations by colonial powers then truncated these local resource use systems, disenfranchising indigenous peoples from their natural patrimony (twentieth-century period).

Table 7.2 Decoupling and recoupling mechanisms between Herero communities and Etosha National Park

Historical decoupling mechanisms	*Prospective recoupling mechanisms*
Forced relocation from the park area to Ombombo, with fences and fines later preventing return.	Complete removal or selective gating of the park fence to permit community access and wildlife connectivity corridors; hence a more porous park boundary for people and wildlife.
Lost reliable water and grazing for livestock inside the park.	Managed emergency grazing for community livestock during periods of drought.
Lost cultural access to Herero birthplaces and ancestral graves inside the park.	Community access to Etosha to visit, tend and commemorate ancestral graves there.
Lost opportunities to hunt wildlife for domestic use and cultural sustenance inside the park.	More park-based wildlife translocations and meat sharing; no need to hunt inside the park with the Ehirovipuka Conservancy in place.
Lost opportunities to gather field foods and medicinal plants inside the park.	Managed community access to the park for harvesting field foods such as mopane worms, wild honey and medicinal plants.
Loss of social memory preserving traditional rules of resource use and environmental knowledge of the park.	Management collaboration between park and conservancy – a real 'voice' in park management; employment in the park.
Increased vulnerability to wildlife due to sedentary lifestyle.	Increased local security around school sites, livestock kraals and water boreholes in collaboration with park managers, mitigating conflict between humans and wildlife.
Loss of community memory and community-based management practices relating to the park area.	Empowerment through the conservancy's common property institutions; collaborations in park and wildlife management, merging science with local and traditional knowledge.
Lost and foregone opportunities for livelihoods inside the park.	Partnerships in ecotourism enterprises in Etosha National Park; wildlife management.
The social injustice of forced relocation from the park.	Social justice through community empowerment, selective and managed access to the park, and collaborative management between conservancy and park.

Source: Adapted from Hoole and Berkes (2010)

Community-based conservation emerged in the late twentieth century in southern Africa in the form of CBNRM programmes, but disconnects persist between CBNRM and protected areas as two solitudes in shared biodiversity space (late twentieth century).

Different futures are needed for the twenty-first century; futures that more fully acknowledge humans as an integral part of ecosystems, the complexity of social and ecological systems, collaborative resources management and increasingly participatory, empowering approaches in conservation. Several such 'futures' are envisioned: a first scenario (Future 1) in which community-based conservation

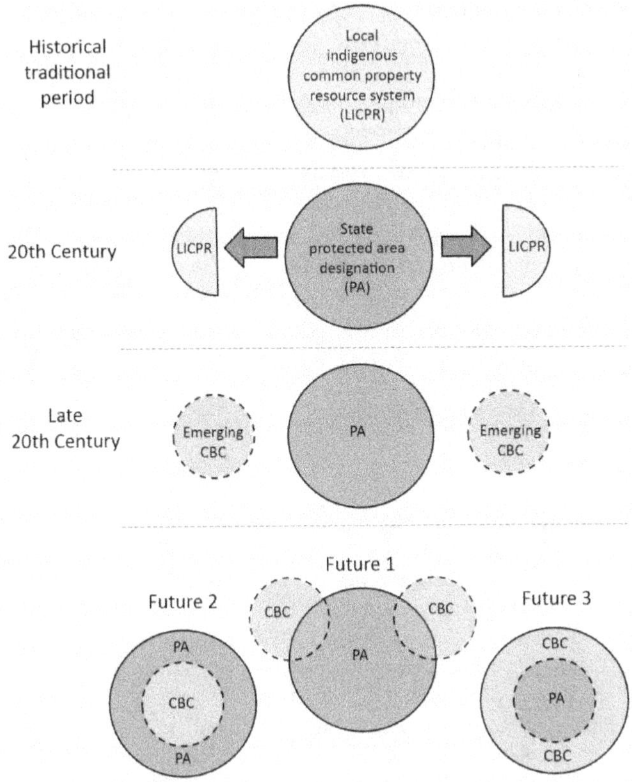

Figure 7.7 Trajectories of community-based conservation (CBC) and protected areas (PA) in southern Africa.

is more directly linked and overlaps with protected areas, much in the manner suggested for the Ehi-rovipuka Conservancy and the Etosha National Park; a second (Future 2) in which community-based conservation is thoroughly embedded within a given protected area, or the community's area is formalized as a protected area in its own right; and a third (Future 3) in which protected areas are more fully integrated with community-based conservation institutions in wider regional landscapes.

These scenarios are not mutually exclusive, nor are they founded on some romantic notion of returning to past indigenous resource management institutions as they may have existed prior to protected areas designations. In fact, while these future scenarios are presented here as theoretical abstractions, they could draw upon and give effect to the most current definition for protected areas and the different protected area management categories now recognized by IUCN (Dudley 2008). They could take advantage of the growing recognition of indigenous and community conserved areas (ICCAs) – that is, indigenous conservation territories and areas conserved by indigenous peoples and local communities (Borrini-Feyerabend et al 2006; Borrini-Feyerabend et al 2010), as well as protected area management categories such as the IUCN Category V (Protected Landscape/Seascape) and Category VI (Protected Area with Sustainable Use of Natural Resources) (Dudley 2008).

The evolution and progress of CBNRM in Namibia, drawing on early experience with a community game guard programme in northern Namibia and lessons learned from regional models such as CAMPFIRE and ADMADE, are very promising for community-based wildlife conservation on the commons. Namibia's maturing communal conservancies, as exemplified by the case of the Ehi-rovipuka Conservancy and Etosha National Park, warrant greater recognition by managers of state-protected areas striving to strengthen collaborative wildlife conservation partnerships in order to achieve greater biodiversity conservation and social justice.

Bibliography

Adams, W. and Hulme, D. (2001) 'Changing narratives, policies and practices in African conservation', in D. Hulme and M. Murphree (eds) *African Wildlife and Livelihoods: The Promise and Performance of Community Conservation*, James Currey, Oxford, pp. 9–23

Agrawal, A. (2002) 'Common resources and institutional sustainability', in E. Ostrom, T. Dietz, N. Dolšak, P. C. Stern, S. Stonich and E. U. Weber (eds) *The Drama of The Commons*, Committee on the Human Dimensions of Global Change, National Research Council, National Academy Press, Washington, DC, pp. 41–85

Armitage, D. R., Plummer, R., Berkes, R., Arthur, R. I., Charles, A. T., Davidson-Hunt. I. J., Diduck, A. P., Doubleday, N. C., Johnson, D. S., Marschke, M., McConney, P., Pinkerton, E. W. and Wollenberg, E. W. (2009) 'Adaptive co-management for social-ecological complexity', *Frontiers in Ecology and the Environment*, vol 7, no 2, pp. 95–102

Berg, B. L. (2004) *Qualitative Research Methods for the Social Sciences*, Pearson Education, Boston

Berkes, F. (2004) 'Rethinking community-based conservation', *Conservation Biology*, vol 18, no 3, pp. 621–30

Berkes, F. (2007) 'Community-based conservation in a globalized world', *Proceedings of the National Academy of Sciences of the United States of America*, vol 104, no 39, pp. 1,5188–93, http://pnas.org/cgi/doi/10.1073/pnas.0702098104, accessed 16 July 2009

Berkes, F. and Folke, C. (eds) (1998) *Linking Social and Ecological Systems: Management Practices and Social Mechanisms for Building Resilience*, Cambridge University Press, Cambridge

Berry, H. H. (1997) 'Historical review of the Etosha Region and its subsequent administration as a national park', *Madoqua*, vol 20, no 1, pp. 3–12

Bond, I. (2001) 'CAMPFIRE and the incentives for institutional change', in D. Hulme and M. Murphree (eds) *African Wildlife and Livelihoods: The Promise and Performance of Community Conservation*, James Currey, Oxford, pp. 227–43

Borrini-Feyerabend, G. et al (2010) 'Biocultural diversity conserved by indigenous peoples and local communities – examples and analysis', ICCA Consortium and Cenesta for GEF SGP, GTZ, IUCN/CEESP, Teheran

Borrini-Feyerabend, G., Johnson, J. and Pansky, D. (2006) 'Governance of protected areas', in M. Lockwood, G. L. Worboys and A. Kothari (eds), *Managing Protected Areas: A Global Guide*, Earthscan, London, pp. 116–45

Brechin, S. R., Wilshusen, P. R., Fortwangler, C. L. and West, P. C. (2003) *Contested Nature: Promoting International Biodiversity with Social Justice in the Twenty-First Century*, State University of New York Press, Albany, NY

Callicott, J. B. (1994) *Earth's Insights: A Survey of Ecological Ethics from the Mediterranean Basin to the Australian Outback*, University of California Press, Berkeley, CA

Cash, D. and Moser, S. (2000) 'Linking global and local scales: designing dynamic assessment and management processes', *Global Environmental Change*, vol 10, no 2, pp. 109–20

Cash, D. W., Adger, W. N., Berkes, F., Garden, P., Lebel, L., Olsson, P., Pritchard L. and Young, O. (2006) 'Scale and cross-scale dynamics: Governance and information in a multilevel world', *Ecology and Society*, vol 11, no 2, http://ecologyandsociety.org/vol11/iss2/art8/, accessed 9 February 2012

Chambers, R. (1997) *Whose Reality Counts? Putting the First Last*, Intermediate Technology Publications, London

Child, G. (2004) 'Growth of modern nature conservation in southern Africa', in B. Child (ed.) *Parks in Transition*, Earthscan, London, pp. 7–27

De la Bat, B. J. G. (1982) 'Etosha: 75 Years', *SWA Annual 1982*, pp. 11–26

Dieckmann, U. (2003) 'The impact of nature conservation on San: A case study of Etosha National Park', in T. Hohmann (ed.) *San and the State: Contested Land, Development, Identity and Representation*, Rüdiger Koeppe Verlag, Cologne, pp. 37–86

Dieckmann, U. (2007) *Hai//om in the Etosha Region: A History of Colonial Settlement, Ethnicity and Nature Conservation*, John Meinert Printing, Windhoek

Dowie, M. (2009) *Conservation Refugees: The Hundred-Year Conflict between Global Conservation and Native Peoples*, MIT Press, Cambridge and London

Dudley, N. (ed.) (2008) *Guidelines for Applying Protected Area Management Categories*, IUCN (The World Conservation Union), Gland, Switzerland

Ehi-rovipuka Conservancy (2000) 'Constitution of the Ehi-rovipuka Conservancy', Otjokavare, Namibia

Fabricius, C., Koch, E., Magome, H. and Turner, S. (eds) (2004) *Rights, Resources and Rural Development: Community-Based Natural Resource Management in Southern Africa*, Earthscan, London

Gibson, C. C. (1999) *Politicians and Poachers: The Political Economy of Wildlife Policy in Africa*, Cambridge University Press, Cambridge

Gibson, C. C. and Marks, S. A. (1995) 'Transforming rural hunters into conservationists: An assessment of community-based wildlife management programs in Africa', *World Development*, vol 23, no 6, pp. 941–57

Hinz, M. O. (2003) *Without Chiefs There Would Be No Game: Customary Law and Nature Conservation*, Out of Africa Publishers, Windhoek, Namibia

Hoole, A. (2008) 'Community-based conservation and protected areas in Namibia: Social-ecological linkages for biodiversity', PhD Thesis, University of Manitoba at Winnipeg, Canada

Hoole, A. F. (2010) 'Place – power – prognosis: Community-based conservation, partnerships and ecotourism enterprise in Namibia', *International Journal of the Commons*, vol 4, no1, pp. 78–99

Hoole, A. and Berkes, F. (2010) 'Breaking down fences: Recoupling social-ecological systems for biodiversity conservation in Namibia', *Geoforum*, vol 41, no 2, pp. 304–17

Hulme, D. and Murphree. M. (2001) 'Community conservation in Africa', in D. Hulme and M. Murphree (eds) *African Wildlife and Livelihoods: The Promise and Performance of Community Conservation*, James Currey, Oxford, pp. 1–8

Infield, M. (2001) 'Cultural values: A forgotten strategy for building community support for protected areas in Africa', *Conservation Biology*, vol 13, no3, pp. 800–2

Jacobson, P., Jacobson, K. M. and Seely, M. K. (1995) *Ephemeral Rivers and Their Catchments: Sustaining People and Development in Western Namibia*, Desert Research Foundation of Namibia, Windhoek, Namibia

Jones, B. (2001) 'The evolution of a community-based approach to wildlife management at Kunene, Namibia', in D. Hulme and M. Murphree (eds) *African Wildlife and Livelihoods: The Promise and Performance of Community Conservation*, James Currey, Oxford, pp. 160–176

Jones, B. T. B. and Murphree, M. (2004) 'Community-based natural resources management as a conservation mechanism: Lessons and directions', in B. Child (ed.) *Parks in Transition*, Earthscan, London

Long, S. A. (ed.) (2004) 'Livelihoods and CBNRM in Namibia: The Findings of the WILD Project', final technical report of the Wildlife Integration for Livelihood Diversification Project (WILD), Directorates of Environmental Affairs and Parks and Wildlife Management, Ministry of Environment and Tourism, Government of the Republic of Namibia, Windhoek

Marks, S. (2005) *Large Mammals and a Brave People*, Transaction Publishers, New Brunswick, NJ

Ministry of Environment and Tourism (2007) *Sandpaper: Etosha Centenary Edition*, issue 6, March 2007, newsletter of the Strengthening the Protected Area Network (SPAN) project, Windhoek, Namibia

Murombedzi, J. (2001) 'Natural resource stewardship and community benefits in Zimbabwe's CAMPFIRE programme', in D. Hulme and M. Murphree (eds) *African Wildlife and Livelihoods: The Promise and Performance of Community Conservation*, James Currey, Oxford, pp. 244–55

Murphree, M.W. (1994) 'The role of institutions in community-based conservation', in D. Western, R. M. Wright and S.C. Strum (eds) *Natural Connections: Perspectives in Community-Based Conservation*, Island Press, Washington, DC

NACSO (2006) *Namibia's Communal Conservancies: A Review of Progress and Challenges in 2005*, Namibian Association of CBNRM Support Organisations, Windhoek, Namibia

NACSO (2010) *Namibia's Communal Conservancies: A Review of Progress and Challenges in 2009*, Namibian Association of CBNRM Support Organisations, Windhoek

NACSO (2012) Summary of conservancies, Namibian Association of CBNRM Support Organisations, www.nacso.org.na/SOC_profiles/conservancysummary.php, accessed 4 January, 2012

Neumann, R. P. (2002) *Imposing Wilderness: Struggles over Livelihood and Nature Preservation in Africa*, University of California Press, Berkeley and Los Angeles, CA

Nott, C. and Jacobsohn, M. (2004) 'Key issues in Namibia's communal conservancy movement', in C. Fabricius, E. Koch, H. Magome and S. Turner (eds) *Rights, Resources and Rural Development: Community-Based Natural Resource Management in Southern Africa*, Earthscan, London, pp. 194–99

Ostrom, E. (1990) *Governing the Commons: The Evolution of Institutions for Collective Action*, Cambridge University Press, Cambridge

Owen-Smith, G. (1987) 'Wildlife conservation in Africa: There is another way', *Quagga*, vol 17, pp. 18–23

Pakenham T. (1991) *The Scramble for Africa: White Man's Conquest of the Dark Continent from 1876 to 1912*, Avon Books, New York, NY

Phillips, A. (2003) 'Turning ideas on their heads: The new paradigm for protected areas', *The George Wright Forum*, vol 20, no 2, pp. 8–32

Protectedplanet.net (n.d.) Etosha National Park, http://protectedplanet.net/sites/Etosha_National_Park, accessed 22 October 2013

South West Africa Administration, undated. *SWAA Nature Conservation and Tourism: iv.* Proclamation of Game Reserves No.'s 1, 2 and 3. National Archives of Namibia. Windhoek

South West Africa Legislative Assembly (1975) *The Nature Conservation Ordinance, No. 4, 1975*, Windhoek

Stern, P. C., Dietz, T., Dolšak, N., Ostrom, E. and Stonich, S. (2002) 'Knowledge and questions after 15 years of research', in E. Ostrom, T. Dietz, N. Dolšak, P. C. Stern, S. Stonich and E. U. Weber (eds) *The Drama of The Commons*, Committee on the Human Dimensions of Global Change, National Research Council, National Academy Press, Washington, DC, pp. 445–89

Stuart-Hill, G., Diggle, R., Munali, B., Tagg, J. and Ward, D. (2005) 'The Event Book System: A community-based natural resource monitoring system from Namibia', *Biodiversity and Conservation*, vol 14, no 11, pp. 2,611–31

Timberlake, L. (1991) *Africa in Crisis*, Earthscan, London

Tobias, T. N. (2000) *Chief Kerry's Moose: A Guidebook to Land Use and Occupancy Mapping, Research Design and Data Collection*, Union of BC Indian Chiefs and Ecotrust Canada, Vancouver, Canada

Turpie J., Lange, G. M., Martin, R., Davies, R. and Barnes, J. (2004) *Economic Value and Financing of Namibia's Protected Areas*, Anchor Environmental Consultants CC, Rhodes Gift, South Africa

Western, D. (2002) *In the Dust of Kilamanjaro*, Island Press, Washington, DC

Western, D. and Pearl, M. C. (eds) (1989) *Conservation for the Twenty-First Century*, Oxford University Press, Oxford

Western, D. and Wright, R. M. (1994) 'The background to community-based conservation', in D. Western, R. M. Wright and S. C. Strum (eds) *Natural Connections: Perspectives in Community-Based Conservation*, Island Press, Washington, DC, pp. 1–12

World Resources Institute, UNDP, UNEP and World Bank (2005) *World Resources 2005 – The Wealth of the Poor: Managing Ecosystems to Fight Poverty*, World Resources Institute, Washington, DC

8 Community-based natural resource management

Micro-governance and face-to-face participatory democracy

Brian Child, Patricia Mupeta, Shylock Muyengwa and Rodgers Lubilo

Introduction

This chapter introduces the concept of micro-governance in community-based natural resource management (CBNRM). The central argument is that wildlife-based CBNRM is good at delivering conservation benefits and generating income. However, we need to understand much more about micro-governance if the goal is also to ensure that CBNRM delivers public participation and equitable benefit sharing. CBNRM is a strategy by which groups of people collectively manage resources that cannot be managed individually because they occur in diverse and scattered configurations or are mobile, or both. Thus CBNRM tends to occur in forests, drylands and mountain environments where individualized agriculture is ecologically difficult.

CBNRM emerged independently and spontaneously for different resources and in different regions of the world in the last quarter of the twentieth century, as rapid demographic growth brought politically and economically marginalized people into increasing interaction with conservation initiatives, while centralized systems of management were failing to provide positive conservation or development outcomes. Participation therefore emerged into the narrative of the development agenda (Chambers 1994). Thus, we see community forestry arising in Mexico in the 1970s (Bray et al 2006; Charnley and Poe 2007), south Asia in the 1980s (Pomeroy 1995; Arnold 2001; Poffenberger 2006), the Amazon and Africa in the 1990s (Charnley and Poe 2007) and community fisheries in south east Asia in the 1990s (Pomeroy 1995; Nasuchona and Charles 2010), with remarkably little cross-referencing or sharing of experiences between these somewhat separate contexts.

CBNRM is complex, but operationally, three issues are critical:

- the economic viability of the natural resources in question
- the willingness of the central government to devolve and protect the rights of local people to manage, benefit from and sell wild resources
- the ability of local people to manage these rights, or micro-governance.

The economics of CBNRM is seldom considered in the literature. In this chapter we take it as given that local collective action can create value in diverse ways. These include regulating use, which includes addressing inefficiencies like the tragedy of the commons, creating synergies and cooperation between local uses and users or specializing in and selling wild resources in a global market as we see in the wildlife sector.

Second, we agree with the literature that CBNRM is often better than what proceeded it (Naughton-Treves et al 2005; Ostrom and Nagendra 2006), even if performance has mostly not lived up to expectation (Hulme and Murphree 2001). Genuine devolution is a necessary condition for CBNRM, and the most important reason for underperformance is the failure to devolve to communities sufficient rights to use, manage and benefit from natural resources, or what Marshall Murphree has termed 'aborted devolution' (Murphree 2004; Ribot et al 2006; Nelson 2010).

The third issue, and the focus of this chapter, is micro-governance: where a resource is economically viable, and most rights have been devolved to the community, what conditions need to be in place to govern it effectively?

Community-based natural resource management, public goods and scale

We define CBNRM as an institutional approach which crafts new economic and political institutions for the purpose of simultaneously achieving both conservation and development. We further suggest that the outcome of CBNRM can be measured in terms of the provision of four sets of public goods (Figure 8.1), namely:

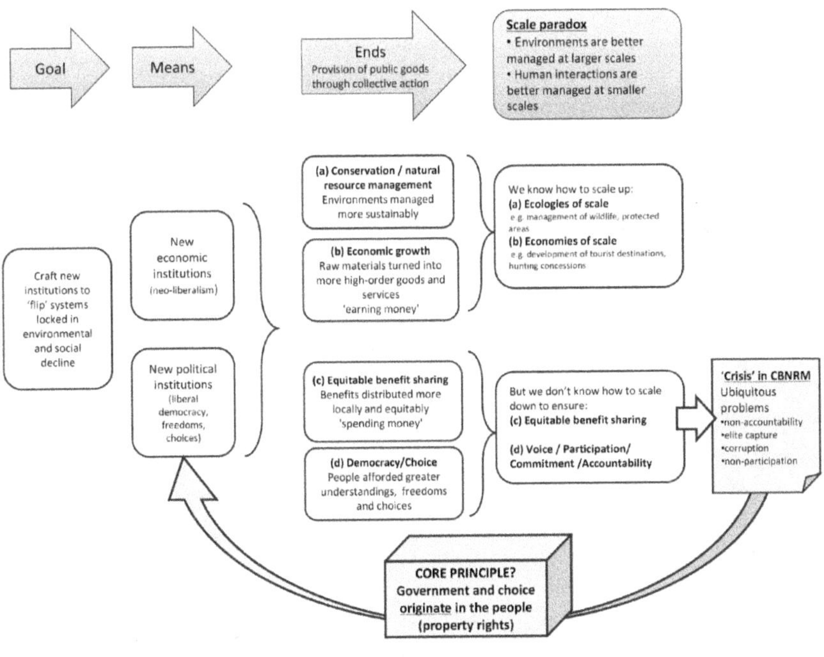

Figure 8.1 CBNRM outcomes and the scale paradox.

- economic growth, by turning environmental raw materials into more highly ordered goods and services
- environmental conservation, via the mechanism of local incentives and empowerment
- more local and more equitable distribution of costs and benefits (i.e. distributive, economic and social justice)
- greater freedoms and choices for people, in the manner of Amartya Sen (i.e. liberal democracy).

Experiences with wildlife-based CBNRM in southern Africa, and especially the well-managed programmes in Namibia and Botswana, show income from tourism and safari hunting in communal areas rapidly increasing, and wildlife populations also increasing or being sustained in the face of population pressure, where before they were rapidly disappearing (NACSO 2008; Rozemeijer 2009; Taylor 2009). A concern is that these benefits often do not reach ordinary people, and that the level of participation in these programmes by ordinary people is low. It appears that CBNRM is often effective at scaling up to generate income through hunting and tourism, and to manage and monitor wildlife and natural resources, but that weaknesses in benefit sharing and participation reflect the challenge of scaling down and the fact that human interactions are better managed on smaller scales (Murphree 2000).

Participatory versus representational governance

Scaling down constitutes micro-governance, and this chapter introduces a framework based on the differences between participatory and representational governance as a way of analysing micro-governance and performance.

Most CBNRM projects, like development programmes generally, work with moderately sized communities on the assumption that this is logistically necessary. These communities typically comprise some 500–5,000 adults. Representational forms of micro-governance are invariably adopted, in which component villages elect or appoint representatives to an overarching management committee with executive responsibility and control of the budget. In these multi-village community-based organizations (CBOs), the primary mechanisms of accountability are an annual general meeting (AGM) and elections held every two or three years.

In the absence of effective top-down oversight, however, these mechanisms may reflect the trappings of democratic accountability rather than its substance, according to the five criteria recognized by democratic theorists: effective participation, voting equality, enlightened understanding, control of the agenda and inclusion of all adults (Dahl 1998; Tilly 2007). In large communities, effective participation is logistically difficult, and most people cannot attend or participate in meetings effectively. More serious problems involve the manipulation of the agenda and information to circumvent accountability. For example, the authors of this chapter have witnessed AGMs at which, after two long days of discussion on non-core issues, the budget is read out in a manner that is too complicated and quick for community members to follow, and in which the haziness of procedures is used to protect people in office that the community clearly wants to remove.

Although AGMs and elections have been adopted ubiquitously, they remain, unless they are deliberately structured to fulfil democratic criteria, more important as ceremonies than as genuine mechanisms of downward accountability. Thus, despite the rubric of CBNRM, multi-village communities are subject to 'committee-based' rather than community-based management. It is not surprising that such processes are associated with low levels of participation and benefit sharing, and high levels of elite control or even elite capture.

Participatory governance exists where the whole community (or at least most of it) meets together to discuss issues and make decisions, and instructs (rather than is instructed by) the committee on matters of implementation. Formal and informal interactions are more frequent, meetings are less hierarchical and information is more egalitarian. Logistically, participatory democracy can only occur where people are able to meet face-to-face – that is, in single-villages or, in the case of larger villages, in village sub-units that are small enough for face-to-face interactions, and to which authority and finances can be devolved.

Superficially, participatory governance in practice may not look too different from representational governance, with anything from 100–300 rural people meeting under a tree, AGM style. However, as described later, there are actually profound differences.

The distinction between participatory and representational governance is not new. According to Madison and de Tocqueville, a democracy is where everyone meets together to represent themselves (as in township government in New England), whereas a republic is where people's interests are represented by elected persons (Mansfield and Winthrop 2000).

Interestingly, the early CBNRM literature emphasizes the importance of scaling down to the level of face-to-face interactions (Murphree 1994; Martin 2009). The cases used to promote CBNRM also typically describe single-village communities, such as Masoka, Mahenye and Chikwarakwara in Zimbabwe (Child et al 1997; Bond 2001; Murphree 2001; Rihoy et al 2007; Taylor and Murphree 2007), even though most CBNRM communities are multi-village. However, the distinction between participatory and representational governance is seldom recognized in theory or in practice: when CBNRM has been implemented programmatically, it has invariably adopted systems of representational governance, perhaps not recognizing the distinction. It is therefore not surprising that CBNRM is said to be in crisis, with ubiquitous challenges of elite capture, financial mismanagement and low levels of participation.

Evidence for the importance of micro-governance

The authors' understanding of micro-governance has grown inductively over more than two decades of managing and researching CBNRM programmes, mainly in Zimbabwe (for the Communal Areas Management Programme for Indigenous Resources [CAMPFIRE]), Zambia (in Luangwa Valley and around

Kafue National Park), Botswana and Namibia (mainly Caprivi). In these situations, two factors have been particularly important in promoting social benefits: face-to-face participatory governance at the scale of single-village CBOs and the power of simple operational procedures that protect the rights of the community to manage their affairs through participatory, accountable and equitable processes. In this section, we present six case studies that demonstrate differences between participatory and representational governance.

Chikwarakwara village and the communal areas management programme for indigenous resource in Zimbabwe

In the late 1980s, private wildlife ranching was expanding rapidly in Zimbabwe. Communal lands had better wildlife than private land, yet the high-value wildlife was disappearing. This led to the establishment of CAMPFIRE (Martin 1981, 1986), which aimed to replicate the economic success of private land wildlife conservation ('game ranching') by devolving similar rights to communities and developing them as 'village companies' to manage their wildlife collectively.

In 1989, a workshop was held on Colin Bristow's Sentinel Limpopo game ranch to discuss conflicts between his wildlife operation and neighbouring communities, whose primary livelihood was cattle, but who were poaching Bristow's grass and wildlife. When Bristow explained why his father, a prominent cattleman, had switched to wildlife, leaders in the Beitbridge District Council became determined to experiment with community wildlife management. Councillors travelled around the district as a team with the wildlife agency: the district's wildlife was depleted, and the challenge was to persuade ordinary villagers that their future lay in managing wildlife sustainably.

A key subject of discussion in these travels was the paradox that livestock, which were unprofitable on private land, were the dominant herbivore in communal lands, whereas wildlife was rapidly disappearing. It was noted that when a farmer sold a cow, he received the income with full discretion as to its use. By contrast, although a buffalo filled a similar ecological niche, the financial flow associated with a buffalo sale was completely different. If it had been hunted by a foreign client, the revenue would go to the government or the district: it would probably be used to build schools or clinics, but no one really knew how it was distributed.

At a critical meeting, the Beitbridge District Council therefore resolved to treat wildlife in exactly the same manner as livestock, which involved two momentous decisions. First, when an animal was shot, that money would go to the village in which it had been harvested, rather than be used more generally for district development. This became known as the 'principle of producer communities'. Second, as with the sale of a cow, people would be encouraged to use the income from wildlife as they saw best, including the option of household cash. In this way people had 'free choice' of how to use wildlife revenues, with the proviso that the decision should be made collectively by the whole village.

The council and wildlife department tested these principles in Chikwarakwara village. First they defined 'household' and compiled a register of households eligible to share the benefits from wildlife. Second, they listed the wildlife shot by foreign hunting clients the previous year and the prices charged. Villagers were very surprised that a warthog, for example, was worth more than an ox, and struggled with large numbers like the Z$10,000 paid for an elephant. At that time, the Zimbabwean dollar was close in value to the US dollar, so this was a lot of money. In 1990, Chikwarakwara earned Z$60,000 (US$1=Z$2.5). There were 149 households, so each household's share was Z$400. However, before the money was handed out, a long debate was held about how best to use it. This meeting to organize the community and allocate wildlife revenues took four days, and the community decided that each household would get Z$400 in cash, after which they would be taxed Z$170 to build a grinding mill and Z$30 for the school.

The council and wildlife agency returned to Beitbridge to collect the Z$60,000 but met strong opposition from the district administrator. He said it was his job, not that of the community, to control the Z$60,000 for district development. The councillors argued against this with some ingenuity, saying that if the cash payment was to be cancelled, the district administrator would have to do so in person because it contradicted the policy of the wildlife agency.

Three days later, the revenue distribution went ahead (Child and Peterson 1991), with the district administrator opening the ceremony. The councillors (Messrs Nare, Modeme and Mulaudzi) then entered the meeting, carrying Z$60,000 in an open-mesh wire basket, and placed this large pile of cash on the table in front of the whole community. All the registered members came up to receive their Z$400, and each then placed Z$170 in a bucket for the grinding mill and Z$30 for the school, and put their signature or mark on a payment voucher, taking home Z$200. This process is illustrated in Figure 8.2.

The philosophy behind CAMPFIRE was to transform wildlife from a nationalized resource into a private good – as on private land, but instead managed collectively and face-to-face at the level of a village (Martin 1986). This argument was used to galvanize the CAMPFIRE programme in over 80 communities and 12 districts, and over the next four years the proportion of revenue allocated to producer communities reached 73 per cent (Child 1993). This happened because the wildlife agency insisted that the producer communities be the primary beneficiaries, and monitored this closely, backing the requirement up by encouraging districts to report the share of revenues getting to communities at national CAMPFIRE workshops. These key financial ratios are still reported by the CAMPFIRE Association (CAMPFIRE 2007).

In retrospect, conformance monitoring was a powerful element of protecting fiscal devolution and the evolution of local collective action, and CAMPFIRE remained quite robust even when technical support dissipated after 2000 (Moinuddin et al 2002). In the early 1990s, the key concern in CAMPFIRE was that district councils were retaining too much income, so financial monitoring focused on the proportion of income devolved from district council to producer community. The allocation within the community was assumed to be non-problematic and was not systematically monitored.[1]

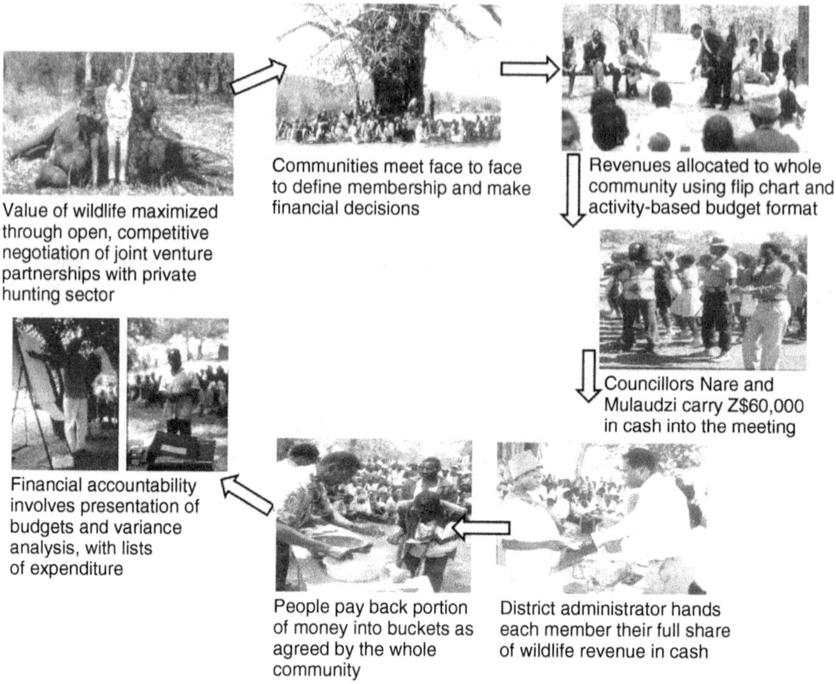

Figure 8.2 Illustration of face-to-face participatory revenue allocation and financial control (photo: Brian Child).

We are able to reconstruct records from three districts to measure equitable benefit sharing, defined as the proportion of the money that reached the community and was used for cash dividends or community projects. In seven wards in Tsholotsho District (i.e. multi-village CBOs), 64 per cent of CAMPFIRE income was shared equitably. In Binga District, 61 per cent was shared equitably in 11 wards within six months (with the rest still unspent in bank accounts). In Hurungwe District, 95 per cent was shared in nine single-village and five multi-village producer communities. This implies that some 60 per cent of benefits was allocated equitably in CAMPFIRE districts, albeit with significant exceptions (such as the highly centralized Nyaminyami District and the disorganized Hwange District).

After 2002, however, fiscal devolution was undermined by Zimbabwe's emerging political culture of leadership impunity (Mapedza and Bond 2006), and by runaway inflation which decimated the value of wildlife accruing to communities until the economy was dollarized in 2011. Fiscal devolution between the district council and producer communities declined to well under 50 per cent from 2002 to 2005, as underfunded district councils coped with Zimbabwe's financial crisis by capturing revenues from communities (Murphree 2005). Importantly, this was reversed through internal political processes facilitated by the CAMPFIRE Association, with district councils accepting that at least 50 per cent of revenue was to be retained by producer communities (CAMPFIRE 2007; Taylor and Murphree 2007).

District councils were always perceived as a double-edged sword, however: while they did extract money from communities, they also provided long-term capacity. In some communities, capacity in areas such as quota-setting and management that was developed through the laudable participatory technology development process of the World Wide Fund for Nature (WWF) in the late 1990s (Goredema et al 2006) is still evident ten years after most technical support ceased.

Community-based natural resource management in the Luangwa Valley, Zambia

The next example reviewed, in the Luangwa Valley in Zambia, was a quasi-experiment: CBNRM was initiated in the form of six representational CBOs, then re-engineered as 43 participatory democracies, and later re-centralized to the same six CBOs.

In the late 1970s and early 1980s, extensive elephant and rhino poaching led Zambian President Kenneth Kaunda, a lover of wildlife but suspicious of his own wildlife agency, to set up the semi-independent Luangwa Integrated Resource Development Project (LIRDP) in his wife's home area in partnership with the Norwegian Agency for Development Cooperation. The project encompassed the 4,500 kilometres squared Lupande Game Management Area (GMA) and the 9,050 kilometres squared South Luangwa National Park. Under the leadership of Fidelius Lungu and Richard Bell, the LIRDP brought poaching under control and embarked on an ambitious integrated rural development project. Forty per cent of the total revenue from tourism in the park and hunting in the GMA was returned to the community in a process controlled by the six chiefs (Dalal-Clayton and Child 2003).

By 1996, it was clear that these revenues were not benefiting ordinary people. Participatory rural analysis showed that ordinary people could hardly name any community projects, and when they did they often raised concerns about mismanagement or worse. At this time, the project was managed as six chieftainships, and decisions were made largely by the six chiefs sitting together with indunas (counsellor of a tribal chief) and women's representatives.

In 1996, the fiscal and governance structure of the project was altered fundamentally. Somewhat surprisingly, this radical proposal was accepted by the five government ministries overseeing the LIRDP in April 1996 with almost no discussion. The transformation is illustrated in Figure 8.3, which replicates LIRDP documents.

The shift to fiscal devolution at the village level was re-emphasized in many village meetings using both role-plays and flip charts, as shown in Figure 8.2. Under the 'old' policy, 40 per cent of revenues were allocated to the chiefs and their area development committees. How this money was spent was never clear. Under the 'new' policy, 80 per cent of wildlife revenues were devolved to 43 village action groups (VAGs) because this was the 'action' level. The area development committees got three per cent for meetings to coordinate the programme. The chiefs each got 1.5 per cent and fought hard for extra money, receiving approximately

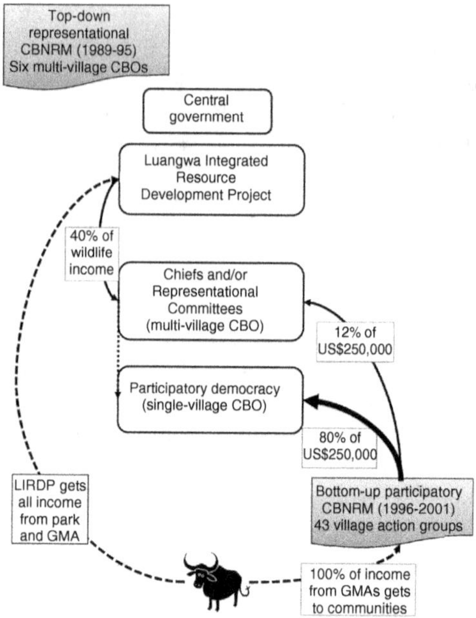

PERFORMANCE METRICS	FIRST GENERATION (downward flow of revenue)	SECOND GENERATION (upward flow of revenue)
Participation (person days/year)	100's	75,000-100,000
Benefits	Few benefits, all public	21,000 people got cash
Community projects	10?	230+
Accountability	40-80% money missing	0.8% money missing
Attitudes to wildlife	–86% negative	+90% positive
Investment in wildlife	0% of income	18% of income
Wildlife trends	Down	Stable/up

Figure 8.3 Comparative performance of CBNRM in Luangwa Valley, Zambia.

10 per cent of total GMA income as personal and non-accountable payments. The remainder was kept in the bank account as a buffer for contingencies.

The chiefs, supported by district and provincial politicians, strongly resisted fiscal devolution, but participatory revenue distribution was initiated in the Malama community in 1996. Following four days of community meetings, Zambian and Norwegian officials and dignitaries travelled from Lusaka to the remote Malama community to witness the first revenue distribution ceremony. Despite their presence and considerable pressure from an excited community, the cash payments were cancelled when the meeting was boycotted by Chief Malama. This nearly ended the policy of fiscal devolution in its tracks. However, the LIRDP was working simultaneously with Chief Msoro, who was often at odds with his fellow chiefs. He allowed revenue distribution to proceed in his area. This was publicized by local radio stations, and popular pressure soon ensured that all 43 VAGs received their benefits, with Chief Malama later asking the LIRDP to facilitate cash distribution in his community.

The process of participatory revenue allocation resembled that developed in Beitbridge, but was managerially more rigorous. A constitution was developed for each community which emphasized that all decision-making, especially on budgets, was to be done face-to-face by the community, and that it was for the community to give instructions to the committee, not the other way around. Committees were to be elected annually, and the constitution clearly specified that finances and progress on projects and wildlife management were to be reported to and approved by the community every quarter.

Two technical procedures strengthened this process. First, the use of a standardized format for activity-based budgeting clarified whether money was to be used for cash or community projects and activities, or wildlife management and administration. This made the second procedure easy, which was to quantify the variance between the budget and actual expenditure at quarterly meetings and present an obligatory financial variance analysis. The constitutions also limited the share for administration to 10 per cent of the budget, to combat the problem of rising and unproductive sitting allowances.

Expenditure data for 43 VAGs over six years (Figure 8.4) shows that communities invested an increasing proportion of their income in wildlife management. On average, 40 per cent was allocated to both projects and cash, but this varied over time and geographically: in drought years people favoured cash, and sometimes communities allocated all their money to complete an important project. But when they did not trust their leaders to conduct their bank accounts properly, they allocated more money to cash to reduce temptation. Most money 'borrowed' by chiefs was, with difficulty, recouped, except in 1998, when the chief in question died suddenly, reportedly having taken community money to treat his HIV/AIDS. Some chiefs continued to try to extract more money from the communities. The project did not intervene in the struggles between chiefs and their 'subjects', despite demands from the latter, but used a strategy of extreme financial transparency, which appeared to work.

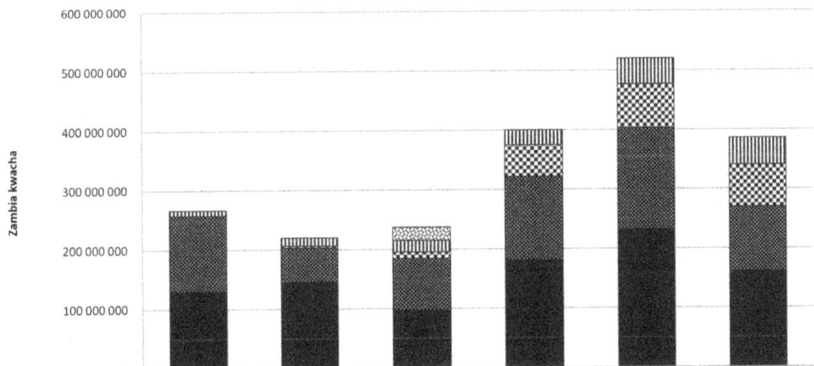

	1996	1997	1998	1999	2000	2001
Chief's extraction			22 498 500			
VAG Administration	9 069 358	13 855 913	19 440 816	25 658 360	43 244 363	45 896 342
Wildlife management		3 082 516	10 413 995	53 243 707	73 287 269	70 612 634
Projects/activities	128 508 800	56 756 138	88 555 519	140 690 623	171 327 922	108 223 251
Member's dividends	129 040 000	146 589 900	96 599 000	180 848 819	232 066 105	161 411 500

Figure 8.4 Expenditure allocation in 43 village action groups in Lupande GMA 1996–2001. During this period the kwacha was unstable, varying from ZK1,207 (1996) to ZK1,862 (1998) to ZK3,848 (2001) to the US$.

As North argues, the structuring of rules – and especially the political institutions and third-party enforcement of such rules – is a serious challenge to the transition from 'big-man' to rule-centred governance (North 2003). In CBNRM, fiscal devolution requires managerial rigour, as well as rigour in the application of participatory principles. In the LIRDP, participatory processes and the rights of ordinary community members were protected by a system of conformance monitoring implemented by the project. A VAG's annual payment was only released to it once it fulfilled procedural requirements aimed at ensuring that the community made all key decisions and had the information it needed to do so. It was a rule that all key decisions (especially budgets) had to be made by the community (not the committee or the chief). Finances had to be properly kept and reported to the community every quarter through a comparison of expenditure and budget. These finances were also audited and had to be formally approved by the community. Elections were held every year, and the membership list was also regularly updated.

These procedures reflect the loose-tight processes in excellent American corporations (Peters and Waterman 1982): the community was strictly accountable for the processes through which they made and monitored their choices, but the decisions on how they allocated their money were entirely theirs.

We learned three useful lessons:

- The bottom-up accountability provided through general meetings was critical for financial accountability.
- It was easy to train all 43 communities to use double-entry bookkeeping, but critical to involve both leaders (i.e. the committee and/or treasurer) and 'followers' in this, so that at least 30 members of any community understood the finances.
- 'Friendly' and formal audits were essential for integrity.

The project tracked key performance indicators on a quarterly basis. Figure 8.3 summarizes the improvement in performance from the top-down representational to the bottom-up participatory phase of the programme. Quarterly audits showed that less than 1 per cent of the money was not accounted for. The 43 communities embarked upon 262 projects that included constructing and repairing schools, clinics, teachers' houses and water wells, food relief in drought years and soccer and sewing clubs. Within three years, 78 village scouts had been employed and were patrolling regularly, and the Msoro community had built two dams to supply water for wildlife. Meetings were generally well attended and dynamic, and gradually conflicts with chiefs were resolved internally. Fully 82 per cent of revenues benefited ordinary people directly through cash payments and community projects and activities. There were also considerable public benefits, including better allocation of scarce revenues, participation, empowerment and other forms of social capital.

Unfortunately for the community, this quasi-experiment entered a third phase when it was re-centralized. The parastatal Zambia Wildlife Authority (ZAWA) replaced the National Parks and Wildlife Service in 2001 and was expected to become financially viable immediately, with the result that even salaries were often not paid. ZAWA 'standardized' CBNRM and retained the majority of hunting revenues from GMAs to fund itself. Formally it retained 50 per cent of trophy revenues and 80 per cent of concession fees, but in the absence of transparent records it is hard to confirm that communities got the money owed to them, which was now about 30–40 per cent of hunting fees. ZAWA also re-centralised CBNRM at the level of chiefdoms by paying revenues to representational Community Resource Boards (CRBs) rather than directly to the five or more VAGs that make up a CRB.

This re-centralized CBNRM in the Lupande GMA (Lubilo and Child 2010) and performance immediately plummeted (Lubilo 2007). The average understanding of community finances declined from 72 per cent in the 1996–2001 period to 20 per cent in 2007 (Lubilo, 2007), and support for the programme declined from over 90 per cent to 39 per cent. In the only community for which we have financial records, the CRB used 63 per cent of income for administration, allocating 28 per cent to wildlife management and only 9 per cent to community benefit, compared to 82 per cent previously.

The one community that runs counter to these trends is Malama. This is the smallest and most remote community, and the switch to CRBs did not change its size. Malama was the weakest community during the bottom-up phase, with members only understanding 50 per cent of finances, compared to an average of 72 per cent in all six areas (Dalal-Clayton and Child 2003). But in the top-down phase, 56 per cent of community members in Malama still understood their finances, while the average in the entire GMA dropped to 20 per cent, suggesting that Malama was also the only community that maintained some level of financial transparency. Support for CBNRM, at 57 per cent of respondents, was much higher than the GMA's average of 39 per cent, despite Malama having the most problem animals.

From this experience, it can be concluded that CBNRM is a highly beneficial and energizing process, with benefits at many levels, including wildlife conservation, economic growth, poverty reduction at household level and rural democratization. Three factors contributed to effective performance here: fiscal devolution to the village level; mechanisms for face-to-face choice, deliberation and performance control; and external support that facilitated experiential capacity creation and held communities procedurally accountable to the principles of CBNRM through conformance monitoring and sanctioning.

Performance of community-based natural resource management according to individual community members

This account now turns from local case studies to a broad analysis of CBNRM in the region. To understand how ordinary people were participating in and benefiting from CBNRM, an action research process called the 'governance dashboard' was developed and applied to more than 1,500 households in CBNRM communities in Tanzania (three communities), Zambia (nine), Namibia (seven), Zimbabwe (two) and Botswana (five) from 2005 to 2010.

The governance dashboard is an adaptive management process, and very deliberately not an extractive research process. Following an introductory workshop with communities, data is collected using normal survey questionnaires administered by local school leavers who have been trained in survey methods. Critically, data is presented and discussed with communities in the same week it is collected, using flip charts and bar charts. Participants are asked if they agree or disagree with what the data says. They then discuss the data, and all comments are written up on the flip chart (i.e. a group memory). Red is used to suggest that an outcome like elections is problematic, amber that it could become so, and green that the community is performing well. Verbal explanatory models are sought from participants to explain the data. Visualization distances communities from the personal or emotional content of the data, encourages them to discuss issues logically (Van der Riet 2008), and often leads to suggestions for corrective action.

After the community feedback meeting, the data is compiled into a report which is given back to the committee and key support organizations. The process requires a considerable commitment to developing relationships with stakeholders, and care and sensitivity in feeding back data as the results can be explosive.

Currently reports are available for more than 20 communities in Zimbabwe, Zambia, Namibia and Botswana, and only an illustrative summary of this data will be presented here for the purposes of argument. Figure 8.5 compares results from two CBOs in the same country: a typical multi-village CBO with representational

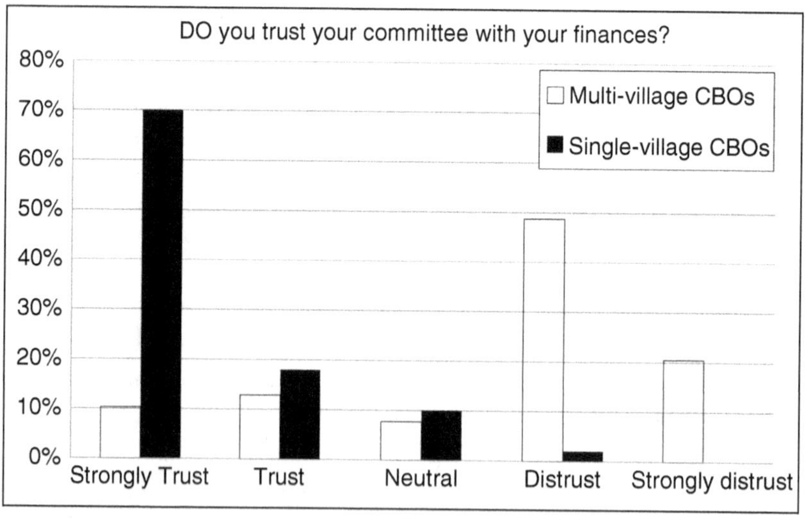

Figure 8.5 Perceptions of performance in single- and multi-village CBOs in Botswana.

governance and a well-functioning, single-village CBO with participatory governance. The variable 'financial trustworthiness' has been selected to illustrate how starkly single-village and multi-village systems differ. However, many other variables – such as satisfaction with AGMs, clarity of decision-making, knowledge of the value of wildlife, receipt of benefits and even knowing the names of elected leaders – also show clear differences between the two systems in levels of understanding and participation. For instance, in a subsample of three single-village and three multi-village CBOs in Botswana and Zambia, 85 per cent of respondents from single-villages attended AGMs, compared to 34 per cent of those from multi-village communities.

This chapter's general conclusions, based on five years of fieldwork, are that wildlife is generating considerable and increasing benefits in all the countries covered here, and is being conserved better than it was, especially in Namibia (Namibian Association of CBNRM Support Organisations [NACSO] 2008) and Zimbabwe (Child et al 2003), but less so in Botswana, for reasons not related to CBNRM, including veterinary fences. The negative exception is Zambia, where poaching is still rife. However, ordinary people, to a great extent, are not benefiting or participating or getting information. CBNRM is therefore earning a lot of money from improved wildlife populations and commercial arrangements, but the spending of this money is problematic. At worst there is corruption, though more often than not money seems to be used wastefully rather than illicitly, and even the best programmes could use money better. For example, in a number of programmes half or more of the expenditure is targeted at a small number of employees relative to the size of the community.

This indicates that CBNRM is performing reasonably well in the dimension of scaling up, providing public or collective goods including improved wildlife management and income (Table 8.1). But in the dimension of scaling down, equitable benefit sharing and participation, CBNRM is underperforming and is disappointing at the household level.

CBNRM is much better at generating revenue than at spending it effectively. Early experiences, such as the Beitbridge and Luangwa cases described above, seem to be exceptions, and only three of the communities surveyed by the authors showed similar attributes. In the small single-village communities of northern Botswana – Sankuyo, Mababe and Khwai – dashboard surveys show much higher satisfaction with governance, and clear evidence of people getting household and

Table 8.1 General conclusion on the performance of 20 CBNRM communities in southern Africa

Provision of public goods through collective action	*Scale dimension*	*General rating of performance*
1. Natural resource management (wildlife populations)	Scale up	Good
2. Gross income ('earning money')	Scale up	Good
3. Equitable benefit sharing ('spending money')	Scale down	Poor
4. Participation/democratization	Scale down	Poor

collective benefits. This suggests that the practice of measuring the performance of CBNRM in terms of wildlife populations and overall income may fail to take account of important deficiencies in household participation and benefit.

Elite capture and low participation are certainly not unique to CBNRM in southern Africa. These challenges are associated with CBNRM globally and for different resources (Arnold 2001; Poffenberger 2006; Charnley and Poe 2007; Frizen 2007; Tacconi 2007; Larson and Soto 2008), and are also widely reported for district decentralization processes (Crook 2003; Frizen 2007; Grindle 2007).

Action research and devolution in Wuparo Conservancy

In the fourth example, action research was used to change the structure of Wuparo Conservancy from a representational to a participatory form of governance, and the impacts of this change were measured. Feeding back the results of governance surveys needs to be handled very carefully, as it creates a demand for change that is not always welcomed by community leaders or even CBNRM support agencies. However, in one community (Wuparo, Namibia), results catalysed a demand for corrective action in the form of more devolved systems of governance. This change was supported by CBNRM support organizations, especially Namibia's Legal Assistance Centre, but also WWF and Integrated Rural Development and Nature Conservation.

Some of the authors of this chapter helped develop new institutions similar to those described in relation to Zambia's Luangwa Valley. Wuparo Conservancy was reorganized through the establishment of seven single-village suborganizations to which half of the wildlife revenues were allocated (Lubilo 2011). Again, the data shows large benefits associated with participatory governance. Thus, in 2010, 1,386 community members received cash dividends of N$50–400 (US$6.50–50) for the first time. Before this, the wildlife income was absorbed by the conservancy committee, including meeting costs and salaries for community game guards. No funds were invested in community projects between 2000 and 2007, and a trifling amount of N$2500 (US$325) was donated to the school in 2008. By contrast, in 2010 and 2011 (the bottom-up phase) 25 community projects were initiated, although not all were successful. Meat from hunted animals was also shared more equitably, according to reports, and specific support was provided to elderly people, orphans and vulnerable children.

Following the money

Data on community income from wildlife generally shows rapid growth, and is a positive aspect of CBNRM that is often highlighted (NACSO 2008; Taylor 2009). However, categorized expenditure (i.e. 'spending money') is difficult to find. A data set was compiled with expenditure information for five single-village and multi-village sites representing 24 CBOs, specifically differentiating between expenditure directly benefiting ordinary community members (i.e. cash dividends or local projects) and money used centrally by committees for management (e.g.

administrative expenses, money not accounted for, payment of a small number of employees and management expenditure). This data is summarized in Figure 8.6. This data admittedly does not reflect considerable complexity or important achievements, especially related to conservation, income generation and capacity-building, but it nonetheless paints a very clear picture.

- In single-village CBOs, some 80 per cent of revenue is converted into community benefit.
- In multi-village CBOs, less than 20 per cent of expenditure benefits community members.
- However, regular monitoring of the adherence of communities to democratic decision-making processes (i.e. 'conformance monitoring' as described on p. 166) is also important.
- Without conformance monitoring, community benefits in single-villages dropped to 60 per cent (e.g. in Sankuyo) and probably less, judged from anecdotal evidence out of Mababe and Khwai.
- With conformance monitoring in multi-village CBOs, the solitary case of CAMPFIRE in the early 1990s suggests that benefit sharing improves from 20–60 per cent of community expenditure.

Overall, the distribution of allocations suggests an 80:20 rule:

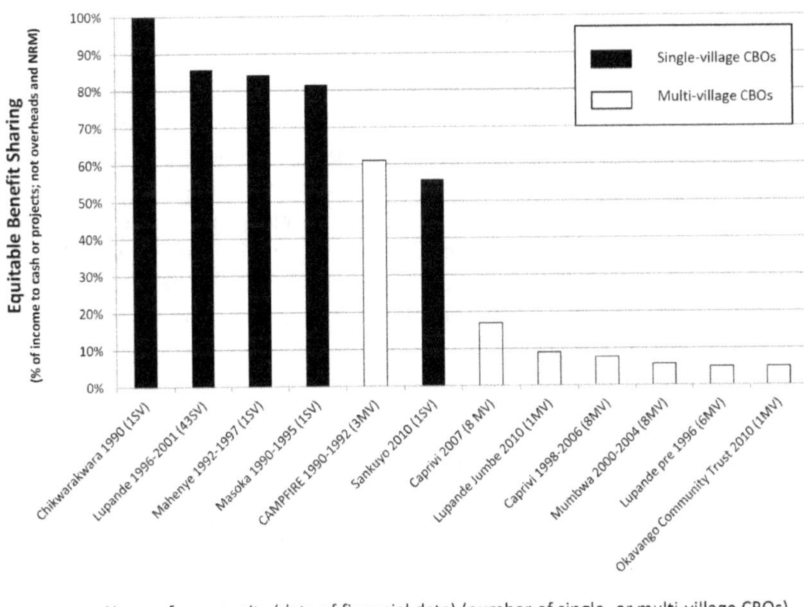

Figure 8.6 Allocation of within-community expenditure to 'equitable benefit sharing'.

- Where sound procedures are implemented in single-villages, 80 per cent of income benefits people (e.g. Lupande, Masoka).
- In multi-village CBOs, in the absence of conformance monitoring, benefit sharing falls to less than 20 per cent of expenditure. The committee uses most of the money for 'management' (e.g. Caprivi, Kafue).
- Single-villages with weaker procedures (e.g. Sankuyo), and multi-villages with strong procedures (e.g. CAMPFIRE), fall somewhere in between.

However, it is important to be cautious in assigning the 'failure' label to programmes that have not benefited people, because judging success or failure depends on one's perspectives and objectives. Thus, CAMPFIRE and the Lupande GMA emphasized multi-dimensional poverty reduction (Sen 1999) as much as conservation, and achieved both (Sen 1999; Dalal-Clayton and Child 2003; Taylor and Murphree 2007). Caprivi originated as a community game guard programme (Owen-Smith 2012). More than half the money is spent employing game guards and resource monitors, and the positive trends in wildlife populations and income reflect success in terms of the original objectives, although we suggest this may ultimately be undermined because participation by, and benefits for, ordinary people are low. In Sankuyo, people were happy with their benefits and rights to participate. Unpublished data suggests they were twice as well off as nearby communities that depended on flood agriculture and livestock, and that their income was more equitably distributed. However, the community had few rights to manage wildlife, and consequently took little action to do so. Recent surveys have shown that wildlife in the Okavango delta was in decline (Chase 2011), albeit for a multitude of reasons largely beyond the control of the community (Fynn and Bonyongo 2010; Child et al 2011).

In the above case, CBNRM achieved some combination of conservation and development objectives. However, in multi-village programmes which are not well managed, CBNRM can fail on all counts. In the eight CRBs near Kafue National Park in Zambia, the combination of representational governance with weak oversight and capacity building left scope for elite capture and mismanagement. Community benefits were negligible (6 per cent), most money was absorbed by the ubiquitous administration and 'sitting allowances' (35 per cent), or was not accounted for (40 per cent). The main added value was the 25 per cent of income used to pay the salaries of a few village scouts, but conservation outcomes were not measured (Malenga 2004).

What workshop games reveal about community budgeting processes and scale

The results of simulations of participatory and representational budgeting processes in training games lends support to findings that there are important differences between the two processes.

At governance training workshops in Maun, Botswana, the participants were divided into groups of 10–15, including community leaders, district officials and

NGO support agencies. First, we simulated a representational budget process. The group elected three or four people to represent them. The representatives constructed a budget on a flip chart in front of the 'community' (but with no input from them). In a simulation of a community AGM, the representatives presented the budget back to the community, who were allowed to make changes. Surprisingly few changes were made, because the 'committee' either defended its decisions or did not change the flip chart budget in response to comments by the community.

Then a participatory budgeting exercise was conducted. The composition of the groups was not changed, but this time the 'community' sat in a circle and made a budget on a flip chart. The budget was messier, with many changes and crossings-out in response to community discussions.

In the representational processes, an average of 36 per cent of the budget was allocated to community benefits (i.e. 20 per cent, 44 per cent, 43 per cent) compared to 72 per cent in the participatory process (80 per cent, 67 per cent, 69 per cent). This approximates the 80:20 rule hypothesized from field data.[2] The social dynamics in the participatory budgeting were more positive than in the representational budgeting exercise, and on a scale of one (terrible) to ten (wonderful), nine participants ranked representational and participatory budget processes as two point zero and seven point seven respectively.

These CBNRM training workshops also included exercises to understand scale. Participants listed on cards the functions that needed to be performed to manage their resources. They pinned these cards hierarchically next to labels that represented scale (individuals, villages, multi-villages, districts and so on), following the instruction that functions should be managed at the lowest level where the costs and benefits were internalized at a single locus of management. Cards for the management of fields, trees, small animals, community projects and finances were usually placed opposite household or village. Those for elephant management, tourism concessions or river management were placed higher, at area, district or even national level. The cards formed a triangle with a large base because most functions were considered to be best managed locally.

In the next exercise, participants placed cards at the level where functions were currently located. The triangle was turned on its head, with more cards at higher levels and relatively few lower down. This supports Murphree's (2000) principles of jurisdictional parsimony. Where scale is built through upward delegation, it is parsimonious in terms of the number of functions aggregating at higher levels. But when scale is built through expropriation from above (rather than delegation from below), many functions accumulate at higher levels of administrations, which become overwhelmed by, and unable to manage, the associated complexity.

Discussion and conclusion

The data presented suggests that processes of representational and participatory governance are profoundly different, with only the latter resulting in equitable benefit sharing and high levels of participation. These, at least rhetorically, are important goals of CBNRM.

There is theoretical support for these results. Subsidiarity is a key principle for building up complex, hierarchical systems of governance (Handy 1994; Meadows 2008), with management theorists and new institutional economists agreeing that complex systems need to be constructed from the bottom-up and to serve the bottom. For example, Douglas McGregor's (1960) observation that people are self-actualizing (Theory Y versus Theory X) lies at the heart of a paradigm shift from the centralized task management of Henry Ford and Frederick Taylor to the devolved, self-managed teams and knowledge economy of modern management practice (Drucker 1973; Peters and Waterman 1982; Micklethwait and Wooldridge 1998).

Similarly, new institutional economists find a strong association between economic performance, individual freedoms, property rights and the unprecedented prosperity of liberal democratic nation states. North makes this point by showing that of 30 countries with a per capita GDP exceeding US$20,000, 23 are liberal democracies (North 2005; North et al 2009). The others are four small oil producing states and three city states. This reinforces our observation that participatory governance both organizes and releases human energy, including the freedom to choose (Sen 1999), and is critical to the multi-dimensional performance of CBNRM.

Given the magnitude of the differences that we have presented between participatory and representational governance, why is participatory democracy so rare in CBNRM and in development more widely? One argument is that small communities face significant diseconomies when it comes to obtaining technical and administrative skills (Overdevest 2000). A more cynical argument is that because participatory democracy can unleash the power and potential of the people, governments and even facilitating agencies are threatened by it. A third reason is logistical: that support agencies prefer to work with fewer, larger communities. Also, the importance of participation and pro-poor benefit may be recognized more in rhetoric than in practice, with implementing agencies favouring clear-cut outcomes like conservation, income and projects over much fuzzier social processes.

Finally, in a world of increasing scale, change and complexity, many commentators wonder if it is realistic to promote face-to-face processes of decision-making and accountability. Our counter-argument is that poverty, mismanagement and elite capture are most closely associated with states and meso-institutions that are democratically hollow, and that, having accepted the logistical challenges, maybe we really do need to build new social institutions from the bottom up and not just from the top down.

Returning to Murphree's (2000) seminal paper on scale, we may be able to have our cake and eat it too, if we think more carefully about the sequencing and structuring of scale, and take especial care not to assume meso-level institutions into place, but rather to build them carefully from the bottom up. Of course many functions need to be managed at higher scales than individuals or single-villages.

The problem is that meso-level institutions like district wildlife or forestry committees are democratically hollow, with weak roots in the community.

Murphree argues eloquently that nested hierarchies should not be constructed from above by usurping rights from landholders. This alienates people who live with wildlife, and it is deterministic about land use outcomes, overwhelms management capacity by aggregating too many functions at higher levels and weakens accountability by lengthening feedback loops. The aggregation of functions should rather occur through a process of upward delegation (Murphree 2000), with the critical Tocquevillian implication that rights and governance should originate in the people (Mansfield and Winthrop 2000).

The common response to this proposal is that participatory democracy is all very well, but at any significant scale, governance has to be representational. The argument put forward in this chapter is that building scale comes later in the sequencing of rural governance. If we are to achieve the socio-politico-economic objectives of equitable benefit sharing, empowerment and participation, we need to build hierarchical systems from the bottom up to ensure that choice and governance originate in the people. Indeed, what if we think of communities as independent jurisdictional units with the same rights as private landholders, rather than as part of bureaucratic hierarchies?

At the beginning of CAMPFIRE, Zimbabwe's chief economist, Norman Reynolds, proposed that the basic unit of governance should be the (single) 'village company'. While most CBNRM initiatives have drifted into structures that resemble district governments, there would be merit in returning to Reynold's suggestion. Ideally, CBNRM should be based on the devolution of rights to benefit, manage and allocate (or sell) wild resources, preferably in the form of collective title. If we were to constitute such communities as village companies, we would be emphasizing the entrepreneurial and productive nature of CBNRM, and not just its regulatory functions. We would also be adding critical protection of the rights of 'shareholders', especially minorities and marginalized groups, using modern and legally enforceable corporate legal documents, not just legislation, policy or administrative arrangements. Perhaps this is a way of addressing North's challenge: that we need to find a way to protect new institutions, including processes for managing, sharing and allocating dividends, information and participation in decision-making.

In conclusion, it should be accepted that the main challenge to CBNRM is failed devolution. However, in places where devolution has occurred, the social performance of CBNRM is closely related to the configuration of micro-governance. Specifically, CBNRM will remain disappointing in terms of equitable benefit sharing and participation unless it is rooted in well-designed face-to-face processes of decision-making and accountability, and unless these conditions are protected by external conformance management and monitoring.

Implementing participatory democracy is probably easier than many people imagine (Child and Dalal-Clayton 2004), but even if it is difficult, do we really have an alternative? Can we succeed in either conservation or development if we do not create robust institutions that empower ordinary people to participate in and receive an equitable share of the benefits from CBNRM?

Notes

1 The data are based on unpublished financial records for 12 CAMPFIRE districts. The principal author of this chapter was responsible for the monitoring between 1989 and 1995.

2 The results probably underestimate differences between participatory and representational processes because most participants were community leaders; much of the training emphasized community benefit and may have biased results in this direction; and the communities concerned were required to employ qualified managers, which imposed fixed costs that might have exaggerated administrative costs in the other direction.

References

Arnold, J. E. M. (2001) *25 Years of Community Forestry*, Food and Agricultural Organization of the United Nations, Rome

Bond, I. (2001) 'CAMPFIRE and the incentives for institutional change', in D. Hulme and M. Murphree (eds) *African Wildlife and Livelihoods: The Promise and Performance of Community Conservation*, James Currey, Oxford, pp. 227–43

Bray, D., Antinori, C. and Torres-Rojo, J. M. (2006). 'The Mexican model of community forest management: The role of agrarian policy, forest policy and entrepreneurial organization', *Forest Policy and Economics*, vol 8, no 4, pp. 470–84

CAMPFIRE (2007) *Annual Report 2006–2007*, CAMPFIRE Association, Harare

Chambers, R. (1994) 'The origins and practice of participatory rural appraisal', *World Development*, vol 22, no 7, pp. 953–69

Charnley, S. and Poe, M. R. (2007) 'Community forestry in theory and practice: Where are we now?' *Annual Review of Anthropology*, vol 36, pp. 301–36

Chase, M. (2011) 'Dry season fixed-wing aerial survey of elephants and wildlife in northern Botswana, September–November 2010', Elephants Without Borders, Kasane, Botswana

Child, B. (1993) 'Zimbabwe's CAMPFIRE programme: Using the high value of wildlife recreation to revolutionize natural resource management in communal areas', *Commonwealth Forestry Review*, vol 72, no 4, pp. 284–96

Child, B. and Peterson, J. (1991) 'CAMPFIRE in rural development: The Beitbridge experience', Branch of Terrestrial Ecology, Department of National Parks and Wildlife Management and Centre for Applied Social Studies, University of Zimbabwe, Harare

Child, B. and Dalal-Clayton, B. (2004) 'Transforming approaches to CBNRM: Learning from the Luangwa experience in Zambia', in T. O. McShane and M. Wells (eds) *Getting Biodiversity Projects to Work. Towards More Effective Conservation and Development*, Columbia University Press, New York, pp. 256–89

Child, B., Ward, S. and Tavengwa, T. (1997) *Zimbabwe's CAMPFIRE Programme: Natural Resource Management by the People*, Canon Press, Harare, Zimbabwe

Child, B., Jones, B., Mazambani, D., Mlalazi, A. and Moinuddin, H. (2003) 'CAMPFIRE: Communal Areas Management Programme for Indigenous Resources, final evaluation report', Zimbabwe Natural Resources Management Program – USAID/Zimbabwe Strategic Objective No. 1, USAID, Harare

Child, B., Barnes, Barnes, G., Nyamoga, G., Odera, E., Pazos, C., Kalala, T., Michaels, L., Diane, T. (2011) 'Sustainable development plan for Sankuyo and management plan for NG 33 and NG 34', Sankuyo Tshwaragano Management Trust, Maun, Botswana

Crook, R. C. (2003) 'Decentralisation and poverty reduction in Africa: The politics of local-central relations.' *Public Administration and Development*, vol 23, no 1, pp. 77–88

Dahl, R. (1998) *On Democracy*, Yale University Press, New Haven, CT

Dalal-Clayton, B. and Child, B. (2003) *Lessons from Luangwa: The Story of the Luangwa Integrated Resource Development Project, Zambia*. International Institute for Environment and Development, London

Drucker, P. (1973) *Management: Tasks, Responsibilities, Practices*, HarperCollins, New York

Frizen, S. A. (2007) 'Can the design of community-driven development reduce the risk of elite capture? Evidence from Indonesia', *World Development*, vol 35, no 8, pp. 1,359–75

Fynn, R. W. S. and Bonyongo, M. C. (2010). 'Functional conservation areas and the future of Africa's wildlife', *African Journal of Ecology*, vol 49, pp. 175–88

Goredema, L., Bond, I. and Taylor, R. (2006) 'Building capacity for local-level management through participatory technology development', *Participatory Learning and Action*, vol 55, pp. 30–6

Grindle, M. S. (2007) *Going Local: Decentralization, Democratization and the Promise of Good Governance*, Princeton University Press, Princeton, NJ

Handy, C. (1994) *The Empty Raincoat: Making Sense of the Future*, Arrow Books, London

Hulme, D. and Murphree, M. (2001) *African Wildlife and Livelihoods: The Promise and Performance of Community Conservation*, James Currey, Oxford

Larson, A. M. and Soto, F. (2008) 'Decentralization of natural resource governance regimes', *Annual Review of Environment and Resources*, vol 33, no 1, pp. 213–39

Lubilo, R. (2007) 'A comprehensive report on the status of governance and participation in Lupande Game Management Area of South Luangwa Area Management Unit in Zambia', unpublished report

Lubilo, R. (2011) 'An assessment of the impact of devolved governance in natural resources management: A case study of Wuparo Conservancy, Namibia', MSc Thesis in Conservation and Rural Development, University of Kent

Lubilo, R. and Child, B. (2010) 'The rise and fall of community-based natural resource management in Zambia's Luangwa Valley: An illustration of micro- and macro-governance issues', in F. Nelson (ed.) *Community Rights, Conservation and Contested Land: The Politics of Natural Resource Governance in Africa*, Earthscan, London, pp. 202–26

Malenga, G. (2004) 'Audit of accounts, financial and management systems of 8 community resource boards in Mumbwa, Namwala and Kafue Flats Game Management Areas for the period covering 1 January 2000 to 31 August 2004', CBNRM-Mumbwa project (Danida), Lusaka

Mansfield, H. C. and Winthrop, D. (2000) *Alexis de Tocqueville, Democracy in America*, University of Chicago Press, Chicago

Mapedza, E. and Bond, I. (2006) 'Political deadlock and devolved wildlife management in Zimbabwe: The case of Nenyunga Ward', *Journal of Environment and Development*, vol 15, no 4, pp. 407–27

Martin, R. (1981) 'Sebungwe region: A planning framework', Department of Physical Planning, joint report, Department of National Parks and Wildlife Management, Harare

Martin, R. (1986) 'Communal Areas Management Programme for Indigenous Resources (CAMPFIRE)', April, revised edition, Branch of Terrestrial Ecology, Department of National Parks and Wildlife Management, Harare

Martin, R. (2009) 'Murphree's laws, principles, rules and definitions', in B. B. Mukamuri, J. M. Manjengwa and S. Anstey (eds) *Beyond Proprietorship: Murphree's Laws on Community-Based Natural Resource Management in Southern Africa*, Weaver Press, Harare, pp. 7–28

McGregor, D. (1960). *The Human Side of Enterprise*, New York, McGraw-Hill

Meadows, D. H. (2008) *Thinking in Systems: A Primer*, Earthscan, London

Micklethwait, J. and Wooldridge, A. (1998) *The Witch Doctors: Making Sense of the Management Gurus*, Times Books, New York

Moinuddin, H., Child, B. Jones, M., Dunham, K., Pope, C., Kemp, I., Sichilongo, C., Simukanze, A. and Mushingo, R. (2002) South Luangwa Area Management Unit, Project Document, Phase V. Lusaka, Royal Norwegian Embassy

Murphree, M. (1994) *Communities as Resource Management Institutions*, Gatekeeper Series 36, International Institute for Environment and Development, London

Murphree, M. (2000) 'Constituting the commons: Crafting sustainable commons in the new millennium', paper presented at Multiple Boundaries, Borders and Scale, Eighth Biennial Conference of the International Association for the Study of Common Property (IASCP), Bloomington, IN, 31 May –4 June

Murphree, M. (2001) 'Community, council and client: A case study in ecotourism development from Mahenye, Zimbabwe', in D. Hulme and M. Murphree (eds) *African Wildlife and Livelihoods: The Promise and Performance of Community Conservation*, James Currey, Oxford, pp. 177–94

Murphree, M. (2004) 'Communal approaches to natural resource management in Africa: From whence to where?', keynote address to the Breslauer Symposium on Natural Resource Issues in Africa, University of California, Berkeley, CA, 5 March

Murphree, M. (2005) 'Congruent objectives, competing interests, and strategic compromise: Concept and process in the evolution of Zimbabwe's CAMPFIRE, 1984–1996', in J. P. Brosius, A. L. Tsing and C. Zerner (eds) *Communities and Conservation: Histories and Politics of Community-Based Natural Resource Management*, Rowman & Littlefield, Oxford, pp. 105–48

NACSO (2008) *Namibia's Commmunal Conservancies: A Review of Progress and Challenges in 2007*, Namibian Association of CBNRM Support Organisations, Windhoek

Nasuchona, N. and Charles, A. (2010) 'Community involvement in fisheries management: Experiences in the Gulf of Thailand countries', *Marine Policy*, vol 34, no 1, pp. 163–9

Naughton-Treves, L., Holland, M. B. and Brandon, K. (2005) 'The role of protected areas in conserving biodiversity and sustaining local livelihoods', *Annual Review of Environment and Resources*, vol 30, pp. 219–152

Nelson, F. (2010) *Community Rights, Conservation and Contested Land: The Politics of Natural Resource Governance in Africa*, Earthscan, London

North, D. C. (2003) 'The role of institutions in economic development', Discussion Paper Series No. 2003.2, United Nations Economic Commission for Europe, Geneva

North, D. C. (2005) *Understanding the Process of Economic Change*, Princeton University Press, Princeton, NJ

North, D. C., Wallis, J. J. and Weingast, B. R. (2009) *Violence and Social Orders: A Conceptual Framework for Interpreting Recorded Human History*, Cambridge University Press, Cambridge

Owen-Smith, G. (2012) *An Arid Eden: A Personal Account of Conservation in the Kaokaveld*. Johannesburg & Cape Town, Jonathan Ball Publishers

Ostrom, E. and Nagendra, H. (2006) 'Insights on linking forests, trees, and people from the air, on the ground, and in the laboratory', *Proceedings of the National Academy of Science*, vol 103, no 51, pp. 19,224–31

Overdevest, C. (2000) 'Participatory democracy, representative democracy, and the nature of diffuse and concentrated interests: A case study of public involvement on a national forest district', *Society and Natural Resources*, vol 13, no 7, pp. 685–96

Peters, T. and Waterman, R. H. (1982) *In Search of Excellence: Lessons from America's Best-Run Companies*, HarperCollinsBusiness, London

Poffenberger, M. (2006) 'People in the forest: Community forestry experiences from Southeast Asia', *International Journal of Environment and Sustainable Development*, vol 5, no 1, pp. 57–69

Pomeroy, R. S. (1995) 'Community-based and co-management institutions for sustainable coastal fisheries management in Southeast Asia', *Ocean & Coastal Management*, vol 27, no 3, pp. 143–62

Ribot, J., Agrawal, A. and Larson, A. M. (2006) 'Recentralizing while decentralizing: How national governments reappropriate forest resources', *World Development*, vol 34, no 11, pp. 1,864–86

Rihoy, E., Chirozva, C. and Anstey, S. (2007) '"People are not happy": Speaking up for adaptive natural resource governance in Mahenye', Occasional Paper No. 31, Programme for Land and Agrarian Studies, University of the Western Cape, Cape Town

Rozemeijer, N. (2009) 'CBNRM in Botswana', in H. Suich and B. Child (eds) *Evolution and Innovation in Wildlife Conservation: Parks and Game Ranches to Transfrontier Conservation Areas*, Earthscan, London, pp. 243–56

Sen, A. (1999) *Development as Freedom*, Anchor Books, New York

Tacconi, L. (2007) 'Decentralization, forests and livelihoods: Theory and narrative', *Global Environmental Change*, vol 17, pp. 338–48

Taylor, R. (2009) 'Community based natural resource management in Zimbabwe: The experience of CAMPFIRE', *Biodiversity and Conservation*, vol 18, pp. 2,563–83

Taylor, R. D. and M. Murphree (2007). 'Case studies on successful southern African NRM initiatives and their impacts on poverty and governance. Zimbabwe: Masoka and Gairesi', USAID-FRAME, Washington, DC; IRG, Washington, DC; IUCN South Africa Country Office, Pretoria

Tilly, C. (2007) *Democracy*, Cambridge University Press, Cambridge

Van der Riet, M. (2008) 'Participatory research and the philosophy of social science: Beyond the moral imperative', *Qualitative Inquiry*, vol 14, no 4, pp. 546–65

9 Fisheries co-management in the Okavango Delta's panhandle

The Okavango Fisheries Management Committee case study

Belda Mosepele, Ketlhatlogile Mosepele, Shadrack Mogotsi and Douglas Thamage

Introduction

Floodplain fisheries are prone to conflict (Charles 1992). Conflict in such natural resource systems is caused by 'rising populations and/or a decreasing natural resource base' (Bennet et al 2001). Generally this becomes an issue when there are competing interests for the same resource and one party tries to assert its 'interests' over those of other parties (Food and Agriculture Organization of the United Nations [FAO] 1998). Fundamentally, these conflicts can be defined at three levels: intra micro-micro conflicts, inter micro-micro conflicts and micro-macro conflicts (Warner 2000). These definitions apply to conflicts occurring respectively within communities (intra micro-micro), among communities (inter micro-micro) and between communities and external groups such as government agencies or private and civil society organizations (micro-macro). Charles (1992) attributes some of the fishery conflict to policy failure, arguing that a management attempt to increase or introduce efficiency in a fishery generally equates to social well-being, as a management objective, to rent maximization from the fishery, which may cause conflict due to conflicting philosophies.

Management of the Okavango Delta fishery was based on a centralized approach (Department of Wildlife and National Parks [DWNP] 2008; Mosepele 2008) which failed to resolve conflict among resource users (Mosepele 2008). Indeed, this top-down management approach, which excludes traditional management (i.e. customary governance) systems, unwittingly contributed to conflict in the fishery (Cassidy et al 2011). While most of the conflict was predicated on allegations of overexploitation of fish stocks by commercial fishing (Haggett 1999), Mosepele (2000) showed that the Delta's key fish stocks (e.g. *Oreochromis andersonii, Oreochromis macrochir, Tilapia rendalli, Hydrocynus vittatus, Clarias gariepinus*) were not being biologically over-exploited. Preliminary investigations revealed that the real source of conflict was competition for access rights (Mendelsohn et al 2010) that is intrinsic to common property resources (Pinho et al 2012). Moreover, Setswalo (2007) found that conflicts increased when floods receded in the upper Delta, before the next flooding season. Understandably, unresolved conflict may ultimately affect the socio-economic livelihoods of the poorest members of society (Bennett et al 2001).

It was against this background that the Building Local Capacity for Conservation and Sustainable Use of Biodiversity in the Okavango Delta [BIOKAVANGO] Project was developed, to mainstream biodiversity conservation into three production sectors in the Delta (water, tourism and fisheries). The project was funded by the United Nations Development Programme and the Botswana government. Because conflict had been identified as a barrier to sustainable development (Warner 2000), which would therefore hamper sustainable biodiversity conservation in the Delta, conflict resolution in the fishery was one of the project's major interventions (Global Environment Facility [GEF] 2006).

Co-management has now been established as a mainstream approach to small-scale fisheries management (Branch et al 2011; Evans et al 2011), and has been implemented in fisheries around the world – for example, Zambia (Ticheler et al 1998), Samoa (King and Faasili 1999) and Laos (Baird 2003; Nsiku 2003) – with varying degrees of success (Gutiérrez et al 2011) and failure (Evans et al 2011). It is increasingly seen as a way to 'move away from past management failures' so that the people and the state can collaborate in management and share responsibilities in the governance of fish resources (Jul-Larsen and Van Zwieten 2002). In what Jul-Larsen and Van Zwieten (2002) call a 'genuinely democratic process', local governance structures and central government can develop and implement an 'adaptive' co-management paradigm. The 'adaptive' aspect is important because it allows for a feedback system that gives local community structures an opportunity to reflect upon, discuss, modify and improve management decisions (Baird 2003).

Adopting a co-management approach to fisheries also promotes (potentially) the utilization of indigenous fisher knowledge, which is an important pre-condition for sustainable fisheries management (e.g. Johannes et al 2000). Indeed, while community-based management of natural resources is not necessarily a panacea for all the ills hampering sustainable resource utilization, it can, where it works, be a vehicle for sustainable development (Charles 2008), and also a conduit for biodiversity conservation, environmental sustainability and social justice. This management paradigm is in line with new trends in resource management that favour the devolution of governance to local communities or resource users (e.g. Jul-Larsen and Van Zwieten 2002; Charles 2008).

The co-management of natural resources in Botswana has been codified within the concept of community-based natural resource management (CBNRM), which started in Botswana in 1989 (Thakadu et al 2005). Its main focus was to promote local participation in natural resource management and tourism development (Mbaiwa and Thakadu 2011). CBNRM has had both successes and failures in Botswana. According to Mbaiwa (2011), the misappropriation of funds and the lack of a culture of entrepreneurship are among the key factors in the failure of CBNRM projects. Notable successes of CBNRM include improved socio-economic benefits to local communities and improved custodianship of natural resources (as evidenced by significant decreases in poaching) by local communities

(Mbaiwa 2011). However, notwithstanding the long history of co-management of natural resources in Botswana, there are no CBNRM activities based on fish, a fact that Mbaiwa (2011) highlights.

The aim of this chapter is therefore to provide an overview of the development and implementation of a fisheries co-management model designed expressly for the Okavango Delta, primarily as a conflict intervention model and secondarily to empower local governance of natural resource management in fisheries. This has been a four-step process: (i) consultations and the identification of key stakeholders; (ii) the formation of a local governance structure, the Okavango Fisheries Management Committee (OFMC); (iii) the adoption of an adaptive system as a model for fisheries management (including conflict resolution); and (iv) negotiation and debate on some key fisheries management issues in the Delta.

Development of a co-management fisheries model: The approach

The upper portion of the Okavango Delta panhandle was the site of the BIOKAVANGO project to pilot a series of interventions in the fish sector from 2007 to 2010. Compared to the rest of the Delta, the panhandle is a well-defined meandering single channel (Wolski and Murray-Hudson 2005; Figure 9.1) which is reached around November by annual seasonal floods that peak between March and April (Murray-Hudson 2009; Mendelsohn et al 2010). The highest population density in the district is found in the upper panhandle because that is where approximately 62 per cent of villages in the entire district with populations of over 1,000 (based on 2001 national census statistics) are found. It is possibly due to this high population density that the highest numbers of fishers are also found in the upper panhandle (Mendelsohn et al 2010). In fact, Mosepele (2001) notes that out of the 3,289 fishers in the entire Delta in 1998, just over 52 per cent were in the upper panhandle.

The Okavango Delta, the world's largest Ramsar[1] site, is a large inland wetland created by the Okavango River (Ramberg et al 2006), where the total flooded area varies annually, depending on the size of the flood (Wolski et al 2006). It is a dynamic ecosystem, characterized by a regular, but highly variable, flood pulse (Ramberg et al 2006; Murray-Hudson 2009), which supports the largest fishery in Botswana (Merron and Bruton 1988; Mosepele 2000). Therefore, while the Delta's waters might be oligotrophic (Cronberg et al 1995), this seasonal flooding enhances aquatic primary production (Mosepele 2008; Mosepele et al 2009) by trapping nutrients of terrestrial origin in the aquatic habitat in a process termed the 'flood-pulse concept', originally in relation to the Amazon (Junk et al 1989).

The seasonality of the flooding has created three major hydro-ecological zones in the Delta (Ramberg et al 2006), which in turn have brought about diverse habitats in the system. According to Smith (1976), the Delta is divided into several main habitats: perennial swamp, perennial and seasonal swamp, floodplains, rivers and water courses, lagoons, islands and mainland edges. This system supports approximately 71 different fish species (Ramberg et al 2006; Mendelsohn et al

2010), distributed heterogeneously throughout the different habitats (Mosepele and Mosepele 2005; Mosepele 2008; Mosepele et al 2009; Mosepele et al 2011). Apart from supporting thriving subsistence (Ngwenya and Mosepele 2008) and small-scale commercial fishery (Mosepele and Ngwenya 2010), the Delta is crucial to Botswana's tourism industry (Mbaiwa 2003) and supports the subsistence of the local population (Wolski and Murray-Hudson 2005).

The species mainly targeted for recreational fishing are the 'trophy sized' *Hydrocynus vittatus* (tigerfish), *Serranochromis robustus* (nembwe), *Clarias gariepinus* (sharptooth catfish), *Clarias ngamensis* (blunttooth catfish) and *Oreochromis andersonii* (threespot tilapia) (Mosepele et al 2003; Mendelsohn et al 2010; Thomassen 2011), for which most fishing is done in the main channel and large lagoons (Setswalo 2007). Subsistence fishers target a wide range of fish species (from smaller ones such as *Barbus afrovernayi* to the larger sizes, including *Clarias gariepinus*), primarily for household consumption (Mosepele et al 2003; Mosepele et al 2006), but sell a small proportion of their surplus catch (Mmopelwa et al 2009). Conversely, the small-scale, yet profitable, commercial fishery (Mmopelwa et al 2005) harvests market-sized *Oreochromis andersonii, Tilapia rendalli* and *Oreochromis macrochir* (Mosepele et al 2003). Both subsistence and commercial fishing are key resources for household food and nutrition security (Mosepele et al 2006; Ngwenya and Mosepele 2008; Mosepele and Ngwenya 2010): children from fishing households are healthier than those from non-fishing households (Nnyepi et al 2007). Essentially, fishing is a social safety net during times of socio-economic hardship (Mosepele 2000; Mbaiwa et al 2011).

These different exploitation regimes, coupled with the different use values of the Delta, resulted in conflict in the fishery (Mosepele 2000; Bokhutlo et al 2007; Setswalo 2007; Mosepele et al 2009). In fact, Bokhutlo et al (2007) revealed that some of the conflict based on access occurred not only between local fishers and tour operators (inter micro-micro conflict), but also among local fishers themselves (intra micro-micro conflict).

Process of developing co-management

Identification of stakeholders

Participation was at the core of the strategy adopted in developing the OFMC model. Key stakeholders were identified through consultative meetings and interviews with selected informants. These included various local governance structures at community level (e.g. village development committees and local headmen), as well as officials in the Fisheries Division of Botswana's Department of Wildlife and National Parks and the Department of Environmental Affairs, which was the major driver of the Okavango Delta Management Plan.[2] As the plan unfolded, the department conducted frequent and intensive *kgotlas* (consultative community meetings) throughout the Ngamiland district, which put it in a good position to identify key issues and actors in the fish sector. The Fisheries Division, on the other hand, is the government agency tasked with the overall

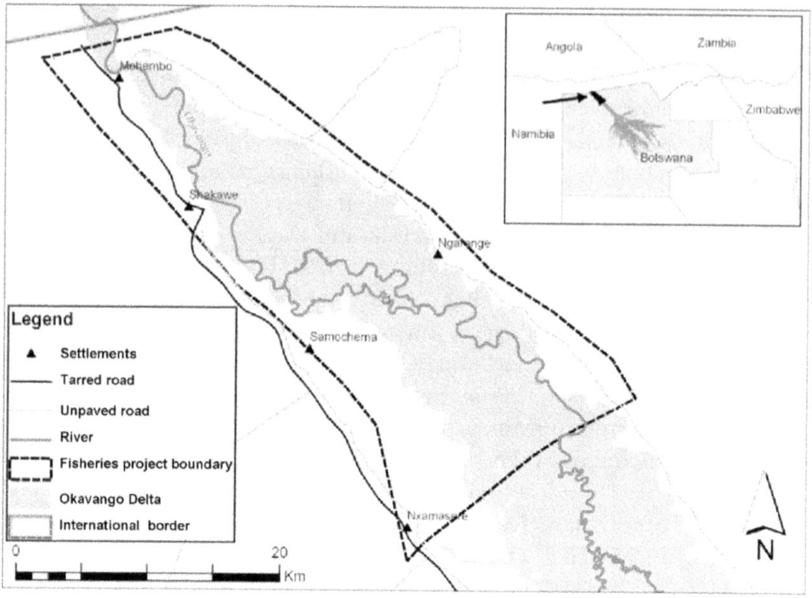

Figure 9.1 Okavango Delta and panhandle with study site marked.

management of Botswana's fish resources and is therefore acquainted with all the main stakeholders in the fish sector.

The philosophy behind this process was to maximize the involvement of all stakeholders and engage all community structures through a series of meetings and workshops. This inclusive participatory approach falls within the framework designed by Welcomme (1998) for the development and management of inland fisheries. Furthermore, a capacity-building approach (based on training needs assessment) was used in a three-pronged strategy to enhance the capacity of the key stakeholders in the sector.

Capacity building

The first strategy entailed capacity building in the Fisheries Division (the regulators) through a series of training programes in fish identification and monitoring. Concurrently, steps were taken to build the capacity of the resource users (fishers) through focused group discussions, indigenous knowledge mapping and training programmes tailored to their needs (e.g. boat maintenance, business and financial procedures, fish monitoring). Existing fisher institutions and community-based organizations were strengthened: for example, the Okavango Fishers Association was helped to revise and amend its constitution, primarily to make it relevant to current issues and concerns. The third strategy was to build the capacity of the private sector, mainly lodge owners, who were engaged in a range of meetings, training courses and workshops (data collection on key fish species targeted by tourists, fish monitoring, etc).

This process of bringing together stakeholders who had previously been at loggerheads, empowering them through, for example, capacity-building interventions, assisting them to identify common issues, sharing information and developing 'local-level, ecologically oriented management initiatives', is similar to the 'Turning the Tide' initiative in Canada's maritime provinces described by Charles (2008). This initiative addressed the efficacy of building community-based fisheries management approaches and creating linkages between the community and non-community actors. The approach corresponds with that of Chuenpagdee and Jentoft (2007), who observe that the success or failure of co-management depends on the pre-implementation processes followed.

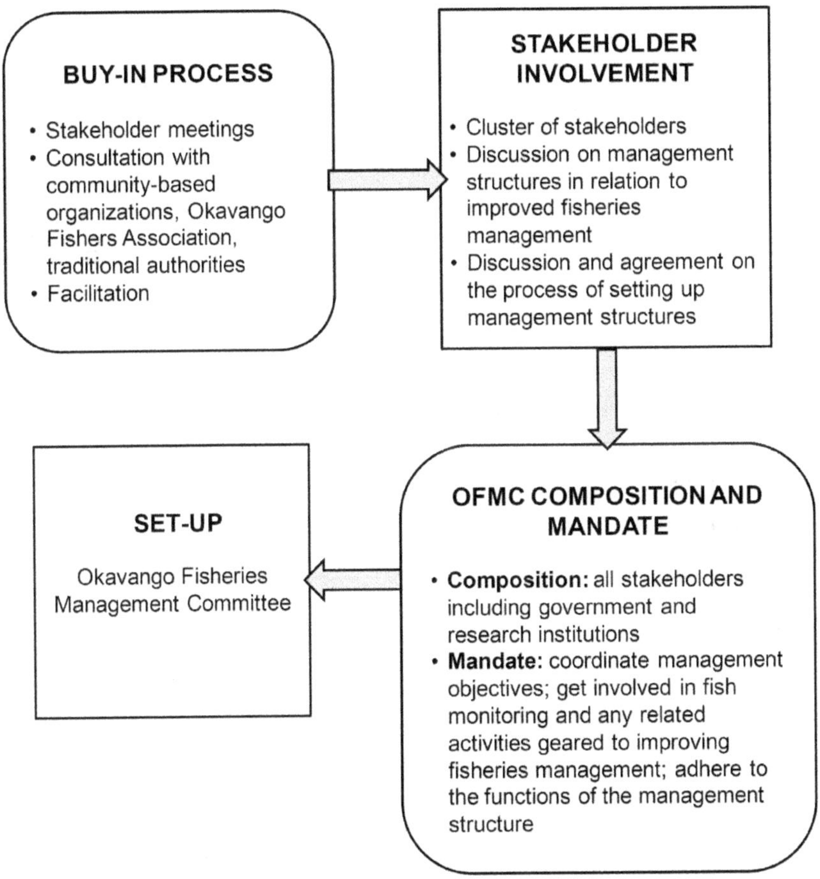

Figure 9.2 Graphic representation of the process followed to form the OFMC.

Formation and composition

The OFMC was formed through the steps depicted in Figure 9.2. The first step involved a series of consultations, which culminated in participants' buy-in of the idea of forming a local co-management structure that was inclusive and that embraced greater understanding of the need to co-manage the fish resources. The next major step was a facilitation process that inculcated greater involvement in and ownership of the process by all stakeholders. A final meeting then discussed the composition and mandate (roles, responsibilities and functions) of the proposed OFMC, whereupon the committee was finally set up.

The OFMC is composed of a diverse group of stakeholders including both state and non-state actors (Figure 9.3). All the state actors are government institutions mandated with implementing sector policies relevant to the management of the Delta fisheries. These include water resources (Department of Water Affairs), environmental issues in general (Department of Environmental Affairs), local government (North West District Council), land use planning and zoning (Tawana Land Board) and tourism activities (Department of Tourism). The Fisheries Division, which is the sole government institution mandated to manage Botswana's fish resources, was allocated the position of committee secretary. The Okavango Research Institute, an institution of the University of Botswana, was given the chair, based on the assumption that it would be impartial with regard to resource

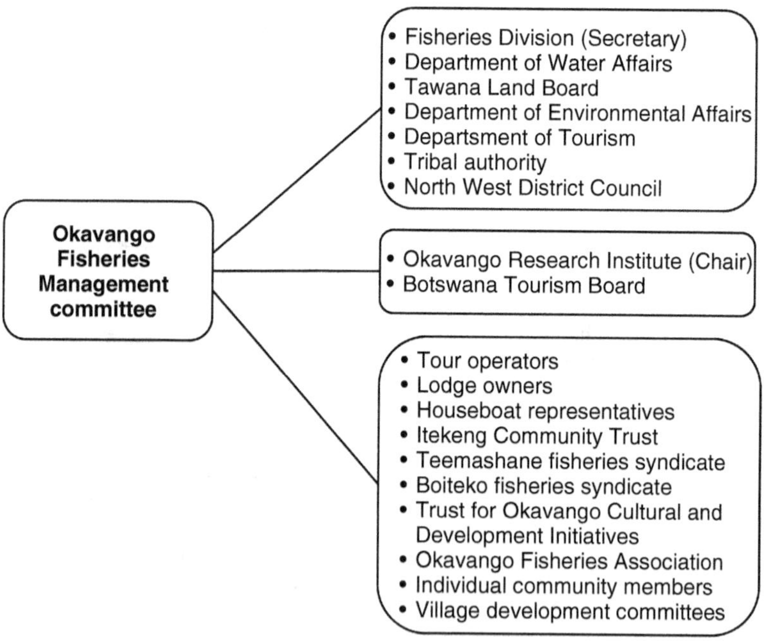

Figure 9.3 Composition of the OFMC indicating state and non-state actors.

management. Effectively, these two bodies have become the anchors of the OFMC. Moreover, it was assumed that both the Okavango Research Institute and the Botswana Tourism Board, as parastatal entities (semi-autonomous because they get funding from government but have a high degree of independence in decision-making), would play a mediating role in the committee.

Non-state actors also cover a wide range, from tour operators and fisher organizations – for example the Okavango Fishers Association and the Teemashane fisheries syndicate, who are key stakeholders and were major opponents in the fishery sector – to village-based government structures such as the village development committees and some individual community members who had strong feelings about fisheries management in the Delta. Many of these individual community members were also embroiled in some of the conflicts observed in the area. The only major group missing from the OFMC are the women basket fishers. However, the representatives of the Trust for Okavango Cultural and Development Initiatives are women and currently represent the interests of women basket fishers. The OFMC's extensive membership is in line with Chuenpagdee and Jentoft's (2007) observation that 'fishers, community members, government officials, scientists and environmental groups' should be included in a participatory process to develop a co-management regime.

Distillation of key management issues

A series of intensive meetings and workshops were held with stakeholders in the study area. The convenors were instructed to follow a strictly democratic and inclusive process in which care was taken, especially in the case of women, to give all participants an equal chance to give their input. The Fisheries Division and Okavango Research Institute facilitated this exercise in order to ensure strict adherence to democratic procedures.

These consultative, participatory meetings were conducted over three years, from 2007 to 2009, their objective being mainly to motivate stakeholders to appreciate the various management issues regarding the fishery and, furthermore, to help them understand the functions of a co-management structure and the concepts of sustainable development and sustainable resource use.

The key issues discussed included the socio-economic dimension (for local fishers and tour operators) of the fishery, indigenous knowledge issues (primarily for local fishers), a biological and ecological understanding of the fisheries, biodiversity conservation (BIOKAVANGO interventions) and a regulatory management approach by the Fisheries Division. This approach to appreciating indigenous knowledge issues is based on the work of experts such as Robin Welcomme (1998), who advise that the cultural dimension of inland fisheries needs to be taken into account as part of the management paradigm, because ignoring fishers' knowledge in fisheries management may put the resource at risk (Johannes et al 2000). King and Faasili (1999) adopted a similar approach in the development of a community-based fisheries management system in Samoa. The basic philosophical construct underpinning the formation of the OFMC was that it would be guided

Table 9.1 Summary of conflict issues raised by stakeholders

Conflict issue	*Explanation or description*
Boats with powerful engines (e.g. 150 h.p.)	These boats create big waves that could destroy nests of endangered birds (e.g. African skimmer) and fish breeding sites.
Houseboats	Stakeholders don't know where these boats come from, whom they belong to and whether they are appropriately licensed for what they do. There are also concerns about the potential impact of the sewage and other waste that they generate.
Fishing competitions	These events bring many boats of different sizes and shapes to the area, some probably not inspected. Stakeholders are concerned that these boats may exceed the limits of what the Delta can support and may be the source of invasive species.
Fishing methods	Gill-net fishing (mostly commercial) is not properly regulated. This practice might be detrimental to fish stocks because there is no daily allowable catch limit.
Disturbance	Noise allegedly caused by night-time fishing affects stakeholders who live or have property near fishing grounds (mostly next to lagoons).
Double netting and drive fishing	These terms describe the practice of beating the water to drive fish into a net and that of placing a second net to catch the fish in case they escape the first one. It is thought that these methods might ultimately be destructive to fish populations.
Legislation	The applicable laws are based on classic fisheries management regulations (e.g. closed seasons, gear restrictions), which do not address conflict issues.
Multiple use	This refers to the use of lagoons by different stakeholders, and fishing in relation to other forms of usage of the lagoons. Different land uses adjacent to the river invariably result in competition for a limited resource – and hence conflict.
Carrying capacity	Concerns about the increased density of boats and lodges in the upper panhandle has created friction between some of the lodge operators. This has spilled over into recreational fishing operations, where tour operators feel there might be too many new entrants into the area, saturating it.
Limited scientific information on fish stocks	This lack creates friction among various fish resource users regarding the status of fish stocks in the Delta.

by the principles of fisheries co-management, focused primarily on conflict resolution. It was also envisaged that the OFMC would introduce a self-policing initiative, which should be both popularly supported and acceptable to the Fisheries Division.

These meetings culminated in a final workshop where all the issues and concerns surrounding the fishery were distilled into the ten problem areas summarized in Table 9.1. These issues were not restricted to overfishing, as had initially been assumed, but rather covered an array of management concerns affecting the Delta's fishery, ranging from general concern about the impact of certain user behaviour on the Delta ecosystem's functioning (e.g. boats with excessively large engines disturbing breeding fish and aquatic birds) to specific concerns about the carrying capacity of the panhandle for tourism-related activities, and the dissemination of scientific information to stakeholders.

Implementation of co-management

Adaptive co-management

Democratic ideals and participatory procedures were advocated and employed as underpinning philosophies of the OFMC. It was envisaged that the committee's decision-making would be highly interactive and democratic, allowing all members the opportunity to contribute their ideas and voice their opinions. During the OFMC's meetings so far, the chair has deliberately solicited opinions from every member on any issue under discussion. This generally results in heated debates, but consensus is always reached eventually.

The decision-making process, which contributes to the overall fisheries management of the Okavango Delta, follows a feedback loop system as illustrated in Figure 9.4. The large arrows follow the committee's decision-making procedure, and the smaller arrows indicate that the OFMC can intervene at any point. The key processes of the feedback system are decision-making, implementation, monitoring and analysis, then revision and modification if and when necessary. This

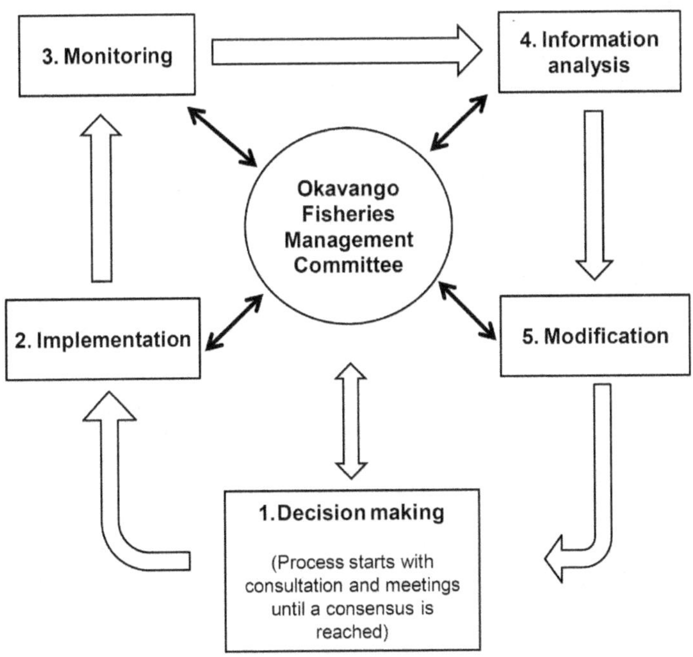

Figure 9.4 Adaptive management strategy used by OFMC in decision-making.

is the process that was followed when the OFMC developed a ten-point code of conduct for responsible fishing (Table 9.2). The first three items are intended to regulate certain practices in the river in order to minimize conflict situations among stakeholders, while the fourth and fifth aim to regulate human behaviour so as to minimize, respectively, the risk of wildfires and the rate of boat collisions. The primary purpose behind items six and seven is to inculcate environmental awareness, particularly in the marking of nets in the water, and hence minimize pollution (thereby increasing the aesthetic value of the system). Items eight and nine endorse government regulations, and the last item requires an affirmation by all parties that they will adhere to the code.

The procedure described here highlights some of the elements of the co-management process and how it is supposed to operate when it is fully functional. In essence, the code of conduct is a blueprint that the OFMC will use in its co-management of the Delta's fishery.

The OFMC also demarcated 'fishing-free zones' in the upper panhandle (Figure 9.5) as one of its management interventions. These zones were primarily designed to address the stakeholders' concern about limited scientific information (Table 9.1). Subsequently, the OFMC instructed the Fisheries Division to initiate a

Table 9.2 Ten-point code of conduct developed by OFMC for conflict resolution

Item	*Code description*
1.	Fishing of any form in front of lodges is restricted. A buffer zone, along the main channels, shall be: 100 metres on either side of the lodge along the channel and 200 metres in front of the lodge.
2.	There shall be 'no wake zones' in the Delta next to vertical banks (which are normally used as bird nesting areas), lagoons, sand banks, in front of lodges and upon meeting traditional craft such as *mekoros* and *rafts*.
3.	For personal safety, *mekoros*, other traditional craft and heavily laden boats shall wait for other faster motorized boats to pass them before they continue on their way.
4.	All campfires must be extinguished after use and when abandoning camps on islands or any part of the Delta.
5.	Anybody driving a boat shall not drink any form of alcoholic beverage.
6.	There shall be no littering in the river, on the banks or on any island in the Delta.
7.	Marking nets with plastic bags shall not be allowed, and only items made out of natural material (e.g. cotton, grass, papyrus) can be used for this purpose. Factory-made floats or any factory-made marker can also be used.
8.	Any fish taken or harvested from the river or water, either by hook, gill net or any other form of fishing gear (by all fisher groups), shall be recorded accordingly to assist the Department of Wildlife and National Parks – Fisheries Division (and other institutions and stakeholders) in medium- and long-term monitoring of fish stocks and fisheries in the Delta.
9.	All stakeholders in the Delta's fishery take it as their individual responsibility to observe and uphold the Fish Protection Regulations of 2008 as promulgated by the Department of Wildlife and National Parks – Fisheries Division.
10.	All stakeholders in the Delta's fishery agree to observe and uphold this code to not only ensure minimum conflict among users but also achieve sustainable fish utilization.

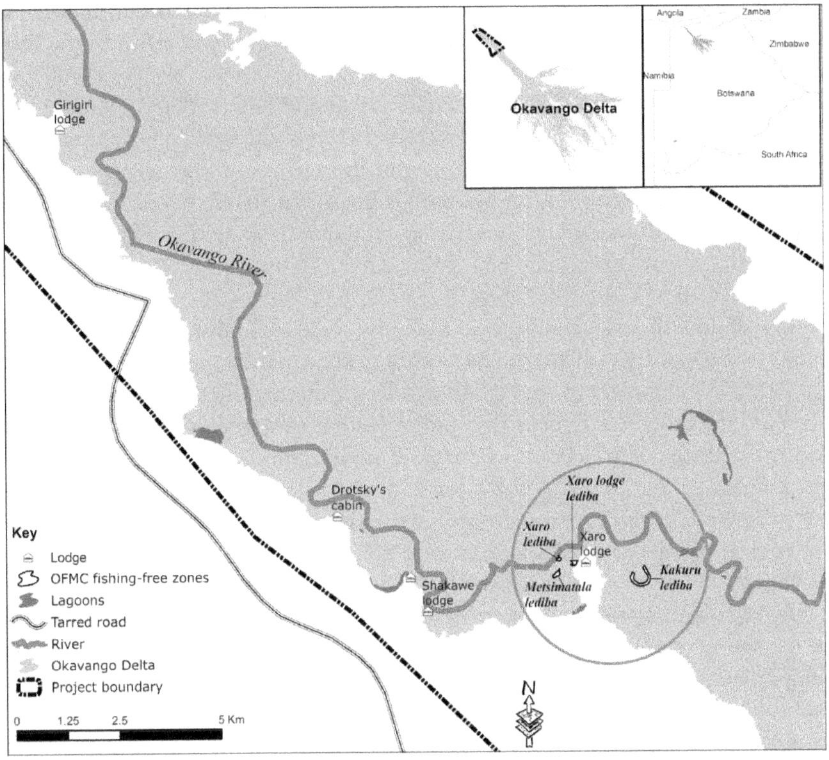

Figure 9.5 Main fishing areas in the upper panhandle and fishing-free zones (lagoons, or ledibas in circle) demarcated by the OFMC as long-term fish monitoring sites.

long-term monitoring protocol in the site so that the committee could be informed regularly on the status of the fish stocks.

It needs to be emphasized that these were the initial issues raised by the OFMC, and more are expected to come up in subsequent meetings. It is quite evident from the code of conduct that the OFMC is concerned about a broad suite of fisheries management issues in the Delta, so it follows that the committee will make further recommendations on the monitoring of issues of concern.

Discussion

This study has proven that the establishment of fisheries co-management in floodplain fisheries like the Okavango Delta is a viable undertaking. While the OFMC was formed primarily as a conflict intervention tool, this study has shown that the issues range well beyond what was documented before (e.g. Kolding 1996; Mosepele 2000; Tweddle et al 2003; Setswalo 2007) and encompass a broad array of management issues in the Delta's fishery. Initially, the main cause of conflict was perceived to be fish stock depletion by commercial 'over-fishing' (Kolding 1996; Haggett 1999; Mosepele 2000), but later analysis revealed that the major conflicts were precipitated by the lack of well-defined property rights in the fishery (Kgathi et al 2005). Indeed, the first documented cases of fishery conflict in the Delta were between commercial and recreational fishers (Bills 1996; Kolding 1996; Haggett 1999). Conflict incidences between commercial and recreational fishers are highest when the water levels in the Delta are low (Setswalo 2007), which suggests that reduced fishing space is the major source of conflict between these two groups. It was against this background that the majority of fishers (67 per cent) felt that the fishery should be regulated by both traditional (local) authorities and government.

This sentiment – that the fishery should be regulated through a collaborative regime by government and traditional authorities (local structures) – laid the foundation for the development of the OFMC as a Delta-centric fisheries co-management model. Ultimately, this information indicates that the process of collaboratively identifying issues of concern can enhance understanding of the fishery system and provide guidance for possible management strategies which can then be jointly devised. In this proposed model, the Fisheries Division and the Okavango Research Institute would play a facilitation role in the management discussions of the OFMC.

The key planning steps – that is, the consultative and facilitation processes – in the formation of the OFMC ensured that there was a high degree of buy-in, ownership, participation and understanding by the key stakeholders. These steps were underpinned by a deliberate decision to initiate consensus-building (Warner 2000) among the previously antagonistic parties. In this respect, special care was taken to ensure that a win-win philosophy was created, so that none of the stakeholders would feel that they had to compromise for the benefit of other groups. This notwithstanding, both sides (i.e. both recreational and commercial fisheries) made some compromises during the negotiations.

The philosophical differences between *consensus* and *compromise* in this context (i.e. fisheries conflict resolution in the Delta) are explicitly defined by Warner (2000), who explains that the latter involves 'trade-offs'. Deliberate steps are then taken to avoid trade-offs by adopting *consensus* as the main conflict resolution strategy. It was perhaps against this background that the 'fishing-free zones' were demarcated in the upper panhandle (see Figure 9.5), where all forms of fishing (commercial, recreational and subsistence) are prohibited. The key objective of this demarcation is to minimize conflict between the recreational and commercial fishers. Generally, the small-scale commercial fishers target their preferred species in floodplain habitats (Mmopelwa et al 2005; Mosepele et al 2006), so this will not have any significant negative effect on their fishing operations. The tour operators also benefit because their 'contentions' against noise from nightly 'drive-fishing' operations by commercial fishers in the periphery of their lodges (Tweddle et al 2003) will now be addressed through the code of conduct.

Moreover, the other reason behind this decision to declare fishing-free zones is the intended implementation of a long-term monitoring programme to monitor fish populations. It is envisaged that this will give the OFMC control and ownership, and will also create a platform for local fishers to integrate their traditional ecological knowledge into the management of the fishery. All these management procedures will then form part of the basis for future assessment of the co-management regime in this fishery.

What is particularly striking about the key conflict issues identified in this study is that they range from biological and ecological concerns to communication problems among stakeholders. The high diversity of management issues in the Delta's fishery is possibly related to that of the stakeholders constituting the OFMC. This is different from previous research (e.g. Tweddle et al 2003), which studied conflict issues only among commercial, subsistence and recreational fishers.

Nonetheless, concerns about 'potential overfishing' through what some OFMC members, especially recreational fishers, describe as 'destructive fishing methods' (e.g. double netting and drive fishing methods as described in Table 9.1), highlight a conservation ethic based on Hardin's (1968) 'tragedy of the commons'. This management approach is the key precursor to the prevailing centralized or top-down governance model of the Delta's fishery (Mosepele 2008). The basic philosophy behind this model is that local communities invariably over-exploit their resources in an open-access regime. This philosophy is also reflected in the 'conservation paradigm' described by Charles (1992), which views fishers as predators who 'act in their self-interest'. Moreover, as noted by Johannes et al (2000), a centralized management paradigm ignores the traditional and indigenous knowledge of fishers.

The formation of the OFMC essentially devolved power from the Fisheries Division and gave the local community a platform on which to participate actively and directly in the management of 'their' resources. Through a consensus-building exercise like that described by Warner (2000), coupled with capacity-building and empowerment through democratic ideals, the ten-point code of conduct was developed.

This co-management approach developed for the Okavango Delta was based upon the FAO Code of Conduct for Responsible Fisheries (FAO 1995), which advocates the inclusion of local communities in the management of their fisheries resources. Co-management is now used extensively to achieve a more holistic approach to fisheries management issues (Jul-Larsen and Van Zwieten 2002), even though Caddy (2002) acknowledges that setting up a 'decision-making process' is the most difficult, and yet important, aspect of fisheries management process. Sipponen and Gréboval (2001) argue that a forum based on a co-management paradigm can be created for commercial and recreational fishers to communicate as a conflict resolution mechanism. One such forum, involving a highly democratic approach, was established in Samoa (King and Faasili 1999) and culminated in a village fisheries management plan that was adopted by most community groups, including women.

Therefore co-management has indeed 'become the buzzword of contemporary fisheries management' (Schreiber 2001; see also Gutiérrez, et al 2011) and has subsequently been adopted or proposed as a major strategy in fisheries management, globally (e.g. Sipponen and Gréboval 2001; Evans et al 2011).

However, Pinho et al (2012) emphasize that cultural and political factors are major barriers to the successful implementation of fisheries co-management. They are certainly potential impediments to the success of the co-management approach in the Delta, where recreational fishers have more economic and political power than the local commercial and subsistence fishers. Moreover, there are cultural distinctions and differences in economic-political power between the commercial and subsistence fishers (Setswalo 2007) which need to be overcome to ensure success and sustainability for this management approach in the Delta.

Jul-Larsen et al (2003) argue that co-management is one of the best tools for conflict resolution in African freshwater fisheries. Moreover, Van Ginkel and Steins (2001) argue that compliance with regulations to solve multiple-use conflicts can be sought and negotiated at the local level before these regulations are eventually codified into law, which agrees with Overby's (2000) observations on the integration of zoning, based on self-regulation, into federal law in the Gulf of Mexico. A more fundamental step towards co-management is recognizing the interests of groups that have been historically disadvantaged, such as women fishers (Haraldsdottir 2000). This observation is relevant in the Okavango Delta, where women basket fishers are not organized into any formal grouping and are therefore not part of any management process. Therefore, while subsistence fishing is a major source of livelihoods in the Delta's communities (Mosepele et al 2006), especially for basket fishers, to whom fishing is also a form of cultural expression (Mmopelwa et al 2009), their concerns still need to be integrated into this co-management paradigm.

It should be emphasized, however, that co-management is not a panacea for all management problems in the Delta's fishery. There is ongoing debate about the efficacy of this approach in fisheries management globally. While Gutiérrez et al (2011) extol the success of co-management in fisheries, they admit that the success of this approach is limited in inland fisheries due to 'weaker social capital and short-term co-management arrangements'. Chuenpagdee and Jentoft (2007) suggest that the success or failure of fisheries co-management regimes depends on the steps taken preceding implementation. Therefore a plethora of issues needs to be taken into account to ensure the success of this management approach in fisheries.

Conclusion

This study has demonstrated the potential of local governance structures with democratic ideals to resolve conflicts through a participatory co-management system in floodplain fisheries, using the case of the OFMC in the Okavango panhandle. The OFMC is multi-stakeholder in character, which makes it appealing to stakeholders. With its democratic approach to dealing with fisheries management issues, it is potentially an effective local governance structure that could, with support (especially financial), become a focal point for fisheries management in the Okavango Delta.

By their nature, floodplain fisheries have diverse management issues (Welcomme 1998) that centralized management alone is generally insufficient to address (Mosepele 2008). The efficacy of multi-stakeholder involvement within a co-management paradigm, initially in conflict resolution and ultimately in overall fisheries management, makes it a more promising approach to floodplain fisheries management.

To enhance the sustainability of the OFMC, especially during its faltering initial steps, it was anchored in two institutions, the Fisheries Division (which appointed the secretary) and the Okavango Research Institute (which provided the chair) as illustrated in Figure 9.3. This was a prudent decision by the committee to ensure the continued existence of this initiative, before it assumes complete financial independence in future. The feedback philosophy of its decision-making process gives the committee an opportunity to reflect upon and assess the relevance and appropriateness of its decisions in relation to certain set goals.

Notwithstanding the initial successes since the establishment of the OFMC, it is not necessarily the final answer, either for conflict resolution or comprehensive fisheries management, in the Okavango Delta. Several challenges remain at the implementation and operational levels that will invariably affect its sustainability. Therefore the long-term sustainability of such structures depends not only on their existence, but also on their functionality.

At the operational level, the multi-stakeholder composition of the committee presents the risk of a lack of common interest, participation and drive in the long run. Currently, the structure is driven by the stakeholders' shared interest in resolving conflict and managing the fish resources. However, this 'collective drive' might reach an optimum level at which conflicts are invariably minimized through effective implementation of the code of conduct, routine fish stock monitoring and judicious adherence to all the management decisions made by the committee. The greatest challenge then would be to maintain the interest of stakeholders in the initiative, and to come up with innovative approaches to fisheries management in the Delta. Other challenges include the need for constant mentorship and capacity-building, the lack of a visible improvement in livelihoods, the absence of a clear understanding of fisheries dynamics and global issues such as climate change, and the failure to include other groups, namely women and young people. A deliberate effort should be made to showcase and integrate indigenous knowledge and cultural values relating to fisheries so that issues of conflict resolution, good governance, long-term monitoring and the management of resources are embraced and sustained by all stakeholders.

Acknowledgements

This study was funded partly by the BIOKAVANGO Project of the United Nations Development Programme, the Global Environment Facility and the Botswana government and partly by the Okavango Research Institute at the University of Botswana.

Notes

1 This designation was made in terms of the Ramsar Convention, whose mission is to ensure 'the conservation and wise use of all wetlands through local, regional and national actions and international cooperation, as a contribution towards achieving sustainable development throughout the world' (Ramsar n.d.).

2 The overall goal of this plan is to 'integrate resource management for the Okavango Delta that will ensure its long-term conservation and that will provide benefits for the present and future well-being of the people, through sustainable use of its natural resources' (Department of Environmental Affairs [DEA] 2008). Subsequently, the BIOKAVANGO project was designed to address some of the knowledge gaps identified through the planning process of the Okavango Delta Management Plan. This resulted in the current study, whose main goal was to find a solution to the conflict that had been identified as a major impediment to sustainable development in the Delta's fishery.

Bibliography

Baird, I. (2003) 'Local ecological knowledge and small-scale fisheries management in the Mekong River in southern Laos', in N. Haggan, C. Brignall and L. Wood, L. (eds) *Putting Fishers' Knowledge to Work, Conference Proceedings August 27–30, 2001*, Fisheries Centre Research Reports, vol 11, no 1, University of British Columbia, Vancouver, BC, pp. 87–99

Bennett, E., Neiland, A., Anang, E., Bannerman, P., Rahman, A.A., Huq, S., Bhuiya, M., Day, M. and Clerveaux, W. (2001) 'Towards a better understanding of conflict management in tropical fisheries: Evidence from Ghana, Bangladesh and the Caribbean', CEMARE Research Paper 159, University of Portsmouth

Bills, R. (1996) *Fish Stock Assessment of the Okavango River*, Investigational Report No. 56, J. L. B. Smith Institute of Ichthyology, Grahamstown, South Africa

Bokhutlo, T., Kootsositse, M.V. and Mosepele, K. (2007) *Okavango Delta Fishery Frame Survey*, Department of Wildlife and National Parks, Gaborone http://scribd.com/doc/37412763/ODMP-Fishery-Frame-Survey-Draft-Report, accessed 24 September 2012

Branch, T. A., Austin, J. D., Acevedo-Whitehouse, K., Gordon, I. J., Gompper, M. E., Katzner, T. E. and Pettorelli, N. (2011) 'Fisheries conservation and management: Finding consensus in the midst of competing paradigms', *Animal Conservation*, vol 15, pp. 1–3

Caddy, J. F. (2002) 'Limit reference points, traffic lights, and holistic approaches to fisheries management with minimal stock assessment input', *Fisheries Research*, vol 56, no 2, pp. 133–7

Cassidy, L., Wilk, J., Kgathi, D. L., Bendsen, H., Ngwenya, B. N and Mosepele, K. (2011) 'Indigenous knowledge, livelihoods and government policy', in D. L. Kgathi, B. N. Ngwenya and M. B. K. Darkoh (eds) *Rural Livelihoods, Risk and Political Economy of Access to Natural Resources in the Okavango Delta, Botswana*, Nova Science Publishers, Hauppauge, NY, pp. 75–98

Charles, A. (2008) 'Turning the tide: Toward community-based fishery management in Canada's Maritimes', *American Fisheries Society Symposium*, vol 49, pp. 569–73

Charles, A. T. (1992) 'Fishery conflicts: A unified framework', *Marine Policy*, vol 16, no 5, pp. 379–93

Chuenpagdee, R. and Jentoft, S. (2007) 'Step zero for fisheries co-management: What precedes implementation', *Marine Policy*, vol 31, pp. 657–68

Cronberg, G., Gieske, A., Martins, E., Prince Nengu, J. and Stenström, I-M. (1995) 'Hydrobiological studies of the Okavango Delta and Kwando/Linyanti/Chobe River, Botswana: Surface water quality analysis', *Botswana Notes and Records*, vol 27, pp. 151–226

DEA (2008) *Okavango Delta Management Plan*, Department of Environmental Affairs, Gaborone, Botswana

DWNP (2008) *Fish Protection Regulations, 2008*. Department of Wildlife and National Parks, Government Printer, Gaborone, Botswana

Evans, L., Cherrett, N. and Pemsl, D. (2011) 'Assessing the impact of fisheries co-management interventions in developing countries: A meta-analysis', *Journal of Environmental Management*, vol 92, pp. 1,938–49

FAO (1995) *Code of Conduct for Responsible Fisheries*, Food and Agriculture Organization of the United Nations, Rome

FAO (1998) *Integrated Coastal Area Management and Agriculture, Forestry and Fisheries.* Food and Agriculture Organization of the United Nations, Rome

GEF (2006) 'Building local capacity for conservation and sustainable use of biodiversity in the Okavango Delta', Global Environment Facility, Washington, DC

Gutiérrez, N. L., Hilborn, R. and Defeo, O. (2011) 'Leadership, social capital and incentives promote successful fisheries', *Nature*, vol 470, pp. 386–9

Haggett, A. R. (1999) 'Is sport fishing in the panhandle of the Okavango in crisis?', *Tight Lines – The Angler's Friend*, pp. 9–13

Haraldsdottir, G. (2000) 'Tradition, co-management, diversity, in small-scale inland fisheries in Africa', in E. P. Durrenberger and T. D. King (eds) *State and Community in Fisheries Management: Power, Policy and Practice*, Bergin and Garvey, Westport, CT, pp. 131–48

Hardin, G. (1968) The tragedy of the commons, *Science*, vol 162, pp. 1,243–48

Johannes, R. E., Freeman, M. M. R. and Hamilton, R. J. (2000) 'Ignore fishers' knowledge and miss the boat', *Fish and Fisheries*, vol 1, pp. 257–71

Jul-Larsen, E. and Van Zwieten, P. (2002) 'African freshwater fisheries: What needs to be managed?', *Naga*, vol 25, nos 3 and 4, pp. 35–40

Jul-Larsen, E., Kolding, J., Overå, R., Raakjær Nielsen, J., Van Zwieten, P. A. M. (2003) 'Management, co-management or no management?' *Major Dilemmas in Southern African Freshwater Fisheries*, 1, Synthesis Report, FAO Fisheries Technical Paper No 426/1, Food and Agriculture Organization of the United Nations, Rome

Junk, W. J., Bayley, P. B. and Sparks, R. E. (1989) 'The flood pulse concept in river-floodplain systems', *Canadian Special Publications of Fisheries and Aquatic Sciences*, vol 106, pp. 110–27

Kgathi, D. L., Mmopelwa, G. and Mosepele, K. (2005) 'Natural resource assessment in the Okavango Delta, Botswana: Case studies of some key resources', *Natural Resource Forum*, vol 29, pp. 70–81

King, M. and Faasili, U. (1999) 'Community-based management of subsistence fisheries in Samoa', *Fisheries Management and Ecology*, vol 6, pp. 133–44

Kolding, J. (1996). *Feasibility Study and Appraisal of Fish Stock Management Plan in Okavango*, University of Bergen, Norway

Mbaiwa, J. E. (2003) 'The socio-economic and environmental impacts of tourism development on the Okavango Delta, north-western Botswana', *Journal of Arid Environments*, vol 54, no 2, pp. 447–67

Mbaiwa, J. E. (2011) *CBNRM Stock Taking in Botswana.* United States Agency for International Development

Mbaiwa, J. E. and Thakadu, O. T. (2011) 'Community trusts and access to natural resources in the Okavango Delta, Botswana', in D. L. Kgathi, B. N. Ngwenya and M. B. K. Darkoh (eds) *Rural Livelihoods, Risk and Political Economy of Access to Natural Resources in the Okavango Delta, Botswana*, Nova Science Publishers, Hauppage, NY, pp. 275–303

Mbaiwa, J. E., Mosepele, K., Bendsen, H. and Kgathi, D. L. (2011) 'Historical trends in livelihoods patterns in the Okavango Delta, Botswana', in D. L. Kgathi, B. N. Ngwenya and M. B. K. Darkoh (eds) *Rural Livelihoods, Risk and Political Economy of Access to Natural Resources in the Okavango Delta, Botswana*, Nova Science Publishers, Hauppage, NY, pp. 17–34

Merron, G. S. and Bruton, M. N. (1988) *The Ecology and Management of the Fishes of the Okavango Delta, Botswana, with Special Reference to the Role of the Seasonal Floods*, Investigational Report No. 29, J. L. B. Smith Institute of Ichthyology, Grahamstown, South Africa

Mendelsohn, J. M., Vanderpost, C., Ramberg, L., Murray-Hudson, M., Wolski, P. and Mosepele, K. (2010) *Okavango Delta: Floods of Life*, Research and Information Services of Namibia, Windhoek, Namibia

Mmopelwa, G., Raletsatsi, S. and Mosepele, K. (2005) 'Cost benefit analysis of commercial fishing in Shakawe, Ngamiland, Botswana', *Botswana Notes and Records*, vol 37, pp. 11–21

Mmopelwa, G., Mosepele, K., Mosepele, B., Moleele, N. and Ngwenya, B. (2009) 'Environmental variability and the fishery dynamics of the Okavango Delta, Botswana: The case of subsistence fishing', *African Journal of Ecology*, vol 47, suppl. 1, pp. 1–9

Mosepele, K. (2000) 'Preliminary length based stock assessment of the main exploited stocks of the Okavango Delta fishery', MPhil thesis, University of Bergen, Norway

Mosepele, K. (2001) *Preliminary Description of the Okavango Delta Fishery*, Ministry of Agriculture, Gaborone, Botswana

Mosepele, K. (2008) 'Flood pulse in a subtropical floodplain fishery and the consequences for steady state management', in O Totolo (ed.) *Proceedings of the Second IASTED Conference on Water Resource Management: Science and Technology Innovation for Sustainable Development*, ACTA Press, Calgary, AB, Canada, pp. 56–62

Mosepele, K. and Mosepele, B. (2005) 'Spatial and temporal variability in fishery and fish community structure in the Okavango Delta, Botswana: Implications towards fisheries management', *Botswana Notes and Records*, vol 37, pp. 280–91

Mosepele, K., Mmopelwa, T. G. and Mosepele, B. (2003) 'Characterization and monitoring of the Okavango Delta artisanal fishery', in T. Bernard, K. Mosepele and L. Ramberg (eds) *Environmental Monitoring of Tropical and Subtropical Wetlands*, University of Botswana, Maun, and University of Florida, Gainesville, FL, pp. 391–413

Mosepele, K., Ngwenya, B. (2010) *Socio-economic Survey of Commercial Fishing in the Okavango Delta, Botswana*, Bay Publishing, Gaborone

Mosepele, K., Ngwenya, B. N. and Bernard, T. (2006) 'Artisanal fishing and food security in the Okavango Delta, Botswana', in Ahmed, A. (ed.) *World Sustainable Development Outlook: Global and Local Resources in Achieving Sustainable Development*, Geneva, pp. 159–68

Mosepele, K., Mosepele, B., Bokhutlo, T. and Amutenya, K. (2011) 'Spatial variability in fish species assemblage and community structure in four subtropical lagoons of the Okavango Delta, Botswana', *Physics and Chemistry of the Earth*, vol 36, nos 14–15, pp. 910–17

Mosepele, K., Moyle, P. B., Merron, G. S., Purkey, D. and Mosepele, B. (2009) 'Fish, floods and ecosystem engineers: Interactions and conservation in the Okavango Delta, Botswana', *Bioscience*, vol 59, no 1, pp. 53–64

Murray-Hudson, M. (2009) 'Floodplain vegetation responses to flood regime in the seasonal Okavango delta, Botswana', PhD thesis, University of Florida http://etd.fcla.edu/UF/UFE0024728/murrayhudson_m.pdf, accessed 24 September 2012

Ngwenya, B. and Mosepele, K. (2008) *Socio-economic Survey of Subsistence Fishing in the Okavango Delta, Botswana*, Bay Publishing, Gaborone, Botswana

Nnyepi, M., Ngwenya, B. and Mosepele, K. (2007) 'Food (in)security and child nutrition in Ngamiland', in A. Ahmed (ed.) *Managing Knowledge, Technology and Development in the Era of Information Revolution*, Fifth International Conference, jointly organised by the World Association for Sustainable Development (WASD) and Griffith University, Australia, pp. 281–91

Nsiku, E. (2003) 'The use of fishers' knowledge in the management of fish resources in Malawi', in N. Haggan, C. Brignall and L. Wood (eds) *Putting Fishers' Knowledge to Work*,

Conference Proceedings August 27–30, 2001, Fisheries Centre Research Reports, vol 11, no 1, University of British Columbia, Vancouver, BC, pp. 148–62

Overby, M. M. (2000) 'Resolving the stone crab-shrimp fisheries conflict: A case of implicit co-management', in E. P. Durrenberger and T. D. King (eds) *State and Community in Fisheries Management: Power, Policy and Practice*, Bergin and Garvey, Westport, CT, pp. 169–86

Pinho, P. F., Orlove, B. and Lubell, M. (2012) 'Overcoming barriers to collective action in community-based fisheries management in the Amazon', *Human Organization*, vol 71, no 1, pp. 99–109

Ramberg, L., Wolski, P. and Krah, M. (2006) 'Water balance and infiltration in a seasonal floodplain in the Okavango Delta, Botswana', *Wetlands*, vol 26, no 3, pp. 677–90

Ramberg, L., Hancock, P., Lindholm, M., Meyer, T., Ringrose, S., Sliva, J., Van As, J. and Vanderpost, C. (2006) 'Species diversity of the Okavango Delta, Botswana', *Aquatic Sciences*, vol 68, pp. 310–37

Ramsar (n.d.) 'The Ramsar Convention and its mission', http://ramsar.org/cda/en/ramsar-about-mission/main/ramsar/1-36-53_4000_0__, accessed 24 September 2012

Schreiber, D. K. (2001) 'Co-management without involvement: The plight of fishing communities', *Fish and Fisheries*, vol 2, pp. 376–84

Setswalo, O. (2007) *Conflict and Conflict Resolution Among and Between Local Fishers and Tour Operators in the Upper Panhandle of the Okavango Delta, Botswana*, Botswana Wildlife Training Institute, Maun, Botswana

Sipponen, M. and Gréboval, D. (2001) 'Social, economic and cultural perspectives of European inland fisheries: Review of the EIFAC symposium on fisheries and society', *Fisheries Management and Ecology*, vol 8, nos 4–5, pp. 283–93

Smith, P. (1976) 'An outline of the vegetation of the Okavango drainage system', in *Proceedings of the Symposium on the Okavango Delta and Its Future Utilization*, National Museum, Gaborone, Botswana, pp. 93–112

Thakadu, O. T., Mangadi, K. T., Bernard, F. E and Mbaiwa, J. E (2005) 'The economic contribution of safari hunting to rural livelihoods in the Okavango: The case of Sankuyo village', *Botswana Notes and Records*, vol 37, no 1, pp. 22–39

Thomassen, C. (2011) Okavango barbel run, *Tight Lines*, vol 617, pp. 22–6

Ticheler, H. J., Kolding, J. and Chanda, B. (1998) 'Participation of local fishermen in scientific fisheries data collection: A case study from the Bangweulu swamps, Zambia', *Fisheries Management and Ecology*, vol 5, pp. 81–92

Tweddle, D., Bills, R., Van der Waal, B., Skelton, P., Kolding, J., Nengu, S. (2003) 'Fish diversity and fisheries in the Okavango Delta, Botswana', in L. E. Alonso and L. Nordin (eds) *A Rapid Biological Assessment of the Aquatic Systems of the Okavango Delta, Botswana: High Water Survey*, RAP Bulletin of Biological Assessment, pp. 97–110

Van Ginkel, R. and Steins, N. (2001) 'Multi-use conflicts in inshore waters', in D. Symes and J. Phillipson (eds) *Inshore Fisheries Management*, Kluwer Academic Publishers, Dordrecht, pp. 257–73

Warner, M. (2000) 'Conflict management in community-based natural resource projects: Experiences from Fiji and Papua New Guinea', Working Paper 135, Overseas Development Institute, London

Welcomme, R.L. (1998) 'Framework for the development and management of inland fisheries', *Fisheries Management and Ecology*, vol 5, pp. 437–57

Wolski, P. and Murray-Hudson, M. (2005) 'Flooding dynamics in Okavango Delta', *Hydrology and Earth System Sciences Discussions*, vol 2, pp. 1,865–92

Wolski, P., Savenije, H. H. G., Murray-Hudson, M. and Gumbricht, T. (2006) 'Modelling of the flooding in the Okavango Delta, Botswana, using hybrid reservoir-GIS model', *Journal of Hydrology*, vol 331, pp. 68–72

10 Shifting gear

A new governance framework for small-scale fisheries in South Africa

Merle Sowman, Serge Raemaekers and Jackie Sunde

Introduction

South Africa's transition to democracy in 1994 precipitated a law reform process that sought to address past injustices and give a voice to marginalized communities. This process was guided by the new South African Constitution (Republic of South Africa [RSA] 1996) and applied to all sectors of society including the fishing industry, following decades of discrimination and exclusion from legal access to marine resources. The new democratic government faced the huge challenge of transforming an industry in which the ownership of marine resources was vested in a handful of large, white-owned companies. Transformation had to take place in a complex policy environment that included balancing the opportunities created by South Africa's reintegration into the global economy and adoption of neo-liberal economic policies with a historically strong conservation agenda, as well as various new social development policies (Van Sittert et al 2006).

However, despite a progressive Constitution that upheld the protection of, and respect for, a range of socio-economic, cultural and environmental rights, and the recognition of 'living customary law',[1] the traditional small-scale fisheries sector in South Africa continued to be marginalized. Decisions regarding rights of access, the use of resources and institutions for the management of marine resources were centrally located, and a powerful, market-based ideology influenced the governing system in favour of commercial fishing interests (Sunde et al 2013).

Although policy reforms from 1998 to 2006, and in particular the promulgation of the Marine Living Resources Act (MLRA) in 1998, resulted in improved access to marine resources for many disadvantaged South Africans (Crosoer et al 2006; Isaacs 2012), these reforms were considered 'window dressing' (Nielsen and Hara 2006), because thousands of traditional fishers still could not gain such access. Therefore, in 2005 a group of traditional fishers took legal action against the minister responsible for fisheries, which culminated in a court ruling that required the minister to develop a policy to address the socio-economic needs of this hitherto excluded group.

In June 2012, nearly five years after the court order, a new small-scale fisheries policy was promulgated in South Africa (Department of Agriculture, Forestry and Fisheries [DAFF] 2012). This chapter offers a critical review and analysis of the new policy and its formulation process. It also examines the new small-scale fisheries governance framework against the backdrop of a set of normative 'good governance' principles derived from international best practice. The final section highlights key implementation challenges and offers suggestions for overcoming some of them.

Governance of small-scale fisheries

Internationally there has been a growing realization that small-scale fisheries, both marine and freshwater, play a critical role in the provision of food, employment and livelihoods for millions of people (Berkes et al 2001; FAO 2005a; Béné 2006; Sharma 2011). Although there are no reliable figures on the actual number of people dependent on small-scale fisheries, the Food and Agriculture Organization (FAO 2005a) estimates that 90 per cent of the 38 million fishers and fish farmers in the world are small-scale, collectively harvesting half of the world's fish catch for consumption, and that 135 million are directly or indirectly employed in small-scale fisheries and aquaculture.

Globally, however, many coastal fisheries resources are declining, habitats are degrading (Pauly et al 1998; FAO 2005b) and food and livelihoods derived from fisheries are as a consequence increasingly at risk (Andrew et al 2007). Given the complexities, uncertainties and challenges prevalent in these coastal fishery systems, conventional top-down, science-based and resource-centred management is no longer considered appropriate: new forms of governance that address both societal and resource sustainability issues are required (Kooiman et al 2005; McConney and Charles 2008; Sowman 2011). These new approaches are systems-orientated, embrace complexity thinking, recognize the plurality of socio-legal systems and are integrated, interdisciplinary, participatory, responsive and adaptive.

As Chapter 1 explains, governance is increasingly being seen as one of the most critical factors in promoting sustainable resource management and human development and in eradicating poverty (United Nations Development Programme [UNDP] 2002; Borrini-Feyerabend 2008). It is concerned with interactions and processes that occur in a diverse group, including non-state actors, in the process of addressing societal issues and needs (Kooiman et al 2005). Central to this governance approach is respect for human rights as well as the participation of all relevant parties in planning and decision-making.

The South African Equality Court's order to the government in 2007 to develop a small-scale fisheries policy that would address the socio-economic needs of traditional small-scale coastal fishers provided an opportunity to embrace this new approach to governance – one that was underpinned by a set of 'good governance' principles, as summarized in Table 10.1.

In this chapter, we examine the extent to which the new small-scale fisheries policy in South Africa – as well as the process through which it was developed – embraces the principles and approaches of good governance. We discuss the implementation challenges and practical changes required to move fisheries onto a different trajectory: one that embraces coastal fisheries in South Africa in all

Table 10.1 Principles of good governance

1. Respect for and protection of human rights
2. Equality
3. Recognition of and respect for the plurality of socio-legal systems
4. Recognition of the rule of law
5. Promotion of intercultural engagement
6. Inclusivity and participation
7. Representivity
8. Free, prior and informed consent
9. Transparency
10. Accountability
11. Transdisciplinarity
12. Embracing of complexity
13. Cross-sectoral coordination, integration and contextualization
14. Integration of indigenous and local knowledge
15. Responsiveness, flexibility and adaptiveness
16. Sustainability
17. Subsidiarity
18. Access to justice and appropriate systems of mediation, conflict and dispute resolution

their complexity and diversity, recognizes the plurality of socio-legal governance systems and seeks to facilitate governance systems that are appropriate to the context and committed to pursuing social justice and environmental sustainability objectives.

Overview of small-scale fisheries in South Africa, past and present

South Africa's productive and diverse marine environment provides food and livelihoods for thousands of coastal dwellers and supports a thriving industrial sector mainly in the south-western region of the country. Extending from Alexander Bay in the north-west to Kosi Bay in the east, the 3,000 kilometre coastline is flanked by the cold, nutrient-rich waters of the Benguela current on the west coast and the warm nutrient-poor waters of the Agulhas current on the east coast (Figure 10.1).

Historical evidence indicates that people have been harvesting resources along the coast for many thousands of years (Parkington et al 1988; Lasiak 1993; Jeradino and Yates 1996). Today, diverse small-scale fisheries operate along the South African coast, ranging from the shore-based harvesting of intertidal resources to the targeting of migratory line-fish stocks using motorized vessels. Based on data from the fisheries authority and research conducted in several small-scale fishing communities along the coast, approximately 110 geographically distinct fishing communities can be identified, varying in size from small villages to larger towns (Sowman et al 2011b) (Figure 10.1). Some of these fisheries are still informal, operate under regulations for recreational fisheries or have only certain components recognized by the fisheries authority (Environmental Evaluation Unit [EEU] 2010).

Figure 10.1 Location of small-scale fishing communities in South Africa.

While the MLRA identified 'subsistence' as a legal category of fishers, the definition only permitted the harvesting of low-value species for local consumption. The new law failed to recognize the range of small-scale fishers that were engaged in fishing for livelihoods and derived an income from these activities. Both government and scholars have agreed that the more encompassing term 'small-scale fishers' should rather be used to cover the range of fishery types taking place along South Africa's coast (Harris et al 2002; Sowman 2006; DAFF 2012).

Small-scale fishing along the west and south-west coastal areas is predominantly boat-based. Generally it is conducted close to shore, boats are launched in local waters and the duration of trips is restricted to one day. The gear used is low-technology and thus does not require large outlays of capital. The fishing activities are labour-intensive. Men usually (but not always) do the harvesting, while women are mainly involved in pre- and post-harvest activities. It is often impossible to disentangle the threads that link one individual to a range of fishing activities spanning the large- and small-scale sectors.

A different scenario has evolved along the east coast of South Africa, where the harvesting of inshore resources by indigenous people was largely regulated by African customary law despite national regulations restricting access and use (Sunde et al 2013). In these areas, there was limited interference by the colonial and apartheid governments in the allocation of access rights and management of resources. This was largely due to the remoteness and inaccessibility of vast stretches of the east coast and the relatively low commercial value of coastal resources there. Furthermore, during the apartheid era large stretches of the eastern seaboard were declared 'homelands'.[2] In these traditional authority areas, rights to access marine resources were inextricably linked to relations of land tenure and social relations within the community (Sunde 2011).

There is not much literature on this, but evidence from recent research on coastal resource use suggests that customary systems constituted the de facto legal dispensation during the twentieth century in many parts of the country (Sunde et al 2013). A consequence of cultural practices involving the ocean, and of resource characteristics, is that coastal communities are engaged in a range of livelihood activities, with fishing contributing to household food and income.

Fisheries reform, 1994–2007

In 1994, South Africa embarked on an extensive law reform programme, guided by the new South African Constitution (RSA 1996), with the human rights principles contained in the Bill of Rights (Witbooi 2006) at its core. All organs of state were required to align their policies, laws and implementation actions to these fundamental rights, as well as to other relevant constitutional provisions.

In the poor coastal fishing communities, expectations were high that the new democratic government would deliver on its promises of an improved life for all. However, although various sectors did participate in deliberations regarding a new fisheries policy in 1996, historically disadvantaged communities were not well represented. Instead, the deliberations were dominated by the fishing industry, which formed an alliance with organized labour to press for the allocation of long-term individual rights. This alliance was strongly opposed to state interference in the redistribution of resources to traditional fishers (Nielsen and Hara 2006).

The three-year law reform process culminated in the promulgation of the MLRA (RSA 1998), which sought to achieve three broad objectives, namely sustainability, equity and economic stability. Transformation of the industry was a further key requirement of the Act, although no clear guidance was given on how this should be achieved (RSA 1998, section 2[j]). The recognition of subsistence fishers as a legal category of fishers and the declaration of coastal areas for their exclusive use, were further indications that a new fisheries dispensation was in the making (RSA 1998, section 19).

Thus, while the restructuring of the fishing industry and the opening up of access to it resulted in a significant increase in new rights holders, and a degree of black economic empowerment in the industry (Isaacs 2006; Nielsen and Hara 2006; Van Sittert et al 2006; Isaacs 2012), there have been serious criticisms regarding the achievement of equity objectives. In particular, many of the quotas allocated to new entrants were considered unviable, and thousands of traditional small-scale fishers were overlooked (Isaacs 2006, 2012; Sowman 2006). A further key criticism has been that the individual rights allocation process largely benefited big industry players and their new black economic empowerment partners (Nielsen and Hara 2006), but failed to address the socio-economic and cultural needs of poor fishing communities (Isaacs 2006; Sowman 2006). Furthermore, there is no evidence to suggest that this approach generated the surplus envisaged for pro-poor development (Isaacs et al 2007).

Although a subsistence category was created, very little progress has been made with respect to the allocation of rights to the subsistence sector, although in the KwaZulu-Natal and Eastern Cape provinces, fishing permits (not rights) have been allocated annually since 2002 (Harris et al 2007). However, this arrangement has been cumbersome and uncertain and, in some cases, has entrenched an individual-rights-based approach that is in conflict with living customary systems which still prevail in many coastal areas of South Africa (Sunde et al 2013). These failures can be attributed in part to institutional inadequacies and the MLRA's narrow definition of subsistence (Sowman 2006; Isaacs 2006, 2012). Certainly the African National Congress' (ANC) objective of 'the upliftment of impoverished coastal communities through improved access to marine resources' was not achieved in the first decade of democracy (African National Congress 1994, 4.5.3.2).

The small-scale fisheries policy development process in South Africa (2007–12)

The failure of the new fisheries regime to formally recognize and cater for small-scale fishers resulted in mass action, including marches to parliament and protests (Sunde 2003; Isaacs 2006), an increasing disregard of formal rules and regulations (Hauck and Kroese 2006; Hauck 2008; Sowman et al 2011a), the holding of Fisher Human Rights Hearings[3] (Masifundise Development Trust [MDT] 2003) and, finally, legal action by a group of fishers from the Western Cape against the Minister of Environmental Affairs and Tourism.[4] The fishers were assisted by the MDT, a non-governmental organization (NGO) supporting fishers' rights, and represented by the Legal Resources Centre (LRC). A key argument in the case was that the government's failure to allocate rights to traditional small-scale fishers had violated their fundamental constitutional rights, resulting in significant socio-economic hardship. According to the Legal Resources Centre, 'these violations were not reasonable and could not be justified in an open and democratic society based on equality, dignity and freedom'. A ruling by the Equality Court in May 2007 required the minister responsible for fisheries to develop a policy that would address the needs of this group of fishers and immediately provide 'interim relief' through access to marine resources until such time as the policy was finalized.

A narrative which emerged as fundamental through the advocacy campaign was that fishers' rights were human rights. Fishing communities began articulating the linkages between their right to access resources and their human rights enshrined in the Constitution, and stating that the upholding of these rights was a precondition for sustainability in its broadest sense. Following the Equality Court Ruling in 2007, a national task team and, later, a technical task team were appointed that included representatives from the government, fisher communities, researchers, NGOs and community-based organizations whose task it was to develop a small-scale fisheries policy. At the same time, the fisheries authority at the time (Marine and Coastal Management, a component of the Department of Environmental Affairs and Tourism), instituted an interim relief system.

The process of formulating the draft policy was lengthy and difficult, due to the widely differing perspectives of the many stakeholders involved. An overriding fear in the government, marine science establishment, industry and recreational sector was that marine resources were insufficient to cater for all those seeking access, and that addressing the socio-economic needs of these fishers would require existing rights to be redistributed, which would destabilize the industry. Yet, although the policy formulation process was hampered by bureaucratic delays, procedural flaws and an inability to reach consensus on a number of 'sticky issues', the process led to a greater understanding among participants of the needs and circumstances of small-scale fishers, as well as the resource opportunities and constraints that needed to be taken into account when decisions were made.

While many of the so-called 'sticky issues' remained unresolved, political pressure from DAFF, which took over responsibility for fisheries management in 2010, led to the submission of a draft policy to the National Economic Development and Labour Council in November 2011. This advisory body, comprising representatives of government, labour, civil society and business, debated various policy issues for four months in pursuit of agreement on contentious issues. A key focus was inserting provisions that would ensure that constitutionally entrenched customary rights would be recognized and catered for in the policy. After a further

Figure 10.2 Fishers and other stakeholders engaged in policy discussions in the Eastern Cape (photo: Jackie Sunde).

six months of public and political review, the policy was eventually promulgated in June 2012 (DAFF 2012).

The new small-scale fisheries policy: A shift to 'good governance'

The process of formulating this policy, the principles and approaches guiding it, and a number of its provisions, suggest a fundamental paradigm shift in the governance of small-scale fishers in South Africa. This is clearly articulated in the introduction to the policy, which states that the aim is 'to provide redress and recognition to the rights of small-scale fisher communities in South Africa . . . in order to fulfil the constitutional promise of substantive equality' (DAFF 2012, section 1). It refers to South Africa's 'obligation in terms of Article 1 of the African Charter on Human and Peoples' Rights to . . . give effect to the protection of the rights enshrined in the Charter' (DAFF 2012, section 1). This call for the realization of human rights in its broadest sense is a theme that recurs throughout the policy, as is the requirement for redress, equity and equitable access (in terms of race, gender and disability) to, and enjoyment of, benefits from living marine resources, as well as participation in decision-making.

The realization of these rights requires the government to change its stance towards the small-scale fisheries sector in certain fundamental ways by adopting

> a developmental approach and an integrated and rights-based allocation system which recognizes the need to ensure the ecological sustainability of the resource; identifies small-scale fishers as a category of fishers for the purposes of the MLRA in law; and provides for community orientation in the management of the marine living resources harvested by these fishers.
>
> (DAFF 2012, section 3.2)

This shift in position is evident from the principles that underpin the policy (Table 10.2) and its strategic policy objectives (DAFF, section 3.2), which seek to give effect to this developmental and human-rights based approach.

In keeping with the Equality Court ruling that the new policy should 'take into account international and national legal obligations and policy directives to accommodate the socio-economic rights of these fishers to ensure equitable access to marine resources', the new policy clearly draws on principles and statements contained in a number of key international instruments such as the United Nations Food and Agriculture Organization's Code of Conduct for Responsible Fisheries (FAO 1995), the Convention on Biological Diversity (United Nations Environment Programme [UNEP] 1992) and the Bangkok Statement (FAO 2008).

The most critical shift from previous policies is that the new small-scale fisheries policy recognizes the rights and needs of small-scale fishers and affords them respect and legal protection. It expands the narrow category of subsistence fishers defined in the MLRA (RSA 1998, section 19) to include a continuum from subsistence fishers to formal small commercial enterprises (Sowman 2006). This clarifies

Table 10.2 Principles of small-scale fishing policy in South Africa.

The state must:

1.	Recognize the existence of any rights conferred by common law, customary law or legislation to the extent that these are consistent with the Bill of Rights.
2.	Recognize rights guaranteed by custom and law, and access to and use of natural resources on a communal basis, to the extent that these are consistent with the Bill of Rights.
3.	Adopt an integrated and holistic approach which is based on human rights principles.
4.	Recognize an approach which contributes to food security, local socio-economic development and the alleviation of poverty.
5.	Promote biodiversity and the sustainable use and management of marine living resources and associated ecosystems.
6.	Recognize that the disturbance of the ecosystem and biological diversity is to be avoided or, where it cannot be avoided, minimized.
7.	Recognize the interdependency of the social, cultural, economic and ecological dimensions of small-scale fishery systems.
8.	Adopt an approach of empowerment through co-management that builds the capacity of fishers through education, training and skills development in all aspects of fisheries.
9.	Develop accountable and transparent structures and mechanisms.
10.	Promote effective participation in policy development, management and decision-making.
11.	Promote equitable access to and involvement in all aspects of fisheries, noting in particular past prejudice against women and other marginalized groups.
12.	Promote preferential access for fishers who are part of a small-scale fishing community, derive their livelihood from the sea and are aligned to a community-based legal entity, to harvest marine living resources.
13.	Ensure that small-scale fishing communities are not denied physical access to infrastructure and amenities that are central to exercising their right to fish.
14.	Incorporate a community-based rights approach to the allocation of marine living resources.
15.	Recognize that resources may be allocated through a multi-species approach.
16.	Recognize the complementary value of indigenous and local knowledge.
17.	Where tenure to coastal land involves coastal communities and affects the implementation of this policy, ensure that there is liaison with the relevant organs of state to resolve such issues.
18.	Contribute to the development of small-scale fisheries by promoting initiatives that improve and support the entire value chain, and improve and support the capacity of small-scale fishing communities to access and optimize the entire value chain.

Source: DAFF 2012 section 3.1

the range of activities that can be categorized as small-scale and includes fishers involved in pre- and post-harvest activities, who are often women. The critical role of women in the fisheries value chain and the need to give them support and protection is highlighted throughout the policy.

Of particular importance for equity and poverty alleviation is the requirement that these fishers be granted preferential access to marine resources, especially where such communities have historically depended on them. Thus, the principles of redress, respect and protection of human rights and equality before the law are firmly embedded in the new policy. Furthermore, the policy affirms the state's responsibility, in terms of the African Charter on Human and Peoples' Rights, to protect the rights of fishing communities to pursue the social and development policies that they have 'freely chosen' (Organization of African Unity [OAU] 1981, article 20). This, together with a commitment to the application of international and regional law on the rights of customary and indigenous communities, gives effect to the principle of free, informed, prior consent which is now recognized as international 'good governance'.

Furthermore, the policy recognizes the contribution of small-scale fisheries to economic development, poverty alleviation and food security and proposes mechanisms to give effect to this contribution through, for example, support for infrastructure development, the provision of facilities in coastal communities, subsidy schemes for fish storage and training and skills development. It also recognizes the important rôle that small-scale fisheries play in food security and sustaining many poor coastal fishing households that rely on fishing, part-time or full-time, as their main source of food and livelihood. Equally important is the role fisheries play as a safety net in poor communities, especially where households do not have access to land or other capital, and also during times of environmental or economic crisis.

While the potential for job creation and value adding in this sector is recognized, the government realizes that incentives will be required to stimulate these activities and that alternative livelihoods need to be explored and developed. The new policy thus offers the potential to locate fisheries in the broader context of local economic development. This is a significant shift from a sectoral approach to fisheries management and calls for greater coordination with other government departments and sectors, the development of other livelihood opportunities – especially in stressed areas – and a recognition of the need to take a holistic, integrated and developmental approach to the fisheries sector. It also recognizes the importance of understanding the fishery context and the linkages between socio-economic, cultural and ecological systems. In this respect, the policy embraces the principles of complexity as well as cross-sectoral coordination and integration.

The policy acknowledges that some communities have retained their customary fishing systems (DAFF 2012, p. 11). The principle that requires respect for and recognition of customary institutions and practices indicates respect for the plurality of different socio-legal systems operating in South Africa. Fishing communities with customary systems of resource use will be able to harvest resources according to customary rules and practices in so far as these practices are sustainable and comply with the Bill of Rights. This applies in particular to customary communities in the Eastern Cape and KwaZulu-Natal, since communities with customary rights are protected by the Constitution,[5] which recognizes customary law as an independent and equal source of law. Precedents set down in the Richtersveld case,[6] to the effect that customary systems can give rise to access and use rights to resources, as well as a recent magistrates' court judgement,[7] confirm that a customary fishing community has a constitutionally protected right to practice its customs in this regard (LRC 2012).

Although not elaborated on in the new policy, the principles described above that give recognition to customary systems and practices, coupled with the principle of recognizing the complementary value of indigenous and local knowledge, create the requirement for a policy implementation approach that promotes intercultural engagement. Where customary rights to marine resources have been restricted due to, for example, the declaration of 'no take' marine protected areas, such rights may need to be reviewed, and in some cases restored, and appropriate management measures will need to be identified and negotiated with local resource users and local leadership structures. In other cases, customary rights to resources may have been lost or restricted because permits have been allocated to other user groups such as the recreational sector. The new policy requires that preferential access be granted to communities that have been excluded in this manner. In this respect, traditional fishers hitherto excluded from access to marine resources will be able to access resources on a just and equitable basis, although ecological sustainability considerations would need to inform the levels of harvesting that could be sustained.

The new policy has been developed in the context of post-apartheid democratic governance, in which a range of mechanisms exist to ensure that policy-making takes place according to a publicly accountable, participatory and transparent process that respects the rule of law. Similarly, the policy itself requires new governance institutions to be representative, transparent and accountable (see DAFF 2012, section 3.1(i)). The involvement of fishers and other stakeholders in planning and decision-making is considered vital to the establishment of participatory governance arrangements (DAFF 2012, section 3.1(j)).

Mechanisms, structures and processes to enable this shift to take place are outlined in section 6 of the new policy (DAFF 2012). It proposes a step-by-step approach to the declaration of a fishing community, the establishment of a legal entity to hold fishing rights and criteria for individuals to gain and exercise their fishing right. Rights will be allocated to a community-based legal entity that comprises individual fishers from a particular fishing community. The policy sets out how to identify such fishers: this includes criteria such as demonstrating a long direct involvement in fishing (ten years) and not being engaged in other permanent employment. DAFF issues the right to fish or to operate a processing facility, but the fishing community determines how these rights will be exercised (DAFF 2012, section 6.2). Most significantly, the policy recognizes that past and existing power relations with respect to the market affect the distribution of benefits from the sector, and hence the policy includes a commitment to introduce mechanisms that will strengthen communities' powers in negotiating with markets, thereby maximizing the redistributive potential of the sector.

The policy recognizes that local contexts and environments differ along the coast, and that management arrangements need to be tailored accordingly. Each fishing community will have to establish a local co-management structure and work with the government to manage local resources. The proposed devolution of decision-making powers to local-level institutions through the community entity and the co-management structure suggest that the powers of the minister and the department have been diminished and real decision-making powers are being devolved downwards (LRC 2012). According to section 5.2.1 of the policy (DAFF 2012), these local co-management structures will be nested within a multitiered institutional system comprising a consultative advisory forum at the national level, a dedicated small-scale fisheries management working group acting as a vehicle for interaction between the sector and the national department, and, at the local level, co-management committees and community-based legal entities to deal with issues affecting the local fishery. The use of local and indigenous knowledge to complement scientific assessments and to inform decision-making is crucial to this new management approach, which is based on the principle of subsidiarity and aims to be locally responsive, context-specific and adaptive.

In accordance with the principles of the rule of law, the policy accommodates access to procedural justice for any persons who may wish to appeal decisions (DAFF 2012, section 6.2.7). Specific mechanisms for protecting the rights of individual members and managing and resolving conflicts are provided for in the process of establishing governance institutions at local level. Where such conflicts cannot be resolved at that level, provisions exist for requesting mediation assistance from the national level.

With respect to the sustainability of resources, the policy sets out requirements for the department to undertake regular assessments of the state of resources, to identify resources which can be allocated to this sector, and to develop holistic management plans, although the scope of these plans will be determined through regulations still to be developed. Although the policy proposes a number of conventional fisheries management tools such as total allowable catch, total allowable effort, closed seasons and bag limits, it requires the participation of resource users in determining appropriate measures and in decision-making processes. Thus concerns for resource sustainability are fundamental to this policy, but resource users will now participate in assessment and decision-making in that regard.

This new policy, which is clearly underpinned by a set of good governance principles developed by a diverse group of fisheries actors, suggests a distinct shift in philosophy and approach to the governance of the small-scale fisheries sector in South Africa. It signals a shift to a human-rights, community-oriented, participatory and developmental approach. The intention is that this new approach will promote local socio-economic development through community-based institutions and benefit-sharing marketing structures such as cooperatives. It will also encourage resource stewardship and the revitalization of the culture of community and sharing that characterized many coastal fishing communities in the past. Finally, the new policy will promote the organizational development, capacity development and empowerment of fishers, equipping them to work collaboratively with the government and other agencies. This will lead to the progressive realization of human rights.

Implementation challenges and prospects

While the policy signals a significant paradigm shift, the extent to which this change in philosophy and approach is actually implemented and leads to more equitable, just and environmentally sustainable outcomes depends on a number of factors. Barriers to implementation and suggestions on how to overcome these are briefly discussed below.

Mismatch between overarching policy frameworks

There are significant tensions between South Africa's neoliberal macroeconomic policy, its conservation policies and the social justice imperatives of fisheries 'transformation' (Sowman 2006; Van Sittert et al 2006; Isaacs 2012). The adoption of a neoliberal economic policy after 1994, reinforced in the recently released National Development Plan (NPC 2012), confirms the government's historical bias towards a capital-intensive, export-driven fishing industry, constraining the state's ability to transform the sector to accommodate small-scale fisheries under a plurality of governance systems. Crosoer et al (2006, p. 23) argue that the market-based approach to fisheries governance will 'favour incumbency and accelerate export orientation in order to maximize the returns for state and capital from the fisheries, both tendencies inimical to the broad aims of populist fisheries transformation'. Without serious economic reform, South Africa's macroeconomic orientation will continue to act as a barrier to the realization of the social justice imperatives outlined in the small-scale fishing policy.

Coexistence of customary systems with statutory systems

The need to develop a legal and policy framework that accords customary law its rightful place as an equal, independent body of law presents a considerable challenge to the fisheries sector. Although the Constitution now protects customary systems in so far as they are consistent with the Bill of Rights, the implications of this provision are not yet understood within the fisheries sector. Customary systems of resource management provide opportunities to give substance to a range of the principles of 'good governance' inherent in the new policy.

In customary systems, rights to access and use resources are nested at the level of local users. Decision-making on and the administration of these rights are usually devolved to this level, which promotes accountability and provides mechanisms for local users to be involved directly in decisions about the resource. These customary mechanisms can be utilized in the implementation of the new policy as the basis for beginning to create a more equitable, bottom-up approach to fisheries management. This, however, requires trust in the adaptive potential of local-level, community-based processes to ultimately evolve sustainable and equitable systems. Closely linked to this principle is the recognition of the value of local knowledge, and of the value of diverse cultural interpretations of the meaning of natural resources.

Responsiveness to different local contexts and needs

The diversity of fishery contexts along the South African coast requires governance proposals to be tailored to suit particular circumstances. DAFF's limited capacity to experiment with and apply new governance models means that it will tend to advocate generic approaches and models. Clearly, a one-size-fits-all approach is not workable in the small-scale fishery sector in South Africa.

Furthermore, the new policy indicates that geographically defined fisher communities will need to identify and co-manage their 'community fishing area', with set boundaries and regulations dependent on traditional fishing practices and sustainable harvesting limits. While these fishing zones can be identified for many rural communities along the coast, at least for the harvesting of sedentary resources such as shellfish and even lobster, the boundaries are not so clear for small-scale fisher groups residing close to urban centres, where there is often competition for resources across fishery sectors. In such cases, agreement will need to be reached between fisher communities who wish to target common resources.

Restoring rights, redistribution and fair allocation of rights

While the policy requires the restoration of rights and redistribution of resources to address the socio-economic and cultural needs of marginalized fishing communities, procedures and mechanisms to achieve these objectives have not been clarified. Where customary rights to resources have been restricted due for example to the presence of a marine protected area, procedures will be required to review the status of such an area, in terms of the rights of the community on one hand and resource sustainability considerations on the other. In cases where customary rights to resources have been lost or restricted because permits have been allocated to other user groups such as the recreational sector, procedures to ensure fair and equitable access will need to be developed. No such procedures and mechanisms have yet been developed and implemented. It is anticipated that there will be resistance to any form of harvesting in 'no take' marine protected areas. However, multiple-use marine protected areas may need to be created to ensure preferential access to local bona fide fishers, as well as mechanisms for enhancing benefit flows to local fishing communities.

The inclusion of women in the fishery value chain

One of the critical challenges facing the sector is how to create opportunities for women to participate in and benefit from the small-scale fisheries sector equitably and without discrimination. There is potential for this sector to empower women in many ways, enabling them to create livelihood opportunities for themselves in the fisheries. This, however, requires a paradigm shift on two levels. First, the value of people's lives cannot merely be reduced to the economic 'value chain'. Rather, policy approaches must recognize the mix of social, cultural and economic values associated with people's enjoyment of human dignity and freedom. For women in the small-scale sector, this means acknowledging the multiple ways in which women contribute towards sustaining life and livelihoods beyond the narrowly defined 'value chain' from 'hook to plate'.

Second, a shift in the current dominant model of development and marketing is required. Poor women will only be able to enter the value chain if the existing control of the value chain by the industrial sector is shifted in favour of small-scale producers, who can increase their share of the market either through the power of collective action or by creating niche markets. This will require substantial state intervention in the market through mechanisms such as subsidies to women for local value-adding activities and infrastructure.

The shift from an individual to a community-based approach

The shift to a community-based approach to fisheries management, with the allocation of rights to a community-based legal entity, may not be practically feasible in all fishing communities. This is especially the case where fishers have gained access to resources through individual rights over the past several years. The intention of the community approach is to include more fishers in the fishery system, enhancing their opportunities, as a collective, to add value to fisheries products. In some contexts, there is significant resistance to the setting up of community-based entities to hold fishing rights, and there are concerns that the elite and powerful groups in the community will benefit as they have done under previous transformation efforts. There are also fears about the difficulties of establishing and maintaining legitimate, transparent and accountable local level institutions without support from the government and NGOs. This raises questions about the capacity of the government and of NGOs involved in supporting small-scale fisheries – as well as that of local institutions – to implement the policy.

Capacity of government and local-level institutions to embrace responsibilities

To realize the ambitious goals of the policy will require considerable institutional capacity at both government and community level. DAFF will need to build capacity among fisheries managers and scientists and restructure its institutional arrangements affecting small-scale fishers. At the same time, fishing communities will need to become more organized so that they can play a central management role and run their own enterprises.

Yet there are already concerns regarding government and grassroots capacity to implement the governance arrangements envisaged in the policy. DAFF is understaffed, under-resourced and bound by an institutional culture that is technocratic, regulatory and based on natural science. These constraints hamper DAFF's ability to embrace the paradigm shift proposed in the new policy. Fishing communities, in many cases, lack the capacity to fulfil their new roles and responsibilities, and there is a danger that neither DAFF nor people in fishing communities will have the capacity to play their respective developmental and co-management roles during policy implementation. This indicates an urgent need for a national training and community development programme.

Small-scale fisheries in the context of broader governance

If the new fisheries governance regime is to be effectively and judiciously implemented, a complete overhaul of the current regime and, in particular, the MLRA is required. Also, governance actors will need to look beyond the fisheries sector for solutions to small-scale fisheries problems. Interaction with other government departments at different levels, as well as with stakeholders from different sectors and contexts, will be needed to address problems and find solutions to the many challenges facing poor fishing communities.

Andrew et al (2007, p. 228) note that 'managing the interface between fisheries and the wider external environment' presents an immense challenge. For example, concerns regarding the overexploitation of coastal resources due to food security requirements may best be addressed by local economic development initiatives or amendments to marine spatial plans, which fall under the mandate of a different government agency. Another solution may be private-sector assistance in the beneficiation and marketing of fishery products – which implies the need to include other private, state and civil society actors in governance. Better communication and coordination are needed across the various agencies and actors with a stake in the issue under consideration.

Conclusions

The lengthy process of challenging the previous fisheries regime on the grounds that it was not participatory, and then of developing a new small-scale fisheries policy that reflects small-scale fishers' needs and rights, has involved the slow and steady inversion of the fisheries governance structure in South Africa. Through the deliberations towards the new policy, both in the policy debates and in the tough and conflict-ridden process of managing the interim mechanisms, fishers and others have started doing governance differently – from the bottom up. The new governance practice that is being developed in villages and coastal towns up and down the coastline is messy, inconsistent, incomplete, complex and fumbling. Superficially it does not look good, and conflicts appear to be increasing as the real power relations of the past are destabilized. This apparent increase in conflict is the downside. The upside is that at last many ordinary men and women from fishing communities are participating in discussions about the future and content of the new policy in a way that has never happened before.

The very first step towards putting into place the conditions necessary to realize the ambitious principles of the policy has thus been taken. The legacy of colonial and apartheid capitalism demands visionary leadership, collective action and an enormous leap of faith to transform the current conditions. Small-scale fishing communities and their civil society partners have taken that leap and, in so doing, are transforming 'government' into 'governance'. Real, good governance is in progress, albeit embryonic right now.

This review has revealed a definite paradigm shift in fisheries governance in South Africa, one that aligns with the principles and approaches of good governance.

However, as Siegele et al (2009, p. 69) conclude, 'good policy is just a starting point – good practice is more difficult to achieve'. Like many policies in South Africa, it is implementation that will bring the severest challenges. However, there is room for optimism when one reviews the progress made in clarifying provisions in the South African Constitution through policy development and implementation, legal challenges and the development of jurisprudence to guide decisions. Similarly, the implementation of this policy will be a matter of trial and error, muddling through, learning by doing, and may well require legal intervention to clarify the meaning of certain principles and provisions.

The development of strategies and administrative procedures to give effect to the principles and provisions in the policy will be an important next step. This will require the involvement of all governance actors, as well as collaboration across sectors, disciplines and actors that have not, in the past, worked together.

Notes

1 'Living customary law' is the term used by South Africa's Constitutional Court to refer to customary law that is 'actually observed by the people who created it', as opposed to 'official customary law', which is 'the body of rules created by the state and legal profession' (Bennett, 2008, p. 138).

2 'Homelands' were the areas to which Africans were given obligatory citizenship under apartheid.

3 Organized by the Masifundise Development Trust, a non-governmental organization, the Fisher Human Rights Hearings were held in 2003 to give fishers an opportunity to tell how existing fisheries policy and legislation were affecting their lives and livelihoods.

4 *George K and others v Minister of Environmental Affairs and Tourism* 2004 (EC1/05).

5 Section 211(3) of the Constitution of South Africa (RSA 1996) states that the courts are obliged to apply customary law when it is applicable, subject to the Constitution and any legislation that deals with customary law.

6 South Africa's Constitutional Court has recognized customary law as a source of law in several cases, for example *S v Makwanyane and Another* 1995(3) SA 391 (CC) and *Alexkor Ltd v Richtersveld Community*, 2004(5) SA 460 (CC) (Wicomb and Smith 2011).

7 *S v Gongqoze*, E 382/10, not yet reported.

Bibliography

African National Congress (1994) *The Reconstruction and Development Programme (RDP): A Policy Framework*, Umanyamo Publications, Johannesburg

Andrew, N. L., Béné, C., Hall, S. J., Allison, E. H., Heck, S. and Ratner, B. D. (2007) 'Diagnosis and management of small-scale fisheries in developing countries', *Fish and Fisheries*, vol 8, pp. 227–40

Béné, C. (2006) 'Small-scale fisheries: Assessing their contribution to rural livelihoods in developing countries', FAO Fisheries Circular no 1008

Bennett, T. (2008) '"Official" vs "living" customary law: Dilemmas of description and recognition', in A. Claassens and B. Cousins (eds) *Land, Power and Custom: Controversies Generated by South Africa's Communal Land Rights Act*, University of Cape Town Press, Cape Town, and Ohio University Press, Athens, OH

Berkes, F., Mahon, R., McConney, P., Pollnac, R. C. and Pomeroy, R. S. (2001) *Managing Small-Scale Fisheries: Alternative Directions and Methods*, International Development Research Centre, Ottawa

Borrini-Feyerabend, G. (2008) 'Governance as key for effective and equitable protected area systems', Briefing Note 8, IUCN Commission on Environmental, Economic and Social Policy (CEESP)

Crosoer, D., Van Sittert, L. and Ponte, S. (2006) 'The integration of South African fisheries into the global economy: Past, present and future', *Marine Policy*, vol 30, pp. 18–29

DAFF (2012) *Small-Scale Fisheries Policy*, Department of Agriculture, Forestry and Fisheries, Pretoria

EEU (2010) 'Small-scale fisheries in South Africa and their potential to obtain Marine Stewardship Council (MSC) certification: A preliminary investigation', prepared for the MSC, Environmental Evaluation Unit, University of Cape Town, Cape Town

FAO (1995) *Code of Conduct for Responsible Fisheries*, Food and Agriculture Organization of the United Nations, Rome

FAO (2005a) 'Increasing the contribution of small-scale fisheries to poverty alleviation and food security', FAO Technical Guidelines for Responsible Fisheries, vol 10, Food and Agriculture Organization, Rome

FAO (2005b) 'Review of the state of world marine fishery resources', FAO Technical Paper no 457, Food and Agriculture Organization, Rome

FAO (2008) 'Report of the Global Conference on Small-Scale Fisheries: Securing sustainable small-scale fisheries: Bringing together responsible fisheries and social development', Bangkok, 13–17 October, Food and Agriculture Organization, Rome

Harris, J. M., Branch, G. M., Clark, B. M., Cockcroft, A. C., Coetzee, C., Dye, A. H., Hauck, M., Johnson, A., Kati-Kati, L., Maseko, Z., Salo, K., Sauer, W. H. H., Siqwana-Ndulo, N. and Sowman, M. (2002) 'Recommendations for the management of subsistence fisheries in South Africa', *South African Journal of Marine Science*, vol 24, pp. 503–23

Harris J. M., Branch G. M., Clark B. M. and Sibiya C. (2007) 'Redressing access inequities and implementing formal management systems for marine and estuarine subsistence fisheries in South Africa', in T. R. McClanahan and J. C. Castilla (eds) *The Process of Developing a Management System for Subsistence Fisheries in South Africa: Recognizing and Formalizing a Marginalized Fishing Sector in South Africa*, Blackwell Publishing, Oxford, pp. 112–28

Hauck, M. (2008) 'Rethinking small-scale fisheries compliance', *Marine Policy*, vol 32, no 4, pp. 635–42

Hauck, M. and Kroese, M. (2006) 'Fisheries compliance in South Africa: A decade of challenges and reform 1994–2004', *Marine Policy*, vol 30, no 1, pp. 74–83

Isaacs, M. (2006) 'Small-scale fisheries reforms: Expectations, hopes and dreams for "a better life for all"', *Marine Policy*, vol 30, no 1, pp. 51–9

Isaacs, M. (2012) 'Individual transferable quotas, poverty alleviation and challenges for small-country fisheries policy in South Africa', *Maritime Studies*, vol 10, no 2, pp. 63–84

Isaacs, M., Hara, M. and Raakjær, J. (2007) 'Has reforming South African fisheries contributed to wealth redistribution and poverty alleviation?', *Ocean and Coastal Management*, vol 50, pp. 301–13

Jeradino, A. and Yates, R. (1996) 'Preliminary results from excavations at Steenbokfontein Cave: Implications for past and future research', *South African Archaeological Bulletin*, vol 51, pp. 7–16

Kooiman, J., Bavinck, M., Jentoft, S. and Pullin, R. (eds) (2005) *Fish for Life: Interactive Governance for Fisheries*, Amsterdam University Press, Amsterdam

Lasiak, T. (1993) 'The shellfish gathering practices of indigenous people in Transkei: Patterns, preferences and perceptions', *South African Journal of Ethnology*, vol 16, pp. 115–20

LRC (2012) 'Proposal to the Department of Agriculture, Forestry and Fisheries (DAFF) for the amendment of the Marine Living Resources Act, 18 of 1998', Legal Resources Centre, Cape Town

McConney, P. and Charles, A. (2008) *Managing Small-Scale Fisheries: Moving Towards People-Centred Perspectives*, Centre for Resource Management and Environmental Studies University of the West Indies, Barbados

MDT (2003) '"We want to be heard, enough is enough": A report on the Fisher Human Rights Hearings 13–14 August 2003', Masifundise Development Trust, Cape Town

Nielsen, J. R. and Hara, M. (2006) 'Transformation of South African industrial fisheries', *Marine Policy*, vol 30, no 1, pp. 43–50

NPC (2012) *National Development Plan: Vision for 2030*, National Planning Commission, Pretoria

OAU (1981) *African Charter on Human and Peoples' Rights*, Organization of African Unity, Addis Ababa

Parkington, J. E., Poggenpoel, C., Bunchanan, B., Robey, T., Manhire, T. and Sealy, J. (1988) 'Holocene coastal settlement patterns in the western Cape', in G. Bailey and J. Parkington (eds) *The Archaeology of Prehistoric Coastlines*, Cambridge University Press, Cambridge, pp. 22–41

Pauly, D., Christensen, V., Dalsgaard, J., Froese, R., and Torres, F. Jr (1998) 'Fishing down marine food webs' *Science*, vol 279, no 5,352, pp. 860–63

RSA (1996) 'Constitution of the Republic of South Africa (Act 108 of 1996)', *Government Gazette*, vol 378, no 17,678, Cape Town

RSA (1998) *Marine Living Resources Act*, Act 18 of 1998, Pretoria, Government Printer

Sharma, C. (2011) 'Securing economic, social and cultural rights of small-scale and artisanal fisherworkers and fishing communities', *Maritime Studies*, vol 10, no 2, pp. 41–61

Siegele, L., Roe, D., Giuliani, A. and Winer, N. (2009) 'Conservation and human rights – who says what? A review of international law and policy', in J. T. Campese, T. Sunderland, T. Greiber, and G. Oviedo (eds) *Rights-Based Approaches: Exploring Issues and Opportunities for Conservation*, CIFOR and IUCN, Bogor, Indonesia, pp. 47–76

Sowman, M. (2006) 'Subsistence and small-scale fisheries in South Africa: A ten year review', *Marine Policy*, vol 30, pp. 60–73

Sowman, M. (2011) 'New perspectives in small-scale fisheries management: challenges and prospects for implementation in South Africa. *African Journal of Marine Science* vol 33, pp. 297–311

Sowman, M., Hauck, M., Van Sittert, L. and Sunde, J. (2011a) 'Marine protected area management in South Africa: New policies – old paradigms', *Environmental Management*, vol 47, no 4, pp. 573–83

Sowman, M., Raemaekers, S., Sunde, J., Schell, N. and Schultz, O. (2011b) 'Integrating the human dimension of an ecosystem approach to fisheries (EAF) into fisheries management in the BCC region, baseline report: South Africa', unpublished report for Benguela Current Commission and FAO, Environmental Evaluation Unit, University of Cape Town, Cape Town

Sunde, J. (2003) 'Access and rights to the "marine commons" in South Africa: Perspectives, lessons and possibilities for future action', Masifundise Development Trust, Cape Town

Sunde, J. (2011) 'Dwesa-Cwebe scoping report', Environmental Evaluation Unit, University of Cape Town, Cape Town

Sunde, J., Sowman, M., Smith, H. and Wicomb, W. (forthcoming) 'Emerging proposals for governance of tenure in small-scale fisheries in South Africa', *FAO Tenure Journal*

UNDP (2002) 'Plan of implementation of the World Summit on Sustainable Development', advance unedited text, 12 June, United Nations Development Programme, http://iisd.ca/2002/wssd/PlanFinal.pdf

UNEP (1992) *Convention on Biological Diversity: Text and Annexes*. Chatelaine, Switzerland: United Nations Environment Programme, CBD

Van Sittert, L., Branch, G., Hauck, M. and Sowman, M. (2006) 'Benchmarking the first decade of post-apartheid fisheries reform in South Africa', *Marine Policy*, vol 30, pp. 96–110

Wicomb, W. and Smith, H. (2011) 'Customary communities as "peoples" and their customary tenure as "culture": What we can do with the *Endorois* decision', *African Human Rights Law Journal*, vol 11, no 2, pp. 422–46

Witbooi, E. (2006) 'Law and fisheries reform: Legislative and policy developments in South African fisheries over the decade 1994–2004', *Marine Policy*, vol 30, pp. 30–42

11 Legal pluralism and the governance of freshwater resources in southern Africa

Can customary governance be embedded within the statutory frameworks for integrated water resources management?

Sharon Pollard and Tessa Cousins

Introduction

Echoing wider political transformations, the last two decades in southern Africa have seen major changes in policies concerning natural resources, especially water, reflecting deepening concerns regarding resource scarcity. Taking account of the interconnectedness of land and water, the emergent view is that freshwater systems need to be managed more holistically: on a catchment basis and in a way that explicitly involves stakeholders. The guiding framework and philosophy for this are captured in the concept of integrated water resources management, which is regarded as a process rather than an end in itself, and one that can be reflexive and adaptable so as to embrace change and learning (Pollard and Du Toit 2011).

Notwithstanding this progress, there are still issues to be addressed. Western norms continue to shape the reform process so that there is almost no recognition of a parallel suite of customary laws that govern water in many areas – and that these generally consider water *part of a system* rather than a separate entity. While statutory frameworks are widely accepted as 'the law of the land' governing a nation's affairs, the reality is that in many areas under communal tenure, locally derived rules and norms shape the day-to-day access, practices and use of the resources (Meinzen-Dick and Nkonya 2007). Such dual governance, or legal pluralism, is widely evident in many countries where customary law exists alongside the statutes usually adopted during the colonial period (Meinzen-Dick and Pradhan 2002). Nonetheless, although this is well documented in the literature on community-based natural resource management, the focus has been largely on terrestrial resources, and there is virtually no meaningful discourse on legal pluralism and water (see Van Koppen et al 2007).

Another issue is that statutory frameworks view different natural resources as separate entities managed under different institutional arrangements, which customary systems generally do not. Despite evidence that it is problematic, this 'silo' approach has continued as government departments struggle to regulate activities that cut across different natural resources (e.g. mining). The 'resolution' has been to emphasize cooperative governance, which, in practice, has proved enormously challenging, given the resource constraints on the state (Murphree 1991; Pollard and Du Toit 2011).

Legal pluralism has a particular bearing on the praxis of sustainable use and management and the norms that shape it. The concept that freshwater systems, especially wetlands, could be used sustainably first entered the literature more than two decades ago (e.g. Heeg and Breen 1982; Claridge and Callaghan 1997), and since then interest has grown (e.g. Ramsar 2007; Wood et al 2013). However, in the case of wetlands, the focus has remained largely on practices and their management, following the reasoning that improved practices lead to improved sustainability, while water resources management has remained largely technocratic. This fails to recognize two critical issues. First, freshwater systems do not exist in a vacuum but are part of broader, mutually influencing systems that also affect sustainability (Pollard and Du Toit 2013). Second, governance, both statutory and 'customary', lies at the heart of *how* freshwater systems in communal areas are accessed, used and managed.

Yet the discourse on the local-level management of freshwater resources is surprisingly inadequate, and, where water has been considered, the focus has been almost exclusively on the management of water supply (e.g. Sithole 2001; Manzungu and Machiridza 2005). Indeed, some authors have pointed to the absence of any discourse regarding indigenous water management experiences in policy, even in postcolonial states with their professed orientation towards stakeholder involvement, or 'indigenization' in the case of Zimbabwe (see for example Mohamed-Katerere and Van der Zaag 2003; Chikozho and Latham 2005; Meinzen-Dick and Nkonya 2007).

The first real attempt to discuss and document customary water law was an edited volume of experiences from different countries (Van Koppen et al 2007). Even there, many of the case studies focused on customary law in the context of water allocations. This chapter aims to step back and examine experiences of the customary management of freshwater systems, such as wetlands and lakes, as a whole. The Lozi in Zambia, for example, manage the Barotse floodplain for multiple uses and not just the allocation of water. Since systemic approaches of this kind are theoretically the central tenet of statutory reforms such as integrated water resources management, we have also briefly examined how and where these can be embraced in evolving arrangements, rather than ignored. Importantly, this should be done judiciously and without a simplistic idealization of customary laws, since those systems can entrench differential power relations, often at the cost of the marginalized or vulnerable, such as women (Bruns 2007).

With that as the background, this chapter seeks to review cases of customary governance of freshwater systems in a number of countries in southern Africa undergoing water reform (Mozambique, Zimbabwe, Zambia and South Africa: see Figure 11.1) in order to broaden the discourse on water governance arrangements. Before moving to the case studies, which demonstrate the dynamic, varied and pervasive nature of legal pluralism, we provide a brief overview of the major

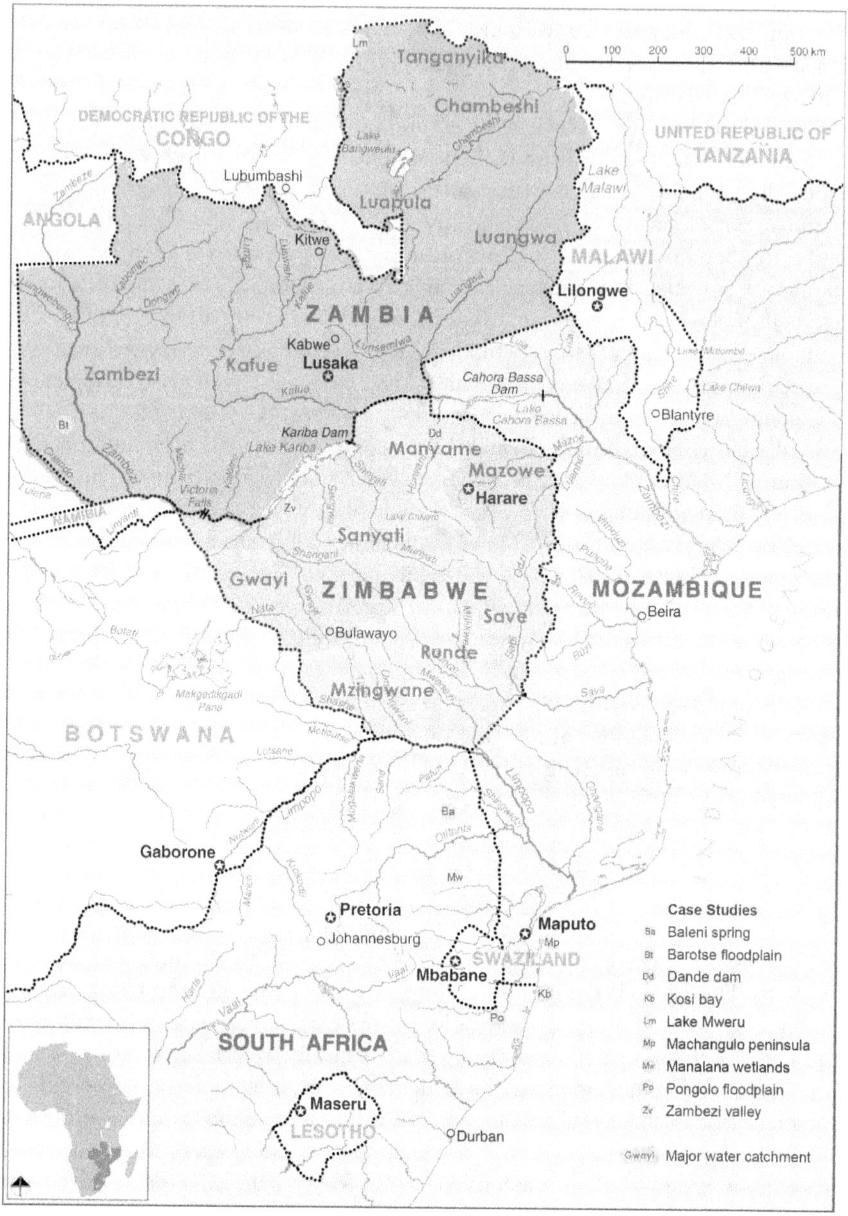

Figure 11.1 Map showing location of case studies.

water-related statutory reforms in the countries under consideration, and of the key conceptual ideas which informed the framework used for the analysis of case studies.

Key concepts in the customary governance of freshwater systems

A number of concepts are central to understanding the customary governance of water resources. While a detailed account is beyond the scope of this chapter (see Chapter 2 of this volume), some of the key points are highlighted here. Although the term 'customary' is adopted, we note that the literature treats it as interchangeable with the terms 'indigenous', 'informal', 'traditional' and 'local'. None of these is entirely satisfactory: 'traditional' can imply a naive and obdurate adherence to unchanging values, and 'informal' is highly pejorative, suggesting a quality inferior to that of the formal system, which carries an implied legality purely through codification. This chapter favours the term 'customary' for its connotation of historical continuity, bearing in mind the concerns described above. Importantly, it does not suggest immobility, but recognizes that governance systems are dynamic in space and time. The term is not synonymous with 'community-based', which can describe systems that are not customary, but the two do overlap to some degree.

Definitions of 'governance' vary, but here it is regarded as 'a socio-political process to manage affairs'. It thus describes the relationships between people and the rules and norms that are set up to guide their interactions (Meinzen-Dick and Nkonya 2007). The nature of the resource informs governance arrangements (Murphree 1991), and freshwater systems such as wetlands are interesting since they represent the nexus between water and land. This is important because a key characteristic of African communal tenurial systems is that they are nested and have multiple and flexible boundaries (Cousins 2007). Hence rights are held at different scales depending on the resource, the user and the season. In wetlands, for example, each resource – reeds, fish or water – has different boundaries that overlap in space and time, and is accessed by different users at different times, illustrating a key difference from Western norms. Moreover, rights are inclusive, flexible and adaptive, and can be renegotiated (Cousins and Claassens 2004). Customary law is based on the notion of community membership rather than that of individual rights, so that customary systems include layered and shared rights of access and use, the institutional nestedness of family, clan and tribe, and the normative values that inform the basis of resource entitlement. The principles are well understood by a local community, but may not conform to the country's legal procedures.

Underlying the discourse on governance is the issue of property regimes: here property is viewed as the bundle of entitlements defining rights and responsibilities for use, decision-making and usufruct – or institutional arrangements (Schlager and Ostrom 1992; Agrawal and Angelsen 2009). Central to these is the concept of *institutions* which have been described in multiple, ambiguous and often conflicting ways. In general we adopt the ideas of Ostrom who describes institutions as 'shared concepts used by humans in repetitive situations organized by rules, norms and strategies' (Ostrom 2007, p. 23). Importantly, institutions are socially constructed; they have normative and cognitive, as well as regulative, dimensions (Jentoft et al 1998), and hence are about the relationship between people. Rules, regulations and sanctions reflect a complex interplay of beliefs, cosmologies and world views and the organizational arrangements for reconciling these. 'Resource tenure' refers to all the ways in which people gain access to natural resources. Lavigne-Deville (2004) defines 'tenure regulation' as 'a set of practical decisions regarding rights including governance, management and operation'.

Finally, in line with a growing body of work that advocates a holistic view (e.g. Cilliers 1998; Berkes et al 2003; Ison 2010; Ostrom and Cox 2010), this chapter uses a systemic lens to examine both the biophysical and socio-political elements of the governance regime (see Pollard and Cousins, 2008, for details). It also examines the literature for reference to issues of learning, flexibility, feedback and risk minimization in local governance systems and practices, since these elements are regarded as critical to management strategies that are adaptive and build resilience (see Folke et al 2002; Janssen and Ostrom 2006).

Case studies

Water reform processes and policies

Although a detailed review of the statutory water reform process is beyond the scope of this chapter, some of the most pertinent changes are summarized in Table 11.1.[1] The principle of equitable access – and, to a lesser extent, sustainability – has guided water reform in all cases, and water is now regarded as a public asset (under the custodianship of a government minister). In most cases, although the current statutes offer opportunities for stakeholder participation, this does not explicitly address opportunities for communities to govern their own freshwater resources, with the possible exception of Zambia.

South Africa

Despite the global interest in community-based natural resource management, the coverage of these issues for freshwater systems in South Africa is negligible. This may reflect the lack of acknowledgement of local-level governance systems in general until recently, as well as the fact that globally the focus of community-based natural resource management studies has tended to be on terrestrial resources. Even in larger systems that have attracted development, community governance arrangements are poorly understood. The broad alluvial Pongolo floodplain close to the border with Mozambique (McCartney et al 2004), which supported an estimated 40,000 AmaThonga people in 1980 (Heeg and Breen 1982; and see Kosi Bay case study) is a case in point. Despite growing contestations around access and use, there is a surprising dearth of substantive work on common property regimes and their transformation in this important floodplain. It must be acknowledged, however, that the colonial interference of the past century and the deep distrust bred then make exploring any of these issues extremely difficult.

Furthermore, in most areas tribal leaders have held little practical power over water use and conservation, with some exceptions. For example, the Baleni hot

Table 11.1 Statutory water reforms in case-study countries

Country	*General reform principles*	*Organizational arrangements*
Mozambique	A 1991 law and a policy document from 1995 regulate water use, although both are under revision. Central is the change from a supply-driven to a demand-driven approach. The 1991 Water Law states that water (surface and ground) is state property and in the public domain. It is based on the principles of unified and coherent management, coordination and participation, and it integrates water policy with national physical planning and environmental policies. It distinguishes two kinds of use: common and private.	The 1995 Water Policy seeks to (a) change the role of the central government from service provider to regulator and (b) decentralize to the watershed level and to local service providers, preferably private. The central body responsible for water management is the National Water Directorate. Intersectoral coordination is guaranteed through the National Water Council, which is an advisory body to the Council of Ministers. The regional water authorities are responsible for integrated water resources management in five regions, and work with sub-basin committees.
Zimbabwe	The Water Act of 1998 brought reforms that were designed to ensure a more equitable distribution of water and stakeholder involvement in water resources management. Water was no longer privately owned, and water permits of limited duration would be allocated by catchment councils. A strong focus was on the reorientation of water from large-scale commercial agricultural to small-scale farmers.	The Zimbabwe National Water Authority Act of 1998 led to the establishment of the Zimbabwe National Water Authority, a parastatal agency responsible for water planning and bulk supply. The authority plans and manages water resources on a catchment basis and is mandated to involve stakeholders. It works with seven river catchment councils to which it is supposed to devolve responsibility for managing river systems and enforcing laws and regulations. Sub-catchment councils are the lowest tier in this system, but these have yet to be established.
Zambia	The Water Act of 1996 is the main statutory legislation dealing with water management. It vests ownership of water in the president, although legal pluralism exists within state law. The Act distinguishes between private and public water.	The Department of Water Affairs is responsible for developing and managing water resources. Most urban water supplies are run by local authorities, while most rural supplies are managed by the department. On customary land, traditional leaders are responsible for administering access to and use of water, in cooperation with local authorities. Stakeholder participation is secured through the Water Sector Advisory Group, which comprises four subsector groups: water supply and sanitation; water resources management; water resources infrastructure development; and monitoring, evaluation and capacity-building.

Table 11.1 Continued

Country	*General reform principles*	*Organizational arrangements*
South Africa	The National Water Act of 1998 is widely regarded as the most progressive water legislation in the world. The statute's overall purpose is to ensure that South Africa's water resources are protected, used and managed in ways which take into account intergenerational equity, equitable access, redressing past racial and gender discrimination, promoting sustainable and beneficial use, facilitating social and economic development, and providing for water quality and environmental protection.	The national Department of Water Affairs has oversight functions with the minister as the overall custodian of the nation's water resources. Certain functions are held by the minister, and others are devolved to the nine catchment management agencies, which are in the process of being established. The democratization of water resources management is supported by representative forums, including catchment management committees, forums and water user associations. Although cooperative governance is seen as central to integrated water resources management, the Act does not mention community governance as such.

spring is in a wetland near the Klein Letaba River under the authority of Chief Muhaumani. The site includes salt-making activities next to the river (Pollard and Cousins 2008). Through myth and legend, salt production has been gendered, so that it is for the exclusive use of certain Tsonga women, and highly ritualized religious and spiritual practices govern the extraction and production of salt (P. Terblanche, curator, Muti wa Vatsonga Museum, personal communication, July 2010). Until recently, Baleni was isolated and subject to little outside influence, but in 1998 it was identified as an African Ivory Route campsite. Although this is seen as a source of revenue for the local community, it has been contentious and is bound to bring change. Theoretically, statutory protection is afforded through the area's proclamation as a national heritage site, although no recognition has been given to the customary system that has afforded protection over the past century.

Craigieburn: Wetland governance in a transforming and plural legal milieu

One of the most detailed systemic explorations of freshwater local-level governance regimes has been on the 365-hectare Manalana wetlands of Craigieburn village in the upper Sand River in north-eastern South Africa (Cousins et al 2007; Cousins and Pollard 2010). Like many other wetlands of the communal area, they play an important role in catchment water security and local livelihoods, where they are used principally for small-scale cropping, reed harvesting and livestock grazing (Pollard ct al 2005). Nonetheless, wide-scale use combined with certain land use practices has resulted in extensive erosion and desiccation, impacting in turn on users' livelihoods (Pollard and Du Toit 2013) The questions raised here regarding norms and rules governing access, use and care catalyzed a detailed examination of community and statutory governance.

Various collaboratively developed conceptual and planning tools were used to examine the local governance systems (see, for example, Figure 11.2). First is the recognition that no discussion on natural resource management would be complete without an examination of history and the influence of apartheid in South Africa. Prior to democratization in 1994, controls were effected through 'chiefs', who were imposed by the white government and dealt with transgressors through a system of 'tribal police'. After 1994 the authority of such individuals, considered to be lackeys of the apartheid state, was contested (this should not be conflated with a contestation of 'traditional systems' per se). Consequently, in many communal areas such as in Craigieburn, this system weakened (Shackleton et al 1995), and today various bodies such as the local government, ward councillors and provincial staff all lay claim, erroneously, to authority over natural resources and their use (Cousins and Pollard 2010).

Today a number of issues are striking. The first is that natural resource management arrangements are fundamentally shaped by plural tenure systems, so that rights and authority derive both from custom and from statutory laws. However, while customary rights are well understood locally, statutory rights are not. The authors have concluded that major local and systemic constraints to effective governance exist, not least of which is that natural resources are not prioritized and hence residents are apathetic about committing time and resources.

Wetlands, as part of the commons, have weaker tenure arrangements than other land uses and, as noted by Lahiff (1997) in the northern Venda-speaking area of South Africa, there is little evidence of 'traditional' practices or institutions concerned with the regulation or conservation of water resources, nor institutionalized forms of cooperation between different tribal areas sharing the same resource. The prevailing sense of individual autonomy around wetland use poses particular challenges for collective, community-based governance (Cousins and Pollard 2010), a situation also reported in other wetlands (Kotze et al 2002; Pollard and du Toit 2013).

Moreover, 16 years on, land tenure reform, which is intimately linked to natural resource use, is regarded by many as being in disarray (Cousins 2007). The insecurity created by such ambiguity, confusion and uncertainty has fostered an environment ripe for opportunism and is likely to reverberate in local efforts to manage natural resources. Ultimately this will impact on livelihoods, especially of the vulnerable people, and on sustainability.

Kosi Bay

Kosi Bay in northern KwaZulu-Natal is well-known for its extensive network of wooden fish traps built by the AmaThonga fishermen, who pass their fish kraals from father to son (Felgate 1982; Webster 1991). They are placed in the shallows between the estuary and lakes, mainly to catch migrating marine fish. The Kosi system falls under the Tembe Traditional Authority, which covers the largest communal area in South Africa (Jones 2006) and is part of the 11,000-hectare Kosi Bay Nature Reserve proclaimed in 1987. Indeed, Kosi Bay's recent history has been largely dictated by the proclamation and expansion of various reserves that did not

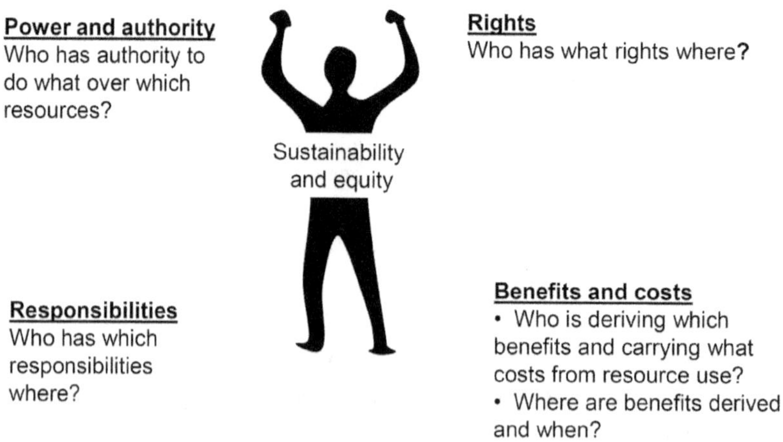

Figure 11.2 Four critical aspects of natural resource governance.

take into account local needs, and relationships have been marked by conflict ever since (McGregor 1997; Guyot 2005; Kloppers 2005). About 40,000 fish are caught for local consumption annually. Kyle (1995) notes that the demand for and value of *Juncus* sedge (or *incema*) are increasing, drawing entrepreneurs from as far as Johannesburg. When sickles came into use, in contrast to traditional hand-picking methods, and exploitation increased markedly, the authorities – later with the support of local users – advocated the ban of sickles in 1992. Aside from this restriction, there are today no limits on the areas of collection, numbers of collectors or season.

Notwithstanding the area's scenic and economic importance, discussions pertaining to traditional property regimes in the Kosi system almost never happen. Any discussion of management focuses on statutory instruments alone, shedding little light on the common property arrangements used by the residents themselves – this despite simmering conflicts in the area over the past three decades. For example, Kyle (1995) notes that the fish trap system is well controlled, but gives no further details about how people gain rights of access. Views on local participation vary widely, from the suggestion that stakeholder participation is central (e.g. Kyle 1995) to the assertion that all moves by the state are duplicitous and hide real political intentions (see Guyot 2005).

Zambia

Even though there has been some sophisticated analysis of legal pluralism in the Zambian context, many studies fail to pinpoint the underlying drivers behind increasing resource depletion and livelihood vulnerability. While these studies recognize the increasing impacts of economic growth and population pressures, they make scant mention of transforming governance regimes (see, for example, International Union for Conservation of Nature [IUCN] 2003; Emerton 2005). In some cases described below, notably those of the Barotse Royal Establishment and Lake Mweru, local-level governance systems are well documented. In others, however, despite the prominence of freshwater systems in Zambia, many of the governance systems, especially in the smaller *dambos* or wetlands, are poorly understood. With over 70 different tribes, these systems are potentially highly diverse.

The Lozi kingdom and the Barotse floodplain

The Barotse floodplain, which covers some 1.2 million hectares (IUCN 2003), is outstanding in its vastness and beauty, and for the long-standing and continued relationship between the communities and their wetland. As noted by Thole and Dodman (1997), 'strong traditional systems have rarely lasted as in the Barotse floodplain'. The four districts of the floodplain are estimated to house about 225,000 people, or about 30 per cent of the population of Zambia's Western Province's (based on figures in Lewanika 2002). Most of the people depend on a mixed livelihood strategy of crop farming, livestock, fishing and natural resource exploitation. More than 75 per cent of the 265,000 cattle are pastured in the floodplain and over half of the floodplain residents are involved in fishing.

Zambia's Western Province is distinct for a number of reasons, not least of which is the role of the Lozi kingdom in safeguarding natural resources. Each natural resource – fisheries, land, water, forests and wildlife – has a designated *induna* (representative of the chief) to supervise its use and management, and they are led by a Lozi paramount chief, or king, the *litunga* (Lewanika 2002). The legal system is based on five cornerstones, which also govern natural resources: *milao* (laws), *liswanelo* (a body of rights and responsibilities), *lituekelo* (rights of a particular position or social status), *mikwa* (methods or ways of doing things) and *mulatu* (offence or wrongdoing).

The Lozi are recognized for their unique traditional methods of wetland cultivation, fishing, hunting and forest use. Most notably, the Lozis continue to practice an annual transhumance, when the entire kingdom undertakes a migration marked by the Kuomboka ceremony, in which, at the height of the floods, the *litunga* relocates from the dry-season to the wet-season capital (Figure 11.3). Importantly, the benefits accrued from natural resources are shared between local people and the authorities (Lewanika 2002). Concerns have been expressed that with increasingly centralized authority, this system is increasingly under threat. Although the Lozi kingdom had occupied the Barotse floodplain since the late eighteenth century, the arrival of the Europeans led to change. In 1890, Barotseland became a British protectorate and enjoyed a certain degree of autonomy. Then, in 1964, the paramount chief renounced the special treaty relationship with the British and accepted integration into the new state of Zambia.

Figure 11.3 Transhumance of Lozi people across the Barotse floodplain.

Before the enactment of the Forest Act in 1974, the Barotse forest reserves were protected under the customary law of 1939, but after 1974 powers of control were placed almost exclusively in the hands of Zambia's forest department.

This sparked the resentment of the local community, which manifested in massive forest destruction and losses of wildlife. Numerous organizations have since attempted to facilitate new governance arrangements, with varying degrees of success. Local people argue that a tenable traditional management system is being eroded by government policies that have largely ignored them (Pollard and Cousins 2008). Undoubtedly recent planned developments outside the Barotse floodplain, including hydropower schemes, reservoirs and dams along the Zambezi River, will affect the wetland status and local livelihoods (IUCN 2003).

Lake Mweru-Luapala: 'Owners of the Lagoon' and changing beneficiaries

Lake Mweru holds some valuable insights into common property theory because of the role of the 'Owners of the Lagoon' (or 'Bamwine Kabanda') in local-level management, and the well-documented conflicts that accompanied colonial interests and the resultant collapse of the lake's fisheries (Gordon 2003). The 4,650 square kilometre Lake Mweru is situated on the border of Zambia and the Democratic Republic of the Congo in the Luapula Valley. The fish of the lake, Luapula River, floodplains, swamps and lagoons provide the basis for food, employment and income for the estimated 400,000 people that live there (Van Zwieten et al 2003). Unlike the Barotse floodplain, this area has a unity determined primarily by geography: it is a valley bounded by escarpments to the east and west.

Gordon (2003) provides a detailed account of changing governance and beneficiaries in a system driven by major external influences. He points out that in contrast to the present day, Mweru-Luapula was not an a priori open-access common resource. Rather, both spiritual and political sanctions were developed and upheld by the Owners of the Lagoon. However, this governance system collapsed with the arrival of two colonial administrations: the Belgians (who regarded the resource as open to all) and the British (who held the resource in trust for Africans).

Widespread exploitation of the fisheries by these newcomers, together with a formal state authority (colonial chiefs, administrators and fisheries officials) that held little standing locally, led to the collapse of the fisheries in the 1950s. Gordon concludes that the colonial administration, apparently increasingly concerned about the growing numbers of fishermen, imposed regulations through a system that sought little coherence with that of the original 'Owners of the Lagoon' – a significant factor, he asserts, in the demise of a tenable governance regime.

Mozambique

Currently, only a very limited body of research work exists for Mozambique, most of it undertaken after the end of the civil war in 1992. Also, in common with most of the countries examined, the focus is mainly on terrestrial or coastal fisheries common-property regimes or on ecological studies (see, for example, Brito 1998;

Boyd et al 2000; Chilundo and Cau 2000; Tique 2000; Anstey and Sousa 2001; Virtanen 2001; Schafer and Bell 2002; Durang and Tanner 2004). Although the demise of common-property regimes in some of the prolific *muchongos* or coastal lakes is mentioned (Chilundo and Cau 2000), there is little discussion of why this is so.

Machangulo Peninsula

The case of the Machangulo Peninsula flanking the bay of Maputo involves a range of resources including the sea, adjacent floodplains and mangroves, forests, farmland and a series of lakes. In his examination of private and common interests, Brouwer (1998) provides a history of the area, which remained relatively obscure until 1995, when proposals for tourism prompted growing economic interest.

Legal pluralism is widely evident, and each of the natural resources is subject to different statutory laws and traditional arrangements (Brouwer 1998). In practice the suite of statutory laws (such as land, forest and environmental legislation) that govern the various resources is given little weight in practice, with locals following their own rules for resource-use rights (Roland Brouwer, Faculty of Agronomy and Forest Engineering, Universidade Eduardo Modlane, personal communication July 2010). For example, a strong local system exists for the tidal plains and mangroves, parallel to the official law. Here locals have mounted *gamboas* (extended nets) which, although creating (temporary) private ownership rights, are constantly encroached upon by the various parties involved.

Brouwer shows not only that the mix of entitlements varies with the nature of resource, but also that the use of these common property resources is subject to two restrictions: it should serve the satisfaction of immediate consumptive needs (i.e. not be for trade), and the user should be a member of the community. For example, access to the lakes for subsistence fisheries is restricted to the inhabitants of a 'cell' (or community group) around the lake. Brouwer points out that up until the late 1990s, when entrepreneurial interests in the area started, the local framework of rules regarding lakes, land, mangroves and the tidal areas existed parallel to, and hardly affected by, the state system. However, growing economic interests mean this situation is likely to change, particularly for the local population, for whom fishing, together with remittances, constitutes the main source of cash and dietary protein.

Zimbabwe

Despite the large body of work regarding community-based natural resource management in Zimbabwe, little of it mentions water resources and even less talks specifically to issues of governance. However, there is mention of customary relationships with the landscape, such as in the case of the Zambezi Valley (see McGregor 2003). Chikozho and Latham (2005) focus on the disjuncture between customary and statutory approaches introduced by Zimbabwe's water reform process and what this means for integrated water resources management. Both cases are discussed below.

The river people: The exclusion of the Tonga from the Zambezi valley

The Zambezi Valley is well known for the removal of 80,000 Tonga people upon the construction of the Kariba Dam in 1958, and the impact that this upheaval had on their livelihoods. The people who lived along this part of the Zambezi referred to themselves as the 'river people' or *bamulwizi*. McGregor (2003) provides a striking account of contested identities, describing how the Tongas' own identity as the 'river people' contrasts with the simplistic and erroneous one constructed by Europeans. The notion of the 'River God' – heavily commodified for tourists at the Victoria Falls – is, she suggests, central to the appropriation of the river and lake for tourism and conservation and to the marginalization of those who lived there.

Even today these communities invoke their relationship with the river. The 19th-century riverine political economy was one of east-west linkage, with the Zambezi as a fearful barrier – and the 'river people' were known for their command of the river crossing. Although the riverine society was not a tightly bounded or centralized political authority, strong sanctions existed for the Zambezi River, which was (and is) viewed as a provider for all, offering sacred places, refuge, fish and gardens for cultivation in retreating floods. Although McGregor does not deal with water resources management per se, she notes that local people mediated their relationship with the river through ancestral spirits. The invocation of sacred places placed limits on activities such as fishing under instruction from those responsible for shrines. The agricultural strategies – using floodplains, riverbanks and rains – provided multiple options for food security which were lost once they were translocated to semi-arid lands, when they became a food-deficit people (Magadza 2003).

The identity of 'river people' is relevant to conflicts over resources evident today. After the construction of the Kariba Dam, the focus on development through tourism and commercial fishing took precedence over local livelihood issues. Today there is increasing anger, especially regarding the breaking of promises that the state would provide new sources of water (McGregor 2003). This never happened, and water shortages are among the most acute problems faced by people. In Hwange the Dombe and Nambya have only just secured rights to have fishing camps, after two decades of prohibition, and are still struggling to cultivate riverbank gardens, something that was deemed illegal.

With reference to the inland fisheries, Jackson (1995) asserts that 'the quest for a long-established customary inland fisheries tenure' is now inappropriate. However, McGregor contests this, pointing out that local communities feel marginalized from a natural resource that was theirs, and so exclusion from the commercialization of the resource is the issue at hand. Residents were ignored when new commercial fishing ventures were established, while whites and other newcomers received preferential treatment. She makes the case that marginalized and dispossessed minorities have invoked a relationship to the landscape that they continue to hold to as part of a claim for rights over resources that they regard as having been taken from them.

The Shona world view and the management of water resources in the Mazowe and Manyame catchments

The Guruve Communal Land, traversed by the Dande and Manyame rivers, which flow into the Zambezi, is home to the Shona, who are thought to have moved there around the second century AD. In recent years the Zimbabwean government has embarked upon an ambitious project to build a dam at the site to irrigate the Zambezi Valley. However, local support for the dam has been less than enthusiastic, with many of the potentially displaced questioning the benefits (Chikozho and Latham 2005). Consultation has been minimal, and the flooding of sacred sites has elicited wide-scale opposition and resistance led by the ancestral spirits (*mhondoros*).

These contestations reflect the hardships already suffered as a consequences of the Mid-Zambezi Rural Development Project, a top-down intervention by the government in the 1980s to rationalize settlement patterns and provide agricultural land to 'outsiders'. The highly controversial forced removal of people away from the riverbanks, where they had lived for centuries, and the prohibition on the use of riverine alluvial soils for cultivation led to a protracted struggle in which the state prevailed (see also Sithole 2001). The use of the sacred Mushongaende Pool as a storage and abstraction point also met with resistance, and the scheme has not materialized due to 'cultural problems' (Chikozho and Latham 2005).

These contestations highlight differences in governance, with the (local) management of natural resources being intimately linked to the Shona world view, encapsulated in dictums such as 'the land is the people' and 'the chief is the people, the people are the chief'. These reflect an institutional reality of governance through consensus and the notion of humans as part of, not separate from, their environment. Traditional Shona religion, still strong today, centres on the belief in a supreme being who is approached through a hierarchy of ancestral spirits called *mhondoros*, whose 'districts and provinces' are concerned with the care and management of the earth and community well-being.

This world view, so fundamental to resource use, is quite contrary to the tenets of Western statutes. Indeed, Chikozho and Latham (2005) make the point that where laws prohibit certain local practices, they are simply disregarded. They suggest that the customary institutional arrangements are better suited to water resource management, thanks to their characteristic flexibility and local sanction. Interestingly, anecdotal evidence suggests that given the current political and economic climate in Zimbabwe, people are relying increasingly on customary systems as statutory arrangements, such as catchment councils (Table 11.1), weaken or collapse (Dr C. Chikozho, International Water Management Institute, personal communication, January 2012).

Implications: Opportunities and challenges for integrating statutory and local-level management of freshwater resources in communal areas

The literature informing this review was written for a variety of reasons, so the treatment of governance issues varies considerably. Nonetheless a number of common characteristics and principles can be discerned. Notably, customary systems are highly dynamic and varied in their own right. For example, Phiri (1999) points out that with over 70 tribes in Zambia, we can expect their associated bodies of 'law' to be very varied. Yet all share a central feature: first and foremost, the customary laws are grounded in the notion of community membership and responsibilities, whereas Western systems place individual rights at centre stage. Moreover, as Derman (1998) notes, the key to understanding customary water resources management lies in grasping the fact that water is part of a social and religious system that is at complete odds with the view that individuals can hold exclusive property rights. These attributes set the customary system apart from Western law. Indeed, the overlapping scales of user rights and responsibilities led Jackson (1995) to conclude that colonial planners found the system confusing and felt the need to 'tidy it up'.

This review finds legal pluralism clearly evident in all four countries, both within the statutory system, where the jurisdictions of multiple laws intersect and overlap, and between the statutory and the customary governance regimes. In all countries, the de jure situation contrasts sharply with the actual circumstances: people are either unaware or unsupportive of state legislation, preferring to use locally derived rules (as the Lozi do) or a blend of statutory and customary (as in Craigieburn).

In areas too remote for state intervention, people rely on local-level common-property arrangements. Much of the work on common-property resources suggests that local-level governance arrangements persist in times of uncertainty and political change. For example, customary systems continued throughout the civil war in Machangulo Peninsula, but are now feeling the impact of external entrepreneurial interests. In areas such as Zimbabwe, where, in the face of political instability and the high transaction costs of statutory institutions such as catchment councils (which are now dysfunctional in many areas), people have fallen back on customary arrangements (Dr J. Latham, Centre for Applied Social Sciences, University of Zimbabwe, personal communication, February 2012). This points to the more resilient nature of locally derived norms, which, as many assert, are better suited to water resource management, given their inherent flexibility, adaptability and congruence (e.g. Cousins and Claassens 2004; Agrawal and Angelsen 2009).

In all cases the dynamic nature of governance is evident – often with local systems weakening as they intersect with Western laws and norms. This is particularly true in the case of Craigieburn in South Africa, where the legacy of apartheid and dispossession, combined with uncertainties introduced by reforms in land, water and natural resources, has led to a vulnerable situation in which opportunism – at the cost of the vulnerable – is rife (Cousins and Pollard 2010).

Another feature of the state's attempts to impose statutory systems is that they may take place on highly contested terrain. The forced removal of the Tonga continues to reverberate even today, and in Zambia, the backlash against state imposition – which took the form of widespread and uncontrolled harvesting of wood and wildlife – had exactly the opposite effect to the one intended in the state's ostensible quest for sustainability. In South Africa, legal pluralism is widely evident in the former Bantustans, but the legacy of apartheid, together with the uncertainties of transformation, has meant that locally based systems, such as they are, are weakening or contested.

Moreover, the contemporary landscape is set to change with the introduction of new laws concerning communal areas. Critics of the process offer some valuable insights, pointing out that these statutes are problematic because of their definition of 'community', their bureaucratization of community rules and the patent lack of capacity to provide meaningful and sustained support for implementation. It is argued by many that this undermines the very basis of common-property regimes that are, by nature, flexible and dynamic.

In none of the cases reviewed here has the central government recognized – let alone made any meaningful attempt to embrace – customary systems, perhaps with the partial exception of the Western Province of Zambia. This is surprising, given the constraints faced by the state in implementing its own laws and the move towards wider stakeholder participation. Nonetheless, the desire to obtain participation should not be confused with a willingness to share governance: the former encourages consultation and involvement, while the latter places the locus of power in the hands of a community and embraces the notion that the power to control rights of access and use should be locally based.

Yet, even mindful of these concerns, opportunities do exist to devolve governance responsibilities within the overall frameworks for integrated water resources management, but mechanisms for this would need to be explored and debated. We suggest, as others have done, that since legal pluralism is a reality, and given the quest for appropriate and lasting governance arrangements in support of sustainability, it is incumbent upon practitioners and academics to consider the potential to embrace customary or local systems within wider statutory frameworks. This is particularly pressing in view of the resource constraints the state faces in managing and regulating water resources alone, since many states have taken on far more resource management authority than they can exercise (Murphree 1991). Indeed, Pollard and Du Toit (2011) argue that given such constraints, and the move to more holistic approaches such as integrated water resources management, the only choice for the state is to embrace self-organization and self-regulation through local-level management actions and governance systems.

Finally, as indicated earlier, no systems are static, and all should be subject to scrutiny against the principles underscoring the democratic reform processes – especially those of equity and sustainability – since, like statutory systems, customary arrangements can entrench power relations at the cost of the vulnerable (Bruns 2007; Derman et al 2007). We must be cognizant of the concerns that have been raised by scholars and practitioners in the field of common-property management, especially in the context of land reform. Thus we should strive to develop, with the requisite simplicity, what is meaningful and appropriate – and no more.

Acknowledgements

The authors wish to thank the Water Research Commission of South Africa for funding this work (WRC grant k8/614). We are grateful to Dr C. Chikozho, Dr P. Chileshe, Dr R. Brouwer, Dr J. Latham and Prof. P. Bernard for their specialist inputs.

Note

1 See Pollard and Cousins (2008) for a review of statutes related to natural resources management and land tenure.

Bibliography

Agrawal, A. and Angelsen, A. (2009) 'Using community forest management to achieve REDD+ goals', in A. Angelsen, M. Brockhaus, M. Kanninen, E. Sills, W. D. Sunderlin and S. Wertz-Kanounnikoff (eds) *Realising REDD+: National Strategy and Policy Options*, CIFOR, Bogor, Indonesia

Anstey, S. and Sousa, C. D. (2001) 'Old ways and new challenges: Traditional resource management systems in the Chimanimani Mountains, Mozambique', in D. Hulme and M. Murphree (eds) *African Wildlife and Livelihoods: The Promise and Performance of Community Conservation*, James Currey, Oxford, pp. 195–207

Berkes, F., Colding, J. and Folke, C. (2003) *Navigating Social-Ecological Systems*, Cambridge University Press, New York

Boyd, C., Pereira, J. and Zaremba, J. (2000) 'Sustainable livelihoods in southern Africa: Institutions, governance and policy processes', SLSA Working Paper 2: Mozambique Country Paper

Brito, L. (1998) 'Case of Study of Santaca Pilot Project 1 – Matutuíne – Mozambique', The World Bank/WBI's CBNRM Initiative

Brouwer, R. (1998) 'Setting the stake: Common and private interests in the redefinition of resources and their access in the Machangulo Peninsula, Mozambique', paper presented at the Seventh Annual Conference of the International Association for the Study of Common Property, Vancouver, Canada

Bruns, B. (2007) 'Community priorities for water rights: Some conjectures on assumptions, principles and programmes', in B. van Koppen, M. Giordano and J. Butterworth (eds) *Community-Based Water Law and Water Resources Management Reform in Developing Countries*, CABI, Wallingford, Oxfordshire

Chikozho, C. and Latham, J. (2005) 'Shona customary practices in the context of water sector reforms in Zimbabwe', paper presented at the International Workshop on African Water Laws: Plural Legislative Frameworks for Rural Water Management in Africa, Johannesburg, South Africa

Chilundo, A. G. and Cau, B. M. (2000) 'Traditional forms of common property rights: A case study in southern Mozambique', paper presented at the Eighth Annual Conference of the International Association for the Study of Common Property, Bloomington, IN, 31 May–4 June

Cilliers, P. (1998) *Complexity and Postmodernism: Understanding Complex Systems*, Routledge, London

Claridge, G. F. and Callaghan, B. O. (eds) (1997) *Community Involvement in Wetland*

Management: Lessons from the Field, Incorporating the Proceedings of Workshop 3: Wetlands, Local People and Development, of the International Conference on Wetlands and Development, Kuala Lumpur, Malaysia, 9–13 October 1995, Wetlands International, Kuala Lumpur, Malaysia

Cousins, B. (2007) 'More than socially embedded: The distinctive character of 'communal tenure' regimes in South Africa and its implications for land policy', *Journal of Agrarian Change*, vol 7, no 3, pp. 281–315

Cousins, B. and Claassens, A. (2004) 'Communal land rights and democracy in post-apartheid South Africa', paper presented at the Conference on the Politics of Socio-economic Rights in South Africa: Ten Years After Apartheid, University of Oslo, Norway, June 8–9

Cousins, T. and Pollard, S. (2010) 'Creating the basis for the governance of wetlands in complex and transforming environments: Lessons from Craigieburn, South Africa'. AWARD internal research paper from the International Development Research Centre, Developing Community Based Governance of Wetlands in Craigieburn, South Africa, IDRC Project No: 103579

Cousins, T., Pollard, S., de Wolf, J., Mabooyi, J. and du Toit, D. (2007) *Developing Community Based Governance of Wetlands: The Tenure Arrangements and Land Management System in Craigieburn*, prepared for the LEAP learning symposium, Cape Town, 6–8 November 2007

Derman, B. (1998) *Preliminary Reflections on a Comparative Study of the Mazowe and Mupfure Pilot Catchments in the Context of Zimbabwe's New Water Act*, CASS, University of Zimbabwe, Harare

Derman, B., Hellum, A., Manzungu, E., Sithole, P. and Machindza, R. (2007) 'Intersections of law, human rights and water management in Zimbabwe: Implications for rural livelihoods', in B. van Koppen, M. Giordano and J. Butterworth (eds) *Community-Based Water Law and Water Resources Management Reform in Developing Countries*, CABI, Wallingford, Oxfordshire

Durang, T. and Tanner, C. (2004) 'Access to land and other natural resources for local communities in Mozambique: Current examples from Manica province', paper presented to the Green Agri Net Conference on Land Registration in Practice, Denmark, 1–2 April 2004

Emerton, L. (ed) (2005) 'Values and rewards: Counting and capturing ecosystem water services for sustainable development', IUCN Water, Nature and Economics Technical Paper No. 1, IUCN (World Conservation Union), Gland, Switzerland

Felgate, W. S. (1982) *The Thembe-Thonga of Natal and Mozambique: An Ecological Approach*, Occasional Publications No. 1, Department of African Studies, University of Natal, Durban

Folke, C., Carpenter, S., Elmqvist, T., Gunderson, L., Holling, C. S., Walker, B., Bengtsson, J., Berkes, F., Colding, J., Danell, K., Falkenmark, M., Gordon, L., Kasperson, R., Kautsky, N., Kinzig, A., Levin, S., Mäler, K.-G., Moberg, F., Ohlsson, L., Olsson, P., Ostrom, E., Reid, W., Rockström, J., Savenije, H. and Svedin, U. (2002) 'Resilience and sustainable development: Building adaptive capacity in a world of transformations', scientific background paper on resilience for the process of the World Summit on Sustainable Development, on behalf of the Environmental Advisory Council to the Swedish Government, Ministry of the Environment, Stockholm

Gordon, D. (2003) 'Technological change and economies of scale in the history of Mweru-Luapula's fishery (Zambia and Democratic Republic of Congo)', in E. Jul-Larsen, J. Kolding, R. Overå, J. R. Nielsen, and P. A. M. van Zwieten (eds) *Management, Co-management or No Management? Major Dilemmas in Southern African Freshwater Fisheries*, FAO Fisheries Technical Paper No. 426/2, Food and Agriculture Organization of the United Nations, Rome

Guyot, S. (2005) 'Political dimensions of environmental conflicts in Kosi Bay, South Africa: Significance of the new post-apartheid governance system', *Development Southern Africa*, vol 22, no 3, pp. 441–58

Heeg, J. and Breen, C. M. (1982) 'Man and the Pongolo floodplain', South African National Scientific Programmes Report No. 56, National Scientific Programmes Unit, Council for Scientific and Industrial Research, Pretoria

Ison, R. (2010) *Systems Practice: How to Act in a Climate-Change World*, Open University, Milton Keynes, UK

IUCN (2003) *Barotse Floodplain, Zambia: Local Economic Dependence on Wetland Resources*, Case Studies in Wetland Valuation #2, IUCN (World Conservation Union), Gland, Switzerland

Jackson, J. C. (1995) *Creating Common Pools in a Lake: Planning for the Community-Based Management of the In-shore Fishery, Lake Kariba Recreational Park, Zimbabwe*, Centre of Applied Social Sciences (CASS), University of Zimbabwe, Harare

Janssen, M. and Ostrom, E. (2006) 'Resilience, vulnerability, and adaptation: A cross-cutting theme of the International Human Dimensions Programme on Global Environmental Change' *Global Environmental Change*, vol 16, no 3, pp. 237–9

Jentoft, S., McCay, B.J. and Wilson, D.C. (1998) 'Social theory and fisheries co-management' *Marine Policy*, vol 22, no 4–5, pp. 423–36

Jones, J. L. (ed.) (2006) *Who Benefits from 'Conservation and Development'? The Political Ecology of Biodiversity in KwaZulu-Natal, South Africa*, Centre for Environmental Studies, University of Pretoria, South Africa

Kloppers, R. (2005) 'Border crossings: Life in the Mozambique/South Africa borderland since 1975', DPhil thesis, University of Pretoria, South Africa

Kotze, D. C., Memela, B., Fuzani, N. and Thobela, M. (2002) *Utilisation of the Mbongolwane Wetland in KwaZulu-Natal, South Africa*, International Water Management Institute (IWMI), Pretoria, South Africa

Kyle, R. (1995) 'Wise use of wetlands by rural indigenous people. The Kosi Bay Nature Reserve: A case study', in G. I. Cowan (ed.) *Wetlands of South Africa*, Department of Environmental Affairs and Tourism, Pretoria, South Africa

Lahiff, E. (1997) 'Land, water and local governance in South Africa: A case study of the Mutale River Valley', Rural Resources Rural Livelihoods Working Papers No. 7, Institute for Development Policy and Management, University of Manchester

Lavigne-Deville, P. (2004) 'Registering and administering customary land rights – current innovations and questions in French-Speaking West Africa', Expert Meeting Group on secure land tenure: New legal frameworks and tools, organized by FIG Commission 7, UN-HABITAT, the Institution of Surveyors of Kenya (ISK) and the Commonwealth

Lewanika, K. M. (2002) 'The traditional socio-economic systems for monitoring wetlands and wetland natural resources utilization and conservation: The case of the Barotseland, Zambia', paper presented at Conference on Environmental Monitoring of Tropical and Subtropical Wetlands, Maun, Botswana

Magadza, C. H. D. (2003) 'Experience and lessons learned: Brief for Lake Kariba (3rd draft)', Department of Biological Sciences, University of Zimbabwe, Harare

Manzungu, E., and Machiridza, R. (2005) 'Economic-legal ideology and water management in Zimbabwe: Implications for smallholder agriculture', paper presented at the International Workshop on African Water Laws: Plural Legislative Frameworks for Rural Water Management in Africa, Johannesburg, South Africa

McCartney, M., Janganyi, J. and Mkhize, S. (2004) 'Comprehensive options assessment: The Pongolo', Dams and Development Project, United Nations Environment

Programme, http://iwmi.cgiar.org/Assessment/files/proceedings/PongoloCaseStudy.pdf, accessed 18 October 2012

McGregor, J. (2003) 'Living with the river: Landscape and memory in the Zambezi Valley, northwest Zimbabwe', in W. Beinart and J. McGregor (eds) *Social History and African Environments*, James Curry, Oxford

McGregor, J. A. (1997) 'Staking their claims: Land disputes in southern Mozambique', Land Tenure Center Paper No. 158, University of Wisconsin–Madison, Madison, WI

Meinzen-Dick, R. and Pradhan, R. (2002) 'Legal pluralism and dynamic property rights, CGIAR Systemwide Program on Collective Action and Property Rights', Working Paper No. 22, International Food Policy Research Institute, Washington, DC

Meinzen-Dick, R. and Nkonya, L. (2007) 'Understanding legal pluralism in water and land rights: lessons from Africa and Asia', in B. Van Koppen (ed.) *Community-Based Water Law and Water Resource Management Reform in Developing Countries*, CABI, London

Mohamed-Katerere, J. and Van der Zaag, P. (2003) *Untying the 'Knot of Silence': Making Water Policy and Law Responsive to Local Normative Systems*, UNESCO, http://webworld.unesco.org/water/wwap/pccp/cd/pdf/history_future_shared_water_resources/indigenuous_approaches.pdf, accessed 18 October 2012

Murphree, M. W. (1991) *Communities as Institutions for Resource Management*, Centre for Applied Social Sciences, University of Zimbabwe, Harare

Ostrom, E. (2007) 'Institutional rational choice: An assessment of the institutional analysis and development framework' in P. A. Sabatier (ed.) *Theories of the Policy Process*, Westview Press, Boulder, CO

Ostrom, E. and Cox, M. (2010) 'Moving beyond panaceas: A multi-tiered diagnostic approach for social-ecological analysis' *Environmental Conservation*, vol 37, no 4, pp. 451–63

Phiri, Z. (1999) 'Water law, water rights and water supply (Africa): Zambia', Study Country Report, Department for International Development, UK

Pollard, S., and Du Toit, D. (2011) 'Towards the sustainability of freshwater systems in South Africa: An exploration of factors that enable and constrain meeting the ecological reserve within the context of integrated water resources management in the catchments of the lowveld', WRC Report No. K8/1711, Final Report to Water Research Commission, Pretoria, South Africa, March

Pollard, S., and Du Toit, D. (2013) 'The emergence of a systemic view for the sustainable governance and use of wetlands in complex and transforming environments: Experiences from Craigieburn, South Africa', in A. P. Wood, A. B. Dixon and M. McCartney (eds) *Wetland Management and Sustainable Livelihoods in Africa*, Earthscan, London

Pollard, S. R., and Cousins, T. (2008). 'Towards integrating community-based governance of water resources with the statutory frameworks for integrated water resources management: A review of community-based governance of freshwater resources in four southern African countries to inform governance arrangements of communal wetlands', WRC Report No. TT 328/08, Water Research Commission, Pretoria, South Africa

Pollard, S. R., Kotze, D., Ellery, W., Cousins, T., Monareng, J. and Jewitt, G. (2005) *Linking Water and Livelihoods: The Development of an Integrated Wetland Rehabilitation Plan in the Communal Areas of the Sand River Catchment as a Test Case*, Association for Water and Rural Development (AWARD), Acornhoek, South Africa, submitted to Water Research Fund for Southern Africa (WARFSA) and Working for Wetlands

Ramsar (2007) *Wise Use of Wetlands: A Conceptual Framework for the Wise Use of Wetlands*, Ramsar handbooks for the wise use of wetlands, 3rd edition, vol 1, Ramsar Convention Secretariat, Gland, Switzerland

Schafer, J. and Bell, R. (2002) 'The state and community-based natural resource management: The case of the Moribane Forest Reserve, Mozambique', *Journal of Southern African Studies*, vol 28, no 2, pp. 401–20

Schlager, E. and Ostrom, E. 1992. 'Property rights regimes and natural resources: A conceptual analysis', *Land Economics*, vol 68, no 3, pp. 249–62

Shackleton, S. E., Stadler, J. J., Jeenes, K. A., Pollard, S. R. and Gear, J. S. S. (1995) 'Adaptive strategies of the poor in arid and semi-arid lands: In search of sustainable livelihoods: A case study of the Bushbuckridge district, Eastern Transvaal, South Africa', unpublished report produced for International Institute for Environment and Development (IIED), Wits Rural Facility, University of the Witwatersrand, Johannesburg

Sithole, B. (2001) 'Participation and stakeholder dynamics in the water reform process in Zimbabwe: The case of the Mazoe pilot catchment board', *African Studies Quarterly*, vol 5, no 3, pp. 19–40

Thole, L. S. and Dodman, T. (1997) 'Traditional and modern approaches to community wetland management in Zambia', in G. Claridge and B. O'Callaghan (eds) *Community Involvement in Wetland Management: Lessons from the Field*, Wetlands International, Kuala Lumpur, Malaysia

Tique, C. (2000) 'Community land and natural resource management in Mozambique: Experiences of pilot community based project: The case of Gondola, Manica province', paper presented at the Eighth *Annual Conference of the International Association for the Study of Common Property*, Bloomington, IN, 31 May–4 June

Van Koppen, B., Giordano, M. and Butterworth, J. (eds) (2007) *Community-Based Water Law and Water Resources Management Reform in Developing Countries*, CABI, Wallingford, Oxfordshire

Van Zwieten, P. A. M., Goudswaard, P. C. and Kapasa. C. K. (2003) 'Mweru-Luapula is an open exit fishery where a highly dynamic population of fishermen makes use of a resilient resource base', in E. Jul-Larsen, J. Kolding, R. Overå, J. Raakjær Nielsen, and P. A. M. van Zwieten (eds) *Management, Co-management or No Management? Major Dilemmas in Southern African Freshwater Fisheries*, FAO Fisheries Technical Paper No. 426/2, Food and Agriculture Organization of the United Nations, Rome

Virtanen, P. (2001) 'Evolving institutional framework for community-based natural resource management in Mozambique: A case study from the Choa Highlands', *African Studies Quarterly* vol 5, no 3, pp. 139–54

Webster, D. J. (1991) 'Abafazi baThonga bafihlakala: Ethnicity and gender in a KwaZulu border community', in A. D. Spiegel and P. A. Mcallister (eds) *Tradition and Transition in Southern Africa*, Witwatersrand Press, Johannesburg

Wood, A. P., Dixon, A. B. and McCartney, M. (eds) (2013) *Wetland Management and Sustainable Livelihoods in Africa*, Earthscan, London

12 Customary governance of baobab in eastern Zimbabwe

Impacts of state-led interventions

Witness Kozanayi, Rachel Wynberg and Frank Matose

Introduction

Zimbabwe has undergone a number of intense governance changes as it has progressed from the colonial to the postcolonial period. These include initiatives to centralize, decentralize, recentralize and democratize the governance of natural resources, as well as economic and land reform policies that have had far-reaching effects. Despite a multitude of statutory laws and policies overlaying existing customary rules, resource degradation continues and pathways towards sustainable natural resource use and the equitable accrual of benefits to local communities remain unclear. This calls for a well-defined resource governance framework that ensures both effective conservation and the enhancement of livelihoods.

The iconic baobab tree (*Adansonia digitata*), reportedly able to live as long as 5,000 years (Sidibe and Williams 2002), has a host of consumptive as well as non-consumptive uses. These uses have a long history, but in the past decade this has changed, with a strong drive to commercialize baobab products which has followed the opening up of new markets in Europe and the USA (Schreckenberg et al 2006; Addy 2009). Most baobab trees in Zimbabwe are located in communal areas which are governed by customary rules and practices (Ribot and Oyono 2005). In response to increased commercialization, however, the Zimbabwean state has intervened by putting in place statutory measures aimed at ensuring ecological and economic sustainability to regulate access, use and trade. Along with new economic and land reform policies, the statutory arrangements governing baobab use have contributed to a somewhat overcrowded institutional landscape.

This chapter describes the coexistence of customary and statutory systems that govern the use and conservation of the baobab tree, and discusses drivers of change in customary rule. In doing so, it examines the conundrum of multilayered authority or governance systems in communal areas of Africa. It draws from broad arguments posited by Ostrom (1990) and Murphree (1991) in relation to common property resources, but goes further to locate these in Mamdani's (1996) theorization about the African state.

At the core of this conundrum is the history of integrating traditional authorities into state institutions during colonial rule. This is argued by Mamdani (1996) to be peculiar to British colonial practice, which was marked by the principle of indirect rule. In this way, the colonial government was able to use customary law as the means for enforcing its own policies of control over the African population resident in the rural periphery. Consequently, the government incorporated customary forms of governance as a central element of its functioning (see also Cousins 2008; Mnisi-Weeks 2011). Postcolonial governments, including Zimbabwe's, have tried to find ways of democratizing governance systems in communal areas, especially in relation to natural resources. The overarching question is whether tinkering with local governance has resulted in enduring, robust and effective systems (Ostrom 1990; Murphree 1991).

Study site

This chapter is based on research undertaken in the Chimanimani District of eastern Zimbabwe, and focused in particular on the people of the Gumbu clan in the Nyanyadzi and Gudyanga wards (see Figure 12.1). The study site has a higher concentration of baobab trees than other areas in the district (Mudavanhu 1998), with between 60 and 70 per cent of adults in Nyanyadzi and Gudyanga directly or indirectly involved in the use of baobab products (Mukamuri and Kozanayi 1999; Veeman et al 2001; Mutasa 2008).

Background to the Gumbu clan

Oral history and archival records show that upon settling in the Nyanyadzi and Gudyanga areas, the Gumbu people spread out to present-day Buhera, across the Save River (Latham 1966; Sinclair 1971). This migration was to provide a vital umbilical cord for the Gumbu people, who would use kinship ties to access scarce baobab products from Buhera. The leader of the Gumbu people was a proficient salt-maker, and he supplied this commodity to neighbouring chiefs, carrying it in a *gumbu* (soft beaten-bark container). Thus the residents of Nyanyadzi and Gudyanga were called VaGumbu, 'those who carry salt in a fibre bag'. The making of *gumbu* from baobab and other trees seems to have been the starting point of what was later to become a lifeline craft-making business for the locals.

Politically, Nyanyadzi and Gudyanga fall under the leadership of ward councillors and a member of parliament for the Movement for Democratic Change, at the time of writing the main opposition party in Zimbabwe. This is important, as politicians from the country's ruling party, the Zimbabwe African National Union – Patriotic Front (ZANU-PF), tend to interfere with how traditional leaders – the guardians of traditional values and rules – discharge their duties, and have been known to coerce traditional leaders into acting as party appendages, making politically partisan rulings (Virtanen 2000; Mtimba 2011).

More people have settled in the study area since 2008, thanks to the discovery of diamonds in the vicinity. People displaced by the diamond fields had the option of being relocated by the government to areas where they would be provided with housing and basic infrastructure, but some families decided to move to neighbouring Nyanyadzi and Gudyanga in the hope of being recruited to work on the mines. The increased population and mingling of peoples and cultures has increased pressure on the resource base and has also led to greater contestations of baobab use, because people have different cultural and historical backgrounds and approaches to resource management.

Data collection methods

Data was collected through interviews with individuals involved in the regulation and use of baobab: local traders and those selling crafts across national borders, sellers at local craft markets and traditional authorities and government officials at local and district level. Group discussions with resource users and other community members were an additional source of information. One of the authors, Witness Kozanayi, grew up in the study area and thus has a profound knowledge of the people involved. Historical records at the National Archives in Harare provided more information. The study also benefited from earlier research in the study area in which two of the authors, Witness Kozanayi and Frank Matose, had participated (Hot Springs Working Group 1995).

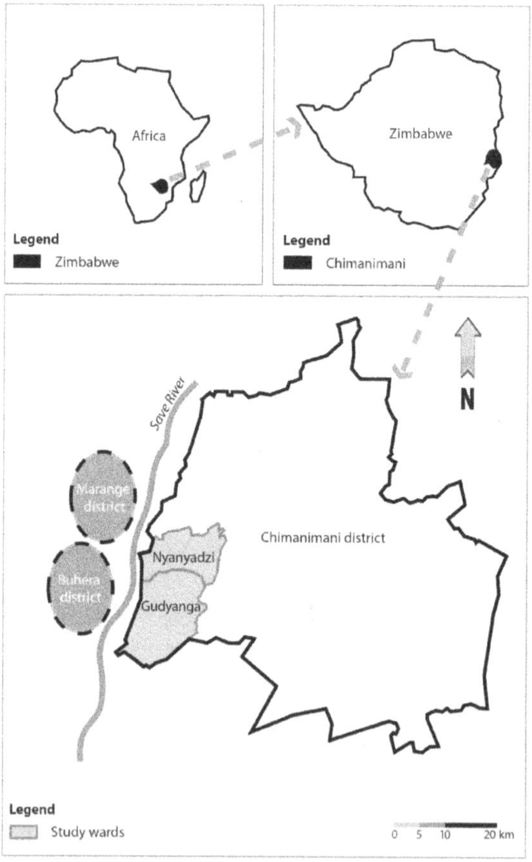

Figure 12.1 Map of Zimbabwe showing the location of the study site.

Baobab use and trade

Much of southern Africa is tropical with infertile soils (Frost and Mandondo 1999), which limits agricultural production. Consequently, the majority of rural people harvest forest or 'veld'[1] resources for subsistence and commercial purposes (Cavendish 2000; Campbell et al 2001; World Resources Institute [WRI] 2005). The baobab tree has gained prominence due to its use for both subsistence and, increasingly, commercial purposes.

The gigantic size of the baobab tree appears to reflect the multiplicity of its uses, of which there are reportedly more than 300 (Sanchez 2011). Not only are components of the tree utilized, such as the fruit, bark, leaves and roots, but the very shape of the tree makes it a good site for water storage, prisons, toilets, burial grounds, sleeping, shelter, ritual and prayer (Mukamuri and Kozanayi 1999; Wickens and Lowe 2008).

Even within the study site, baobab trees are used for many purposes, both non-consumptive and consumptive. The most frequently observed non-consumptive uses relate to traditional rituals and worship, and some trees are used as burial sites, because symbolically the tree offers shade or rest for the spirits of the dead. Most parts of the tree can be used consumptively: the fruit are eaten raw, the leaves are cooked and served as a vegetable and the roots are taken as a snack. The pulp can be added to maize or sorghum flour to make porridge. Some residents collect the fruit and sell it to local makers of ice lollies[2] and in urban areas. A key product is the bark fibre, which is used in handicrafts such as mats, hats and bags for the urban and export markets.

Apart from these handicraft products, which have been exported mainly to South Africa since the 1990s, the main baobab exports are the fruit pulp, which is marketed as a novel food ingredient, and the seed oil, used in the cosmetics industry. The commercialization of baobab has long been touted as an opportunity for rural development. The Southern African Development Community Protocol on Forestry, for example, recommends commercial utilization of natural products in the region, based on the premise that the 'promotion of innovative forest resource utilization technologies' will 'generate income for smallholder farmers' and 'reduce forest destruction' (Mubaiwa 2004, p. 31).

The existence of a ready market has contributed significantly to the fast growth of the baobab industry. The granting of 'novel food' status for baobab in the European Union in 2008, and 'generally regarded as safe' status in the USA in 2009, opened up new and bigger markets for baobab products (*Official Journal of the European Union* [OJEU] 2008; Addy 2009). The Overseas Development Institute has projected that the European market could generate more than US$750 million annually from baobab products for producer countries in southern Africa, making it the highest earner of all traded non-timber forest products in the region (Schreckenberg et al 2006). With increased volumes, this annual income could rise to an estimated US$1 billion, benefiting more than 2.6 million people along the marketing chain (Bennett 2006; Regional Trade Facilitation Programme [RTFP] 2007).

A review of baobab governance frameworks

A closer look at the governance arrangements for baobab reveals the legal pluralism frequently found in the regulation of non-timber forest products. As a 'low-value' non-timber forest product, baobab is regulated by customary systems, but at the same time increasing commercial use labels it a 'high-value' resource, meaning that statutory systems also apply to its governance (Laird et al 2010).

Unpacking customary systems

Anthropologists, sociologists and policy researchers have written extensively on customary systems (Weinrich 1971; Ranger 1992; Chanock 1998; Piot 1999). Such a system is understood to be 'a code of conduct approved by tribal tradition, the hereditary body of established conduct, taboos and belief systems which has been observed, recognized and enjoyed from time immemorial, and handed down by the forefathers' (Mohamed-Katerere and Van der Zaag 2003, p. 2).

Many early anthropological studies romanticized customary systems, viewing them as unique to local communities. However, as noted by scholars of complex systems (see, for example, Scheffer et al 2001; Mitchell and Newman 2002), no system exists on its own, and what is considered customary at a local level might well be a 'hybrid' of different cultures resulting from the networks between the local and the outside world (Anderson 1996; Piot 1999).

Although the role of customary systems in the governance of natural resources is, in most cases, confined to domestic and not commercial purposes (Mohamed-Katerere and Van der Zaag 2003; International Institute for Environment and Development [IIED] 2009), that role is central in regulating the use of the baobab tree in many parts of sub-Saharan Africa. As the governance of the baobab in eastern Zimbabwe shows, however, customary systems can be used in combination with statutory laws – which then falls within the realm of legal pluralism (Bavinck 2003). Such systems of dual governance can both strengthen and weaken efficacy. For example, Murombedzi (1998) argues compellingly that the state's attempt to elbow out customary systems and usurp the governance of natural resources in Zimbabwe has resulted in resource degradation in communal areas.

The role of customary systems in managing the baobab tree

A range of customary systems regulate access to baobab products in eastern Zimbabwe. Such systems include use of taboos, belief in autochthonous spirits and ascribing the value of sacredness to the tree. As a cardinal rule, anyone wishing to collect baobab products for commercial purposes first has to seek permission from traditional leaders and, when the baobab trees are located in crop fields or around homesteads, from the field or homestead owners themselves. Different rules regulate the use of the different baobab products, but most rules regulate bark and fruit collection.

Depending on the tenure system, different rules may apply to the same product. For example, fruit collection for domestic use is generally allowed, but is forbidden in the case of sacred baobab trees. The belief is that the area around a sacred tree and the products of such a tree are also sacred and therefore untouchable. However, high demand for baobab products and disobedience on the part of outside harvesters have resulted in the opportunistic redefinition of 'sacredness' so as to enable harvesting to take place. Some argue that only the tree is sacred, not its products or the land around it, or that the degree of sacredness differs from tree to tree: for example, trees in the burial grounds of the ruling family are deemed more sacred than trees in the burial grounds of commoners.

In one village in Nyanyadzi, trees in a cemetery were heavily debarked and some harvesters even stood on top of graves to debark the higher parts of baobab trees (Mukamuri and Kozanayi 1999). This is believed to have been the work of fly-by-night bark harvesters from outside the area (P. Matute, former craft maker, personal communication, 2 November 2011). Local leaders are adamant that such behaviour must not go unpunished and that the spirits will strike in ways that are not readily comprehended by mortals. The local adage '*Chisi hachieri musi wacharimwa*' ('Punishment for working on the field on a traditional resting day is meted out on a later day') aptly encapsulates this belief system. However, sometimes the traditional leaders punish errant behaviour, rather than waiting for the local spirits to do so. Local elders, for example, noted the case of a villager in Gudyanga ward burning down a sacred muucha tree (*Xanthocersis zambeziaca*) in his field to get wood to fire his kiln of bricks: he was fined a goat and had the firewood annexed by the village head. Much as the spirits of the land protect the sacred trees, it is also incumbent upon the traditional leaders to do so.

Role of autochthonous spirits

Native inhabitants of the study area strongly believe that if residents abuse resources, the spirits of the land will mete out punishment at two levels, and more so if such resources are sacred. The first punishment will be targeted at the offender and the second at the community. The offender may face misfortune such as the death of livestock or the mysterious illness of a family member. At community level, punishment can come in the form of droughts or marauding wild animals such as baboons destroying crops. The fact that the whole community can be punished by an avenging spirit forces residents both to restrain themselves and to monitor their peers.

Adult informants claim that the onset of black soot disease, frequent droughts and spates of cyclones are examples of enraged spirits expressing their anger over something extremely wrong, such as the spilling of innocent blood during local and national elections, the wanton destruction of natural resources and people working in crop fields on *chisi*.[3] During field work in the Gudyanga area in 1997, an account was given of how a construction company that had uprooted sacred baobab trees ended up brewing beer to appease local spirits after the company faced many 'unexplained' accidents involving machinery and the death of staff (Mukamuri and Kozanayi 1999). A second example, from Nyanyadzi, was the hacking down by the United Methodist Church of the sacred tree *Xanthocersis zambeziaca* to make way for a crèche in 2010. The roof of the church was allegedly blown off by mysterious winds on two occasions, forcing the church elders to approach the local leaders for purposes of rapprochement with the angry spirits of the land (M. Mukute, Church Elder of the Nyanyadzi United Methodist Church branch, personal communication, 15 June 2011). Residents thus tend to respect the spirits of the land more than any kind of customary or state authority. The nearest representative of state authority is located 120 kilometres away, whereas it is believed that the spirits of the land are omnipresent and witness all offences.

Other cultural practices that have been useful in managing the baobab trees include moral self-restraint on the part of harvesters. For example, those caught harvesting fibre without permission voluntarily surrender all the harvested material to local leaders. Some may also offer an apology and pay fines. This tends to be the case with culprits from within the ward or village, whereas those from other wards usually run away. Residents are under a lot of pressure to conform to local customs in the name of good neighbourliness. Those who are quarrelsome 'with the community' find themselves punished by losing the cooperation of the community in critical times, for example when residents refuse to attend the funeral of a troublesome neighbour. Having neighbours boycott the funeral of one's family member is a serious disgrace to a family and can oblige it to migrate to a far-off village.

Use of kinship ties

There has been widespread use of social relations to access resources outside the Nyanyadzi and Gudyanga areas. In some villages in Gudyanga and Nyanyadzi, baobab bark is now difficult to get for two main reasons: first, trees have been severely debarked and no longer produce good-quality bark; second, some traditional leaders have banned debarking, especially in grazing lands, which are the worst-affected areas. As several village heads have become very strict about who can access baobab products, craft makers now travel outside their ward and district and go to collect baobab fibre in Buhera and Marange districts, where some of the Gudyanga people have settled. Local residents argue that the two districts are just an extension of their own area, because they are closely related to the people there, so they have a right to access those resources.

This results in a disjuncture between administrative and traditional boundaries, however, as such an arrangement contravenes the provision of the Communal Land Forest Produce Act, which stipulates that natural resources in a community may only be harvested by local residents for their own use. Although the definition of 'community' has been contested (Mamimine and Mandivengerei 2001), the Chimanimani and Buhera Districts fall under different local government authorities and their identification as a community may therefore be difficult. Trans-district sharing of resources can only be explained and facilitated with reference to kinship ties, which can be crucial in times of critical resource scarcity (Nemarundwe and Kozanayi 2003).

Guarding traditional values

Positions of authority give their incumbents leverage to influence events and assert their interests (Kozanayi 2007). This accounts for the fact that traditional leaders or their next of kin tend to occupy all key influential positions in the study area. For example, the ward councillors for both Gudyanga and Nyanyadzi are the sons of local village heads. Bringing this about was a tactical move by traditional leaders to make sure that local customs were well articulated even at rural district council (RDC) level, where policies are formulated. As one traditional leader aptly put it, '*Kana unewako ari kunzou, haushai mapapu*', literally meaning, 'If you have a member of your family at the hunting front, at least your meat needs are well taken care of'. In the face of dramatic political changes, this strategy appears to work, especially considering that the councillor for Gudyanga ward occupies the powerful post of deputy council chairperson in the Chimanimani RDC. Nonetheless, issues of fiscal responsibility remain the preserve of the RDC, and traditional leaders do not have the power to handle levies from natural resources on behalf of the RDC (see Table 12.1).

Table 12.1 Key laws that potentially affect customary systems and baobab use patterns

Act	*Key provisions*	*Implications for customary systems and baobab use*
Environmental Management Act, Chapter 20:27 (Act No. 13 of 2002)	As the principal law, provides a broad framework for the conservation and sustainable use of all natural resources. It makes explicit the rights of communities to have access to, and enjoy the benefits of, the country's biodiversity. Obliges every council to have an environment management subcommittee. The Environmental Management Agency, the implementing arm, has arresting powers.	This Act has the potential to support local management of baobab. However, implementation of this Act is weak due to a lack of government resources.
Traditional Leadership Act, Chapter 29:17 (Act No. 25 of 1998)	Empowers traditional leaders to participate actively in the conservation of all natural resources that fall within their jurisdiction. Confers a wide range of powers on traditional leaders. Provides for the appointment of traditional leaders, from village head to chief, based on lineage.	Very few traditional leaders are conversant with this law. The Act has been ignored by politicians in the appointment of traditional leaders, so appointment is sometimes done on a partisan rather than lineage basis.
Rural District Councils (RDC) Act, Chapter 29:13	Vests RDCs with powers to raise revenue from any business activity in the district.	Resource users feel that the Chimanimani RDC is taking its fiscal power too far by collecting annual levies – especially from impoverished traders of baobab products.

Table 12.1 Continued

Act	*Key provisions*	*Implications for customary systems and baobab use*
(Act No. 8 of 1988)	Sections 56 and 59 mandate RDCs to set up environmental management committees at village and ward levels which have powers to arrest anyone found violating district environmental by-laws.	Village and ward environmental management committees are sometimes in conflict with traditional authorities about which authorities should grant permission for baobab harvesting.
Communal Land Forest Produce Act, Chapter 19:04 (Act No. 20 of 1987)	Regulates the exploitation of timber resources in communal areas by restricting the utilization of natural resources by local communities to own use. Commercial harvesting is regulated by RDCs through the issuing of concessions. Provides for the gazetting of certain tree species, largely on the basis of extinction threats.	Criminalizes commercial use of natural resources. Sometimes used by the Forestry Commission to prosecute those who harvest baobab products for commercial use. The baobab tree is not gazetted despite its importance and the threat of overharvesting in some areas.
Forest Act, Chapter 19:05 (Act No. 37 of 1949)	Deals with all forestry resources in the country, regardless of tenure. Sets up the Forestry Commission as the regulatory authority which grants permits to all timber and non-timber resource harvesters.	The Forestry Commission is seen as usurping the fiscal role of the traditional authority by collecting levies from baobab craft vendors and offenders.
By-laws at ward level	Residents in Gudyanga and Nyanyadzi, with the help of EAfrica (a local non-governmental organizations [NGO]), the RDC, the Environmental Management Agency and the Forestry Commission, have crafted by-laws that govern the management of baobab resources at the local level. Among other provisions, these stipulate a resting period of two years after each bark harvesting session.	Enforcement of the by-laws is problematic because there are few incentivized personnel on the ground. Logistical problems include the need for the ward councillor, village head and ward environment management committee to approve the harvesting of baobab products, especially fibre, yet these institutions are usually located far from each other.

Sources: Mamimine and Mandivengerei 2001; Mandondo 2001

Use of threats by the marginalized

Some women say that if collectors stole bark from trees owned by them, they would consider enlisting the assistance of a particular popular magician. The women, mainly widows, allege that when they go to traditional leaders for arbitration, they sometimes do not get a fair trial, because some bark collectors bribe the traditional leaders, or are their relatives, and are promptly exonerated. This has resulted in an unusual practice of 'forum shopping', in which the aggrieved parties consult various arbitrators in pursuit of a fair trial. Though no one has consulted the magician over environment-related conflicts, the mere mention of his name in a society that believes deeply in the world of spirits is enough to deter would-be offenders. Thus it is not uncommon to find baobab trees around widows' homesteads untouched by the voracious bark collectors.

The statutory landscape

The use of natural resources in Zimbabwe is regulated by a range of laws that impinge on customary systems governing natural resources in general and baobab products in particular. Table 12.1 summarizes these laws and their implications for baobab use.

As Table 12.1 indicates, a multiplicity of statutory interventions have introduced a variety of laws, institutions and rules which sometimes compete with traditional authorities and are often difficult to implement. For instance, many of the laws in the framework for natural resource management in Zimbabwe recognize the role of local people and make a case for the involvement of traditional authorities. However, as Table 12.1 illustrates, statutes such as the RDC Act and the Forest Act vest considerable powers in rural district councils and the Forestry Commission as regulatory bodies – in effect excluding traditional authorities from involvement in decisions such as the granting of permission for harvesting. Additionally, certain technical officials view local people as agents of resource degradation (Matose 2002), not as potential co-managers, thereby thwarting the umbrella Environmental Management Act's objective of involving local people in natural resource management.

A further example of tension between a statutory intervention and its implementation is the possibility of a complete ban on craft making in terms of the Communal Land Forest Produce Act. In an attempt to relieve pressure on the resource base, the state may use this instrument to prohibit craft making, while trying to identify alternative livelihood options for forest-dependent residents. However, such options are limited in the study site, as rain-fed agriculture, the most likely option, is not feasible due to frequent and prolonged droughts, and irrigated farming is threatened by old and dilapidated infrastructure. Moreover, craft making is an integral part of the Gumbu people's cultural identity and, thanks to co-mingling, has spread to other tribes too – something the government would have to take into consideration before embarking on such drastic action.

The state has used an approach best described as 'walking softly but carrying a big stick'. In the early 2000s, this took the form of banning trade in baobab products, which entailed mounting roadblocks and arresting and fining all those found with baobab fibre and fruits. Not much progress was made, as offenders simply did not pay their fines. For example, a review of RDC ticket books for environmental fines showed that of 140 fines issued by April 2011, fewer than 20 had been paid six months later. This kind of approach led local people to view the state with mistrust, in the belief that its main aim was to extract fines and levies from them.

In 2004 the state changed its tactics, and instead reached out to the residents of Nyanyadzi and Guyanga, involving them in formulating by-laws and creating village environment management committees (VEMECs) and ward environment management committees (WEMECs).

To facilitate monitoring, the state encourages resource harvesters to form groups or cooperatives, but traders claim that there is an ulterior motive, in that these structures make it easy to collect levies. Meanwhile, the state still makes sporadic raids on those who do not comply. Some traditional leaders believe the RDC is attempting to usurp their legitimate roles in baobab management. Others welcome the state's involvement, but only if the RDC shares part of the proceeds from fines with them.

Purportedly good policy has thus resulted in unintended consequences – and it is sometimes difficult to disentangle the various effects of regulation. For example, Gudyanga by-laws restrict bark collectors from harvesting a debarked portion until at least two years have passed, giving the tree time to recover. Local people have therefore traditionally debarked the lower part of the tree, leaving some vertical strips intact to allow for nutrient and water uptake. Increasingly, however, the upper parts of the baobab tree, including branches, are being harvested, stressing the tree and predisposing it to disease (Figure 12.2).

Figure 12.2 Over-harvesting of baobab bark can severely stress the tree and predispose it to disease (photo: Witness Kozanayi).

Drivers of changes in customary systems and use of the baobab tree.

In addition to the statutory laws described above, a number of macro- and micro-level processes have shaped customary systems over time. Below we discuss some of the key processes that have taken place and had a bearing on the efficacy of customary systems of baobab governance.

Government interference drives population growth

There has been a marked increase in population in Gudyanga, from 874 adults (tax-payers) in 1966 (Latham 1966) to over 3,500 households (about 10,000 adults) in 2002 (Central Statistical Office [CSO] 2002). Nyanyadzi's population has shown a similar trend. Some of the people who settled in the area came from former commercial farms which were annexed by the government under its land reform programme from 2000, ostensibly to correct historical imbalances in landownership between the minority of whites and the majority of blacks (Masiiwa and Chigejo 2003).

The Economic Structural Adjustment Programme in the early to mid-1990s saw many people being retrenched and going to reside in rural areas, increasing their reliance on the environment to support their livelihoods (Mukamuri et al 1998). The influx of retrenched people in rural areas resulted in the weakening and sometimes breakdown of local rules and customs for natural resources, as those who were retrenched often did not respect established traditional rules and customs (Braedt and Standa-Gunda 1998). Local leaders allege that unsustainable bark harvesting practices such as the debarking of branches started during the period of retrenchment. With more people displaced from former commercial farms that have been annexed, and newly opened up diamond fields, it is likely that local systems of natural resource governance will weaken even further.

Operation Murambatsvina ('Remove the Filth'), a government programme initiated in 2005 to remove people forcibly from slum areas, led to over 700,000 people being displaced from urban areas into rural areas (Tibaijuka 2005). Murambatsvina also saw the demolition of former baobab marketing stalls along the portion of the highway from Mutare to Beit Bridge that passes through Nyanyadzi. Market surveys in 2011 showed that of the 41 marketing stalls that had been located between Nyanyadzi and Gudyanga (Veeman et al 2001), only three remained after Operation Murambatsvina. Such unpopular policies and actions by the state have led to deep resentment on the part of local residents. Therefore the residents view any attempt by the state to engage locals in the management of the baobab tree, however well meaning, with disdain. The impact of these macro-level processes has thus put pressure on the resource base and, at times, weakened the traditional leadership.

Macro-economic pressures

Trade liberalization was a key World Bank recommendation as part of the broader Economic Structural Adjustment Programme implemented during the early to mid 1990s. A positive spin-off was a tourist boom and concomitant rise in handicraft marketing in Zimbabwe (Braedt and Standa-Gunda 1998; Campbell et al 2001). Data collected for Nyanyadzi and Gudyanga over a 24 kilometre stretch of highway on 12 occasions between 1995 and 1999 show that of the total of 41 marketing stalls for baobab products (Veeman et al 2001), the majority (64 per cent) entered the baobab business during the trade liberalization years (Veeman et al 2001).

The harvesting of baobab fibre products increased as the state was withdrawing funding for the forestry sector as part of the structural adjustment programme. Traditional authorities took over the management of natural resources, including the baobab tree, and tightened rules regulating the collection of fibre. Anyone wanting to harvest fibre had to first report to the village head and get instructions on debarking before being allocated trees to harvest. While most residents obeyed instructions from the traditional leaders, those from urban areas often did not.

Zimbabwe experienced unprecedented inflation of over 500 billion per cent and unemployment of over 90 per cent from 2007 to 2009. At the height of the economic meltdown in 2008, many unemployed people migrated to neighbouring countries to seek employment. The net effect of this was a reduction in the harvesting and processing of baobab products. At the same time some of the former city dwellers who had settled in rural areas started to see the economic potential of baobab and to question – or disregard – the customary systems governing its use.

Disempowerment of traditional leaders

Traditional leaders, supposedly the guardians of customary systems, have undergone episodes of empowerment, disempowerment and re-empowerment at the hands of colonial and postcolonial governments (Mohamed-Katerere 2001). Upon the attainment of independence in 1980, the new Zimbabwean government stripped traditional leaders of their powers because of their alleged cooperation with the colonial government. That period was also characterized by a 'freedom frenzy', as people misconstrued freedom to mean that they could now do whatever they wanted without any interference from the state or traditional leaders (Nemarundwe et al 1998).

Although the enactment of the Traditional Leadership Act in 1998 returned powers to traditional leaders, the 'rights' agenda also had unintended negative consequences. As one village head remarked to researchers:

> [T]hese days everyone is obsessed with rights: they tell you, 'I have a right to the environment, I have a right to this and that.' It is like the 1980s, when everyone could do anything in the name of exercising their freedom!
>
> (Muzviziyi, village head, personal communication, 1 April 2011)

Rights are claimed by the locals regardless of their attendant duty to care for the environment and the regulatory powers of traditional leaders.

Some traditional leaders in Gudyanga and Nyanyadzi have remained sympathetic to the Movement for Democratic Change, the opposition political party that dominates elected positions in administrative structures. Consequently such leaders tend to gain the allegiance of the majority of the residents who, going by the results of national elections since 2000, support the opposition political party. Winning the allegiance of the residents helps traditional leaders when it comes to the governance of natural resources, as residents are then more cooperative.

Politicization of institutions

Political challenges in Zimbabwe over the past decade have resulted in the state 're-orienting'– a euphemism for 'politicizing' – traditional authorities. This is driven by an ulterior motive: to make the re-oriented traditional authorities functionaries of the state rather than guardians of local customary systems. Generally such politicization has impaired the integrity of traditional leaders, and residents have reacted with rebelliousness, manifested by, for example, not observing customary ways of harvesting baobab.

Another challenge is that elected leaders such as ward councillors have to make sure that they do not undermine their political capital by implementing policies that are unfavourable to the electorate, especially if they anticipate campaigning for re-election in future. 'If we were to implement the local by-laws to the word in such times of distress, we would literally choke people to death,' said one such leader (Mr. C. Dirikwe, councillor in Nyanyadzi ward, personal communication, 6 June 2011). As an elected leader, Mr Dirikwe will give the welfare and survival of his subjects high priority, even if it means slackening the enforcement of some by-laws.

The RDC and the Forestry Commission also explain that at times they have to show a human face by 'looking the other way' in order to let people survive: 'All we require of the harvesters is that they do it sustainably and plant more trees' (Mr. W. Sibanda, Forestry Commission district officer for Chimanimani, personal communication, 3 February 2011). However, the state's failure to take action could be due more to a lack of resources than to moral obligation. For example, the RDC, the Environmental Management Agency and the Forestry Commission have only one old truck between them.

Changing practices

Customary practices have changed in a number of ways in response to the effects of legal pluralism and increased pressure on the resource base.

Conflict over fiscal responsibility

WEMECs are supposed to operate side by side with traditional leaders. In theory the two are complementary, but in practice they are often antagonists. Traditional leaders believe they have absolute authority over resources under their jurisdiction, while WEMECs claim that they are empowered by the RDC Act and local by-laws to oversee the utilization of the baobab tree. All the WEMEC chairpersons are young men, whereas the traditional leaders are mostly older men, so the 'battle' between the two assumes an ageist dimension – young versus old – with the young accused of being responsible for the violation of local customs. This tends to weaken the role and credibility of WEMECs, the grassroots representatives of the state.

The Achilles heel of effective joint management of the baobab tree by traditional and state authorities is fiscal responsibility. Both want to control the revenue or fines accruing from the use of baobab trees. The WEMECs ticket offenders and the RDC collects the imposed fine. Local by-laws, on the other hand, refer to a trust fund in which all fines and revenue from baobab use accrue. This is in direct contrast with the RDC Act, which states that the RDC is the responsible authority for all fiscal issues related to use of natural products.

Traditionally, all fines and revenue accrue to the traditional leader and his aides. Such fines include *marime* (money for the convenience of the court), which is shared by the council of elders who try the case. Anyone who is found guilty pays *matenda ndaa* (admission of guilt) and then the fine, which, depending on the gravity of the case, can be a goat or the equivalent. In cases of arrest by the state police and prosecution by a magistrate, the village head can impose a second fine for 'holding the name of the village in disrepute'. Once someone has been fined, the case is publicized at the community court in the spirit of 'naming and shaming', to deter any future offenders.

The 2004 ward by-laws usurp some of the fiscal responsibilities which were the preserve of the traditional leaders in that they provide for the creation of a trust fund into which all fines have to be deposited. The village head and headmen hold the funds on behalf of the community. While the object of this arrangement is to democratize fiscal accountability, it has promoted corruption among some traditional leaders. Former offenders report that they had to buy their freedom by purchasing 'scuds'[4] for the village heads after being caught collecting baobab bark without prior permission. Though it is not publicly admitted, there are reports of resource harvesters who pay 'expedite fees' to village heads in order to get permission to harvest baobab bark.

> [W]hy should I not pay the village head if the Forestry Commission forces me to pay them for selling my crafts by the road side? At least the village head manages the baobab tree that is important for me to remain in business.
>
> (Peter Sithole, a craft maker in Gudyanga ward, 11 April 2010)

Commercialization of baobab around homesteads

Another changing practice that is gaining prominence is that of harvesting trees near and at homesteads. This is a negotiated process, with the homestead owner normally setting the terms of harvesting. Usually it entails the bark harvester ceding 50 per cent of the harvest to the homestead head. There is a ready market for bark at US$0.50 per bundle. Preliminary results from an ecological survey in the area show that in some Nyanyadzi villages, trees around homesteads are increasingly becoming debarked, especially since the creation of local by-laws for baobab trees and the election of the WEMEC in 2004.

Population pressure and the increased value of baobab products have led to other changes in harvesting practices, such as the use of sticks and stones to dislodge fruit, whereas traditionally only fruit that has fallen to the ground is supposed to be collected. There is also the practice, described above, of collecting fruit and fibre from trees in cemeteries, something that historically would have been seen as taboo.

Discussion and conclusion

As customary systems for the governance of the baobab tree in eastern Zimbabwe weaken, largely due to unfolding socio-political and economic developments and the increased commercialization of baobab products following the opening up of export markets, there has been a spate of statutory interventions aimed at regulating the use of the tree. While the state's involvement could be viewed as a welcome development for the sustainability of the tree, the way in which the state has intervened has created mistrust among residents – a mistrust arising from the fact that the state wants to achieve many goals at the same time: ecological sustainability, levy collection, co-managing the resource base with the local people and penalizing offenders.

Mamdani (1996) concludes that postcolonial states perpetuate the colonial legacy that created a bifurcated state and decentralized despotism. Empirical evidence from this case study, however, challenges that conclusion. Traditional authorities have integrated themselves into the structures of statutory governance systems (for example, by becoming ward councillors) for purposes of safeguarding and articulating their customary systems, and several key natural resource laws acknowledge their role. In Mamdani's conundrum of authority of governance under indirect rule, there are two distinct systems that govern the rural populace (subjects) and the masters. But despite efforts to democratize and modernize customary systems, the rural Zimbabweans who use the baobab tree and other communally owned natural resources still remain sceptical about the state's sincerity in trying to create an enabling environment for the sustainable and equitable utilization of baobab products. In their view, the state is more interested in collecting levies.

Creating an enabling environment for the baobab remains a challenge, as dual governance of the tree and its components has weakened the efficacy of pre-existing customary systems, while degradation of the resource continues. Traditional leaders have controlled use of baobab for decades, using systems that include sacredness, taboos and supernatural beliefs. Different parts of the tree are governed by different rules, but permission to collect baobab products has to be granted by traditional authorities and homestead owners. Disregard for these rules is rising, driven in part by the escalation in population in the study area, economic hardship and a growing export market for baobab goods. Concomitantly, the state's intervention in the form of a host of laws and by-laws has been hindered by a lack of capacity and a myriad of unpopular policies embarked upon as part of the broader political agenda. Such interventions include the politicization of traditional authorities for political expediency, an apparent latter-day version of Mamdani's 'indirect rule' by British colonialists. Locals have resisted all interventions by the state, however well intended.

Several challenges remain. Firstly, diverse and conflicting administrative and traditional boundaries result in customary systems conferring user rights that overlap with those offered by the state, resulting in confusion and a disregard of rules. Secondly, re-empowering traditional authorities through instruments such as the Traditional Leadership Act will remain ineffective as long as traditional leaders are unaware of these provisions. Monetization of the industry has brought with it corrupt tendencies among some of the leaders, and has led to the emergence of new local arrangements such as the privatization of trees around homesteads. Invoking kinship ties to access resources in neighbouring areas is another development that arises from governance systems that fail to allocate available resources equitably, and from the sheer over-utilization of baobab products. Use of such ties, if unchecked, may facilitate resource degradation in unaffected areas.

Customary and statutory systems of governance are both crucial in the management of the baobab tree. The argument for respecting customary systems hinges on the fact that these systems have been among the foundational elements of the laws of all states in Africa (McAuslan 2005; Wynberg and Laird 2007), but whether it is workable to have these two governance systems operating together depends on many factors, not least of which are the timing and degree of intervention by the state (Mandondo 2006). The roots of the inefficiencies in the customary system of governance, or its amalgam with statutory systems, can also be found in the postcolonial state's desperate recent attempts to rope in traditional authorities to buttress waning political support. The traditional authorities' prime mandate is to safeguard customary systems, not to be an extension of political parties or, as Mamdani puts it, 'for indirect rule' (Mamdani 1996: 18). Such a change in focus compromises the integrity of the traditional leaders and their ability to influence resource governance.

Despite its weaknesses, the state could still play a key role in the management of the baobab tree – for example, by carrying out sound ecological surveys and helping communities plant baobab trees. The state would also be performing an important function if it harmonized the roles of traditional leaders with those of WEMECs, because it is imperative for the two to complement each other.

The revered baobab tree offers more than economic opportunities to local people; it also has profound religious meaning. It forms a link between the living and the spiritual or ancestral. It is believed that when the living mismanage the baobab tree, the spirits of the land punish the living; on the other hand, when the living behave well, the spirits take care of them. The tree also defines the nature of the multiple political and administrative links between the state and the peasants (natural resource users). When baobab users are unhappy with particular government policies, they do not cooperate with the state in the management of the baobab tree. The survival of the tree thus signifies the survival of these important links between the living and the spiritual and between the state and local baobab users.

The state of Zimbabwe should take note of powerful connections such as these between local communities and the natural resources in their vicinity. With the value of baobab products likely to rise in the near future, a well-defined governance framework to guide sustainable use and enhance livelihoods, without eroding effective existing customary practices, is critical.

Notes

1 Grassland or grazing.
2 These confections are made of sugared baobab pulp frozen on a stick. Milk can be added to improve the flavour.
3 *Chisi* is a traditional rest day when everyone is expected to stop working in their fields or tilling the land. However, with the advent of NGO-funded community gardens, some residents no longer observe *chisi* on the pretext that such days for residents do not apply to community projects.
4 Traditional opaque beer is sold in containers called 'scuds'.

Bibliography

Addy, R. (2009) 'Baobab fruit approved as food ingredient in US', *Nutra Ingredients-USA.com*, William Reed Business Media, http://nutraingredients-usa.com/Industry/Baobab-fruit-approved-as-food-ingredient-in-US

Anderson, N. E. (1996) *Ecologies of the Heart: Emotion, Belief and the Environment*, Oxford University Press, New York

Bavinck, M. (2003) 'Understanding fisheries conflicts in the South – A legal pluralist perspective', *Society and Natural Resources*, vol 18, no 9, pp. 805–20

Bennett, B. (2006) *Natural Products: The New Engine for African Trade Growth. Consultancy to Further Develop the Trade Component of the Natural Resources Enterprise Programme (NATPRO)*, NRI

Braedt, O. and Standa-Gunda, W. (1998) 'The woodcraft industry in Zimbabwe: An overview', IES working paper, Institute of Environmental Studies, Harare

Campbell, B. M., Luckert, M., Mutamba, M., Zindi, C. and Kozanayi, W. (2001) *Household Livelihoods in Semi-Arid Regions: Options and Constraints*, CIFOR, Bogor, Indonesia

Cavendish, W. (2000) 'Empirical regularities in the poverty-environment relationship of rural households: Evidence from Zimbabwe', *World Development*, vol 28, no 11, pp. 1979–2003

Chanock, M. (1998) *Law, Custom and Social Order: The Colonial Experience in Malawi and Zambia*, Heinemann, Portsmouth, NH

Cousins, B. (2008) 'Characterising "communal" tenure: Nested systems and flexible boundaries', in A. Claassens and B. Cousins (eds) *Land, Power and Custom: Controversies Generated by South Africa's Communal Land Rights Act*, UCT Press, Cape Town, pp. 109–37

CSO (2002) *National Census Profile*, Central Statistical Office, Government of Zimbabwe, Harare

Frost, P. and Mandondo, A. (1999) 'Improving rural livelihoods in semi-arid regions through management of catchments', Institute of Environmental Studies, University of Zimbabwe, Harare

Hobsbawm, E. and Ranger, T. (eds) (1983) *The Invention of Tradition*, Cambridge University Press, Cambridge

Hot Springs Working Group (1995) 'Local-level economic valuation of savannah woodland resources: Village cases from Zimbabwe', Hidden Harvest Project Research Series 3, International Institute for Environment and Development, London

IIED (2009) 'Protecting traditional knowledge from the grassroots up', IIED Briefing, June, International Institute for Environment and Development, London, http://iied.org/pubs/display.php?o=17067IIED

Kozanayi, W. (2007) 'Mapping stakeholders for tropical ecosystem functions assessment: The case of Gola forest', unpublished MSc thesis, University of Cranfield, UK

Laird, A. S., Ingram, V., Awono, A., Ndoye, O., Sunderland, T., Fotang, L. E. and Nkuinkeu, R. (2010) 'Integrating customary and statutory systems: The struggle to develop a legal and policy framework for NTFPs in Cameroon', in S. A. Laird, R. J. McLain and R. P. Wynberg (eds) *Wild Product Governance: Finding Policies that Work for Non-Timber Forest Products*, Earthscan, London

Latham, C. J. K. (1966) *Delineation Report: Tribal Trust Lands in Melsetter District*, National Archives, Harare

Mamdani, M. (1996) *Citizen and Subject*, Princeton University Press, Princeton, NJ

Mamimine, W. P. and Mandivengerei, S. (2001) 'Traditional and modern institutions of governance in community-based natural resource management', Commons Southern Africa Occasional Paper series no 5/2001, University of Zimbabwe, Harare

Mandondo, A. (2001) 'A critique of by-law development and implementation in Chivi District, Zimbabwe', IES Working Paper no 19, Institute of Environmental Studies, Harare

Mandondo, A. (2006) 'Dialogue of theory and empirical evidence: A weighted decision and tenurial niche approach to reviewing the operation of natural resource policy in rural southern Africa', CASS/PLAAS Occasional Paper Series no 10, Centre for Applied Social Sciences (CASS), University of Zimbabwe, Harare, and Programme for Land and Agrarian Studies (PLAAS), School of Government, University of the Western Cape, Cape Town

Masiiwa, M. and Chigejo, O. (2003) *The Agrarian Reform in Zimbabwe: Sustainability and Empowerment of Rural Communities*, Department of Agrarian and Labour Studies, Institute of Development Studies, University of Zimbabwe, Harare

Matose, F. (2002) 'Local people and reserved forests in Zimbabwe: What prospects for co-management?', unpublished DPhil thesis, University of Sussex, UK

McAuslan, P. (2005) 'Legal pluralism as a policy option: Is it desirable? Is it doable?', in E. Mwangi and E. Patrick (eds) *Land Rights for African Development: From Knowledge to Action*, Capri CGIAR System-Wide Program on Collective Action and Property Rights, UNDP, International Land Coalition, www.capri.cgiar.org/pdf/brief_land.pdf

Mitchell, M. and Newman, M. (2002) 'Complex systems theory and evolution', in M. Pagel (ed.) *Encyclopedia of Evolution*, Oxford University Press, New York

Mnisi-Weeks, S. (2011) 'Layers of authority, boundaries of decision-making: Controversies around the Traditional Courts Bill', paper presented at the 13th International Association for the Study of the Commons (IASC) Biennial International Conference, Sustaining Commons: Sustaining Our Future, Hyderabad, India, 10–14 January

Mohamed-Katerere, J. (2001) 'Participatory natural resource management in the communal lands of Zimbabwe: What role for customary law?', *African Studies Quarterly*, vol 5, no 3, pp. 115–38

Mohamed-Katerere, J. and Van der Zaag, P. (2003) *Untying the 'Knot of Silence': Making Policy and Law Responsive to Local Normative Systems*, UNESCO, Paris, http://webworld.unesco.org/water/wwap/pccp/cd/pdf/history_future_shared_water_resources/indigenuous_approaches.pdf

Mtimba, G. (2011) 'We do not belong to anyone: Chiefs', *Dailynews*, 1 October

Mubaiwa, L. (2004) 'The Southern African Development Community Protocol on Forestry: Can it stop the mounting threats to the region's forest?', *Unasylva*, vol 55, no 218, pp. 27–33

Mudavanhu, H. T. (1998) 'Demography and population dynamics of baobabs (Adansonia digitata) harvested for bark in south-eastern Zimbabwe', unpublished MSc thesis, University of Zimbabwe, Zimbabwe.

Mukamuri B. B. and Kozanayi, W. (1999) 'Institutions surrounding the use of marketed bark products: The case of *Berchemia discolors*, *Adansonia digitata*, and *Warburgia salutaris*', IES Working Paper no 17, Institute of Environmental Studies, Harare

Mukamuri, B. B., Campbell, B. M. and Kowero, G. (1998) *The Impacts of Macro-Economic Changes on Rural Institutions*, Center for International Forestry Research, Bogor, Indonesia

Murombedzi, J. (1998) 'The evolving context of community-based natural resource management in sub-Saharan Africa in historical perspective', plenary presentation, International Workshop on Community-Based Natural Resource Management, Washington DC, 10–14 May

Murphree, M. W. (1991) 'Communities as institutions for resource management', Centre for Applied Social Sciences (CASS), University of Zimbabwe, Harare

Mutasa, B. K. (2008) 'Exploitation of the baobab tree (*Adansonia digitata*) in Zimbabwe: Products, markets, income and impact on household livelihoods in Nyanyadzi Communal Area', unpublished MBA thesis, Africa University, Mutare, Zimbabwe

Nemarundwe, N. and Kozanayi, W. (2003) 'Institutional arrangements for water resource use: A case study from southern Zimbabwe', *Journal of Southern African Studies*, vol 29, no 1, pp. 193–206

Nemarundwe, N., Mutamba, M. and Kozanayi, W. (1998) 'An overview of woodland utilization and management in three communal areas in Zimbabwe', Special Report no 16, Institute of Environmental Studies, University of Zimbabwe

OJEU (2008) 'Commission decision of 27 June 2008 authorising the placing on the market of baobab dried fruit pulp as a novel food ingredient under Regulation (EC) No 258/97 of the European Parliament and of the Council', *Official Journal of the European Union*, 11 July, Brussels, Belgium

Ostrom, E. (1990) *Governing the Commons: The Evolution of Institutions for Collective Action*, Cambridge University Press, Cambridge, UK

Piot, C. (1999) *Remotely Global: Village Modernity in West Africa*, University of Chicago Press, Chicago, IL

Ranger, T. (1992) 'The invention of tradition in colonial Africa', in E. Hobsbawm and T. Ranger (eds) *The Invention of Tradition*, Cambridge University Press, Cambridge, UK

Ribot, J.C. and Oyono, R. (2005) 'The politics of decentralisation', in B. Weisner, C. Toulimm and R. Chitiga (eds) *Towards a New Map of Africa*, Earthscan, London

RTFP (2007) *News Brief No 2*, Regional Trade Facilitation Programme, http://www.rtfp.org

Sanchez, C.A. (2011) The status of baobab tree populations in southern Malawi: Implications for further exploitation, *Forests, Trees and Livelihoods*, vol 20, no 2–3, pp. 1–17

Scheffer, M., Carpenter, S., Fohley, A. J., Folke, C. and Walker, B. (2001) 'Catastrophic shifts in ecosystems', *Nature*, vol 413, pp. 591–96

Schreckenberg, K., Marshall, E., Newton, A., Te Velde, D. W., Rushton, J., Edouard, F. (2006) *Forestry Briefing no 10*, March, Overseas Development Institute, London

Sidibe, M. and Williams, J. T. (2002) 'Baobab. Adansonia digitata', International Centre for Underutilised Crops, Southampton, UK

Sinclair, S. (1971) *The Story of Melsetter*, M. O. Collins, Salisbury, Harare

Tibaijuka, A. (2005) 'Report of the Fact-Finding Mission to Zimbabwe to Assess the Scope and Impact of Operation Murambatsvina by the UN Special Envoy on Human Settlements Issues in Zimbabwe', United Nations, Harare, Zimbabwe

Veeman, M., Cocks, M., Muwonge, A., Choge, S. and Campbell, B. (2001) 'Markets for three bark products in Zimbabwe: A case study of markets of *Adansonia digitata, Berchemia discolor and Warburgia salutaris*', IES Working Paper no 18, Institute of Environmental Studies, Harare

Virtanen, P. (2000) 'Community-based natural resources management: Defining the community in Mozambique', in P. Virtanen and M. Nummelin (eds) *Forests, Chiefs and Peasants in Africa: Local Management of Natural Resources in Tanzania, Zimbabwe and Mozambique, Silva Carelica 34*, Faculty of Forestry, University of Joensuu, Joensuu, Finland

Weinrich, H. K. A (1971) *Chiefs and Councils in Rhodesia: Transition from Patriachical to Bureaucratic Power*, Heinemann, London

Wickens, G.E., Lowe, P., (2008) *The Baobabs: Pachycauls of Africa, Madagascar and Australia*, Springer

WRI (2005) 'The wealth of the poor: Managing ecosystems to fight poverty', NEP/UNDP/WRI? World Bank, http://wri.org/biodiv/pubs

Wynberg, R. P. and Laird, S. A. (2007) 'Less is often more: Governance of a non-timber forest product, marula (*Sclerocarya birrea, subsp caffra*) in southern Africa', *International Forestry Review*, vol 9, no 1, pp. 475–90

13 Partnerships are not forever

The limits of collaborative governance in diamond mining in Namaqualand

Ralph Hamann

Introduction

The semi-arid, diamond-rich coastal plain of Namaqualand in the north-western corner of South Africa is the stark setting of conflicting views on the potential for collaborative environmental governance. A desktop study in 2010 funded by the United States Agency for International Development (USAID) identified Namaqualand as one of four case studies in Africa demonstrating that collaboration between a mining company and a conservation non-governmental organization (NGO) could contribute to significant conservation outcomes (Smuts 2010). The report argued that a partnership between Conservation International (CI), a Washington-based NGO, and De Beers, the world's largest diamond mining and marketing company and owner of the largest mine in the area (called Namaqualand Mine), had made a 'significant contribution . . . towards conservation planning in an internationally recognized biodiversity priority area (namely the Succulent Karoo hotspot)' (Smuts 2010, p. 27). Furthermore, a 'multi-pronged approach to the partnership, which assisted in rehabilitation, job creation and protected area establishment, assisted in realizing the conservation vision. CI engaged the local and provincial government authorities in tandem to engaging the mining company' (Smuts 2010, p. 27).

This partnership thus seemed to illustrate the potential of collaborative governance, more generally, to proactively combine contributions from business – motivated in part by corporate social responsibility (CSR) – with those of government and civil society in the pursuit of common or compatible objectives and in support of sustainable development (Business Partners for Development [BPD] 2002; Donahue 2004; Esteves and Barclay 2011; Huxham and Vangen 2000; Moon 2002).[1] Yet starkly opposing assessments exist. Van Wyk et al (2009) lament the social and environmental legacies of 80 years of diamond mining along this coast, highlighting local communities' high levels of poverty, exacerbated by the more recent downscaling of on-shore diamond mining and resulting retrenchments.[2] They argue that 'CSR driven development is rendered ineffective unless the national and provincial governments take the necessary steps [to enforce improved compliance with relevant legislation]' (ibid: vii).

Similarly, Blair (2011) finds that neither De Beers nor the government have lived up to the ideals of the South African Constitution or notions of corporate social responsibility, and that both have shown scant regard for commitments to environmental rehabilitation and procedural or substantive justice for local communities. With reference to Swyngedouw (2005) she argues that 'governance-beyond-the-state' may promise to be inclusive and democratic on the basis of more proactive governance contributions from business and civil society, but actually contributes further to entrenched power relations and the marginalization of the poor and disenfranchised.

In the wake of efforts by De Beers to sell the Namaqualand Mine in 2010 and 2011, even CI (now known in South Africa as Conservation South Africa, or CSA) has become critical of De Beers, arguing in a press release that 'De Beers' claim that the sale is part of their commitment to a "lasting contribution to the communities in which they live and work" is unsubstantiated given their lack of meaningful engagement with key stakeholders in the process of selling the mine' (CSA 2011, p. 1). CSA's point person argues that there had been a positive partnership between CSA and De Beers in developing a plan for post-mining land use and economic development opportunities, but 'there was a change of management and a change of heart' in the company, which led to a tenser relationship and even threats of legal action (Frazee, personal communication, 24 August 2011; Centre for Environmental Rights [CER] 2011).

The shift from partnership to acrimony was not limited to corporate and NGO offices. A visit to this remote area made the significant conflict potential quite apparent, with local community representatives distrusting the willingness or capacity of either De Beers or the government to respond effectively to their concerns and grievances. One interviewee gave vent to a common refrain: 'De Beers is so mighty – the community is in despair' (Markus, personal communication, 26 September 2011).

Is this a story of a collaborative relationship gone sour in the wake of changing management or economic circumstances, suggesting practical limits to cross-sector partnerships? Or does it demonstrate more fundamental constraints to collaborative governance, supporting critics such as Swyngedouw (2005)? This chapter discusses both the initial CSA–De Beers partnership and the more recent controversy surrounding the sale of the mine, as well as related concerns pertaining to post-mining economic development, land rights and environmental rehabilitation.[3] For our purposes, the Namaqualand Mine is an illustrative case (Flyvberg 2006) that is used here to aid understanding of collaborative governance and define its opportunities and boundary conditions. In setting all this out, the chapter emphasizes that partnership between role-players can be an important and effective governance mechanism, but only if certain conditions are met. If these conditions change beyond a certain point, a partnership may become untenable.

Specifically, the chapter argues that collaborative governance based on common interests is constrained in the context of historical legacies of social injustice and environmental degradation. A particularly important constraint is the lack of a clear and predictable framework of government regulation, which is vital in keeping open the 'negotiation space' among potential collaborators. For local communities, a more accommodating space needs to be created, with recourse to rights (including in particular the right to threaten or take legal action) and power (such as protest action, alliance building and media campaigns) as legitimate engagement options.

Such engagement can also function to increase the scope of the company's definition of self-interest.[4]

The Conservation South Africa–De Beers partnership and its demise[5]

The Namaqualand mine is situated in the north-western corner of South Africa (see Figure 13.1), an area characterized by very low rainfall and entrenched poverty. Most residents eke out a living farming livestock or working on the mines. Ecologically the region is known as the Succulent Karoo: it has the richest variety of succulent plants in the world, with nearly one-third of its floral species found nowhere else, and it is also a centre of endemism for reptiles and many groups of invertebrates. It is under significant pressure from overgrazing, illegal collection of fauna and flora, climate change and mining: in particular, it overlaps with De Beers' 80-year-old mining operations in Namaqualand.

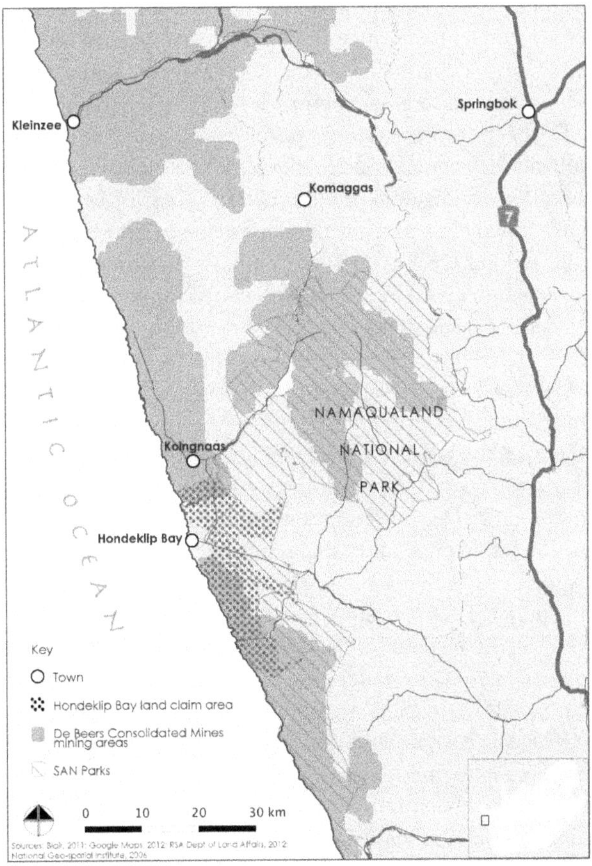

Figure 13.1 Case study area.

In 2001, CSA facilitated the development of the Succulent Karoo Ecosystem Programme, a 'long term, multi-stakeholder bioregional conservation and development partnership programme' with government and non-government partners.[6] De Beers staff members from corporate offices and the Namaqualand Mine participated in these discussions, with a particular focus on the transfer of De Beers coastal properties (between the Groen and Spoeg rivers, south of Hondeklip Bay) for the purpose of extending the existing Namaqualand National Park conservation area on land donated as a 99-year lease to the South African National Parks Board in 2008. These conversations also considered environmental rehabilitation aspects of the planned closure of the Namaqualand Mine, which had been envisaged for some time already. (University of Cape Town students had conducted environmental and social impact studies on mine closure as early as 1992 (Gosling 1992).) Environmental rehabilitation methods were developed in collaboration with researchers at the University of Cape Town's Namaqualand Restoration Initiative, and subsequently applied by De Beers.

Recognizing that the success of conservation would rely on the creation of viable socio-economic development and job creation opportunities, CSA organized a five-day 'charrette' (collaborative design exercise) in Namaqualand in 2006. In this design workshop, experts and stakeholders were invited to discuss options for linking conservation and socio-economic development in the area, with a focus on ecotourism. De Beers representatives participated, and the relationship between the two organizations strengthened on the basis of a shared recognition of the need to link conservation and livelihood opportunities.[7] Inspired by the Eden Project in Cornwall, UK, in which a post-mining landscape had been transformed into an ecotourism attraction,[8] CSA proposed a similarly ambitious programme of developing ecotourism and 'green industry' projects in Namaqualand under the banner of the Living Edge of Africa Project (LEAP).[9]

Two further charrettes were held in September 2007 and February 2009 to develop 'sustainable livelihoods' projects, the second part-funded by De Beers. CSA was contracted by De Beers to develop the LEAP concept further. Strictly speaking, therefore, the CSA–De Beers relationship was a contractual one, rather than a collaboration based on common interests (Wickens, personal communication, 16 September 2011), but common interests clearly were a vital motive for both parties. The charrettes and CSA's subsequent work led to a pre-feasibility report published in May 2009, suggesting a number of projects focused on mariculture, wind energy, environmental rehabilitation and land art.

There are divergent accounts of the degree to which local communities were represented in these charrettes, or indeed the LEAP process generally. Frazee (personal communication, 24 August 2011) mentioned that many community members had been involved, especially representatives of local and provincial government, but De Beers interviewees highlighted the role of experts and architects in these workshops and suggested that 'local communities were not really involved' (Wickens, personal communication, 16 September 2011). The local community members interviewed by the author also expressed some uncertainty regarding community involvement in LEAP. Markus (personal communication, 26 September 2011) noted, 'I've heard about LEAP, but no one from Hondeklip Bay was involved . . . it is very vague, we don't know what is going on.' But later on, he argued that CSA had made an effort to involve the community in issues such as environmental rehabilitation. In this case there may have been some confusion regarding the definition of LEAP. Pienaar and other community members from Komaggas (personal communication, 27 September 2011]) claimed that they had not gone to LEAP workshops, nor had other community members, and they noted that language barriers generally made participation in such processes difficult for them.

Community participation in the LEAP process thus seems to have concentrated on representatives of local and provincial government 'at the end, at the closing function' (Pisane, personal communication, 27 September 2011) of the workshops. This reliance on elected representatives and government officials was made more problematic by changes in local government resulting from the elections of May 2011.[10]

Frazee (personal communication, 24 August 2011) emphasizes the excitement about the proposals in the LEAP pre-feasibility plan and the subsequent disappointment when De Beers ostensibly decided to abandon the plan in early 2011. This was triggered, she argues, by a change in corporate leadership, with Nicky Oppenheimer stepping down as chair of the board, and also changes in mine-level leadership. The sale of the mine, discussed below, is seen by Frazee (personal communication, 24 August 2011) as a convenient means for De Beers to avoid making the more committed investments in the social and natural environment envisaged in the LEAP document. Furthermore, while the land uses proposed by the company in its amended environmental management plan (EMP) in preparation for the sale of the mine do include some of the LEAP proposals, they also include arguably incongruent proposals such as correctional services (i.e. a prison) and hazardous waste facilities (Frazee, personal communication, 24 August 2011).[11]

De Beers staff, however, describe the situation differently, denying that LEAP has been abandoned. Pisane and Wickens (personal communication, 16 September 2011) emphasize that the process of developing a pre-feasibility report for LEAP projects was a success and that some of the proposals were considered feasible in the subsequent De Beers investigations (specifically the wind farm and the abalone mariculture project) and are currently part of the amended EMP that awaits approval from the government (at the time of writing). Some, such as the wind energy plant and a mariculture project, are already being implemented (see Creamer 2011). They argue that other options discussed in the pre-feasibility plan, such as the land art and tourist accommodation proposals, were not considered feasible because of the remoteness of the area, among other reasons, so they have not been taken further.[12] Furthermore, Wickens (personal communication, 16 September 2011) points out that the responsible government department, the Department of Mineral Resources, will only approve the post-mining land use options if their feasibility is proven, a process for which no precedent exists in South Africa, so there is still significant regulatory uncertainty involved. Generally, however, Wickens (personal communication, 16 September 2011) characterizes the partnership between CSA and De Beers as a positive one that has achieved its purpose: specifically, CSA had been able to provide a vision for the area and 'made us think differently'.

Mine sale controversy

As noted, the closure of Namaqualand Mine had been planned for many years. More recently, it was decided to sell the mine instead, based on the premise that a smaller company with lower overheads might yet be able to operate the mine profitably. A further incentive is likely to have been the desire to respond to the South African government's requirement to see mines operated by black economic empowerment companies. (The sale conditions emphasized the 'empowerment' credentials of preferred bidders.)

A bidding process managed by one of the country's big banks attracted bids from a significant number of applicants. The shortlist consisted of six companies, out of which Transhex was eventually chosen.[13] This choice in itself was criticized by CSA and others, because the company was seen as having a bad track record in the Namaqualand region (CSA 2011; Pienaar, personal communication, 27 September 2011). More fundamentally, however, the concern has been that the sale of the mine will dilute the social and environmental commitments and liabilities associated with the mine.

CSA (2011) argues that the sale of the mine has not involved sufficient stakeholder consultation and, in particular, that pertinent information about the social and environmental commitments associated with the sale has not been made available. Specifically, the financial commitment required by law for environmental rehabilitation has not been publicized, in the absence of which, argues Frazee (personal communication, 24 August 2011), the inclusion of broad principles and commitments in the amended EMP for the mine is futile. Some of the substantive conditions themselves have also been criticized, such as the reportedly insufficient density of proposed netting (Frazee, personal communication, 24 August 2011).

Blair (2011: 92) also suggests ulterior motives in the mine sale:

> What mine sale might mean in practice is that De Beers will attempt to dispense with the social and environmental responsibilities they had initially agreed to take on, selling on instead to a smaller company without pretensions to, or capacity to achieve high corporate responsibility standards.

Wickens (personal communication, 16 September 2011) and her De Beers colleagues (personal communication, 27 October 2011), however, reject this possibility because all remaining liabilities associated with the mine, as specified in the EMP, are transferred to the new owner.[14] She does accept that limited government capacity to enforce adequate rehabilitation may mean that liabilities are, in practice, diluted, but maintains that this is not so here, with the Department of Mineral Resources having adopted a 'very diligent approach to approving the EMP'.

Furthermore, De Beers interviewees emphasized that the financial commitments for environmental rehabilitation could not be made public because the departmental review of the EMP was still in progress. Indeed, De Beers was reportedly asked by the department not to make public the financial commitments associated with the amended EMP (Wickens, personal communication, 16 September 2011).

De Beers interviewees also noted that there had been engagement with local communities regarding mine closure, going back as far as the studies conducted by University of Cape Town students in the early 1990s. In addition, the mine closure plan and social and labour plan required by mining legislation were developed according to pertinent regulations, which meant that they would respond to the priorities identified in the local integrated development plans (IDPs) published by the municipalities (which ought to have been compiled on the basis of local community participation).[15] These interviewees mentioned that there was continuing interaction between De Beers and local communities with regard to the company's corporate social investment spending in the area, specifically in the form of a 'local area committee' established for this purpose (Pisane, personal communication, 16 September 2011). Furthermore De Beers representatives are reportedly participating in local municipalities' local economic development forums.

Yet company interviewees emphasized the difficulties in identifying and communicating with legitimate and recognized community representatives, a task apparently made more difficult by local government elections in May 2011, which saw control of the local council shift from the African National Congress to a rival political party, the Democratic Alliance. 'Now previous agreements are questioned all over again . . . . We cannot be seen to be favouring any particular political party' (Pisane, personal communication, 27 October 2011).

Local resentments

Notwithstanding these structures and mechanisms ostensibly for communication between local communities and De Beers, company interviewees' remarks regarding the priorities of local community groups (e.g. 'They want a new boat launch . . .') cannot be reconciled with the level of frustration and the fundamental nature of the grievances among community members interviewed. In particular, local community members spoke of long-standing grievances with regard to land grabs before and during the apartheid period, which had benefited the mining companies but left local communities disenfranchised, poor and dependent on wage labour. The retrenchments of recent years had made these grievances more acute (see also Bregman 2010).

Since the transition to democracy in 1994, there have been attempts to use the statutory land restitution process to reclaim some of this land, spurred on by the legally successful land claim lodged by communities in the Richtersveld, some 150 kilometres north of the study area (Ramutsindela 2003; Barry 2004; Blair 2011). According to Markus (personal communication, 26 September 2011), a land claim that included five farms owned by De Beers was lodged on behalf of 40 families in Hondeklip Bay in 1998, was gazetted in 2008, and is still under review by the regional land claims commission (this was confirmed by the land claims commission based in Kimberley). The land is estimated by De Beers to contain nine million carats of diamonds (Jordan 2011). Quite apart from the absence of substantive benefits for the local people from the diamonds mined from these areas, Markus (personal communication, 26 September 2011) emphasizes grievances related to the loss of grazing land and livestock, as well as harassment and beatings by security guards in instances of trespassing. The poverty of residents in Hondeklip Bay, with unemployment estimated at 80 per cent, is contrasted with the wealth that has been extracted by De Beers (Markus, personal communication, 26 September 2011).

Concerns regarding the land claim have become more pressing in the context of De Beers' attempts to sell the mine. As noted by Markus (personal communication, 26 September 2011), 'De Beers never engaged the community with regard to the sale – you cannot sell land that is in dispute; the land claim has been totally ignored.' A De Beers spokesperson was quoted in March 2011 as saying that the company had not received any notification from the Land Claims Commission, 'with the result that De Beers has not been placed in a position to properly evaluate the claim. De Beers, accordingly, has placed on record with the Land Claims Commission that it disputes the merits of the claim' (quoted in Jordan 2011). The same stance was reported by company interviewees in this research. Markus (personal communication) noted that De Beers representatives (including one of the interviewees) had reportedly come to talk to him about the company's corporate social investment projects in the town, but had refused to talk about the land claims as these were not within the ambit of their responsibilities, as 'their focus is on CSI [corporate social investment]'.

The De Beers interviewees subsequently emphasized that the land claims needed to be seen in context, and that the company was willing to engage with local communities about their concerns. They pointed out that De Beers had granted three farms each to the towns of Hondeklip Bay and Komaggas, through the municipality, as a positive response to community members' requests for land (see also Bregman 2010). The company's opposition to the Hondeklip Bay land claim, they argue, needs to be seen in the context of a Chamber of Mines resolution to oppose land claims as a matter of course, and is motivated by a concern that 'just giving the land . . . will lead to a precedent that will make other communities around the country say, what about us, we also have a claim?' (Ngcobo, personal communication, 27 October 2011). However, the existence of such a resolution was denied by the Chamber of Mines (van Achterbergh, personal communication, 10 July 2012).

In Komaggas, a town with a particularly close relationship to the Namaqualand Mine because the majority of the mine's workers were from there, Pienaar and fellow community members[16] (personal communication, 27 September 2011) also emphasized the importance of historical injustice and their attempts to regain some of their land through a land claim. Their claim, which is under development and has not been lodged (at the time of writing), is particularly challenging because it relies on the principle of aboriginal title, which, though it contributed to the legal victory of the Richtersveld community mentioned above (Barry 2004), is difficult to prove and furthermore can create a problematic legal precedent given South Africa's history (Blair 2011; though more supportive arguments are given by Bregman 2010 and Smith, personal communication, 7 July 2012). The claimed land includes many farms currently owned by De Beers, as well as land transferred by De Beers to the national park, which has further contributed to the exclusion of community land use (Pienaar, personal communication, 27 September 2011; Blair 2011). Pienaar (personal communication, 27 September 2011) said they have been trying to discuss their claim with De Beers, but 'they just say, "show us your title deeds"'.

But the relationship between De Beers and Komaggas community members has not always been tense. Referring to the time of a certain former mine manager, Pienaar (personal communication, 27 September 2011) notes, 'we started to have neighbourly relations'. In response to the community's high court challenge against De Beers' plans to sell two farms in the late 1990s, De Beers reportedly yielded and sold the land to the government, and the land is currently being used for grazing by the community (see also Bregman 2010). However, a change in the management (particularly that of Kleinsee itself)[17] and the planned mine sale have reportedly led to a more tense relationship. The very fact that the retrenchment packages were reportedly twice as valuable as the legal requirement has been interpreted as reflecting the company's desire to 'entice' employees to accept them. More time and support was expected to prepare the town for the shock of the retrenchments (Pienaar, personal communication, 27 September 2011).

Komaggas interviewees emphasized that they had lost hope of De Beers or the government making substantive commitments in response to their concerns related to their land claim, as well as broader issues including unemployment and lack of infrastructure. They therefore embarked on a public campaign to air their grievances, similar to the efforts by Hondeklip Bay activists prominently featured in media reports in October 2011 (e.g. Blaine 2011). In addition, the socio-economic hardships and a sense of injustice are reportedly contributing to illicit diamond mining.[18]

> [C]ommunity members are now up in arms; some of our people are starting to defy the rules and taking the law into their own hands; there are allegations that some of the former De Beers employees are involved in illegal digging, as well as others who come from further away.
>
> (Pienaar, personal communication, 27 September 2011)

Such illegal mining is very dangerous, both in immediate physical terms, as demonstrated by the deaths of two illegal miners when their tunnel collapsed in July 2011 (Nicholson 2011), and in relation to the associated criminal marketing activities. The risk is that an escalation of illicit mining, coupled with increasingly forceful action by security guards, may lead to further deaths.

Discussion and conclusion

In a previous version, this chapter started off as a 'good news story' of how a multinational corporation and an international conservation NGO could collaborate in the interests of both conservation objectives and local community interests. It was framed as an example of how collaborative governance could contribute to addressing governance gaps created by challenges faced by the state.

However, recent controversy and more detailed investigations have turned this into a more critical account. One lens through which to interpret these developments is negotiation theory, recognizing that negotiation based on interests takes place in the context of critical rights and power relations (Lytle et al 1999; Covey and Brown 2001). Each collaborative relationship thus relies on a 'negotiation space' based on a certain set of converging or at least reconcilable interests, and that space is created and maintained by a system of rights and power-based relationships.

In the erstwhile De Beers–CSA partnership, the relationship was built upon a convergence of interests specifically involving conservation objectives, with links to community development opportunities (Smuts 2010). De Beers' motivation in this regard was based on its explicit and relatively ambitious conservation policies, as well as compliance with government law and regulations, specifically with regard to mine closure.

However, even though the conservation–development link was an explicit objective of the partnership, it is not clear to what extent the partnership actually involved local communities either procedurally or substantively. Procedurally, it is not apparent that they were a prominent and influential role player in the LEAP process. Substantively, the environmentally friendly development options were clearly meant to encourage much-needed employment generation, but community interviewees' primary concerns relating to land claims were not a prominent feature of the LEAP discussions. The negotiation space within which both the NGO and the company could engage thus relied on a relatively specific set of converging interests, which could not substantially include the land claims.

This negotiation space contracted significantly and rapidly as a consequence of the company's decision to sell the mine. It could be argued that the recession in 2008–10 also had a significant influence, but although this is likely to have diminished resources available to the company for corporate responsibility activities in that period, it ought not to have had long-term, strategic implications, considering the positive market outlook for the company.[19]

The company's interest calculus thus moved from a mine closure scenario to one of mine sale. The claim by critics that this was part of an effort to minimize or even avoid existing social and environmental commitments and liabilities cannot be fully substantiated, though crucial information is still outstanding while the final government ruling on the sale (and the associated EMP) is awaited.[20] In particular, the financial commitment to environmental rehabilitation, which will be part of the new owner's commitments, needs to be commensurate with the substantive content of the EMP, and the government will have to enforce that conscientiously. De Beers' decision to sell the mine has arguably not been the only or even primary cause for tension between the company and CSA. Another has been the problematic regulatory framework governing the sale, specifically with regard to the lack of disclosure requirements for financial rehabilitation commitments.

This is yet another compelling illustration of why collaborative governance arrangements cannot function as a replacement for state regulation, but rather require a clear regulatory framework to operate effectively and legitimately (see also Hamann 2004). Clear rules about the 'basics' – in this instance, with regard to environmental liabilities in the case of mine sales and corresponding transparency requirements – need to be established and enforced in order to create a negotiation space for partnerships. In the absence of such enforced rules, the negotiation space contracts and no negotiation in search of converging interests can take place.

So the negotiation space between company and NGO with regard to conservation and post-mining land uses has changed as a consequence of the company's decision to sell the mine, and the state's governance framework has been unable to keep it open. On the other hand, it was never really large enough to include local communities' land claims, which seem to have been more fundamentally at odds with the company's interests (notwithstanding some interviewees' comments to the contrary). This could be interpreted as an illustration of the fundamental limits to collaborative governance in the context of historical injustices.

However, again it is the regulative framework that ought to clarify the status of such rights – in this instance with regard to the land claims process, hampered by significant bureaucratic inefficiencies. Implemented more effectively, this regulatory framework would create a new negotiation space, this time including local communities in the negotiations more substantively.

It is clear that the adjudication of land claims is a complicated and controversial process, but nevertheless much of the uncertainty and tension in the case study area could probably be attenuated if the Land Claims Commission implemented a more structured and transparent process. Again, therefore, the negotiation space in which the company is motivated to find more amenable outcomes for local communities is effectively closed down by weak or haphazard state regulation.

In the absence of compelling interests inducing the company to respond to local community grievances, and in view of uncertain government regulation and regulatory enforcement, the local communities face the greatest risks in the mine sale. Recognizing that they are likely to have greater influence over De Beers, a company with a significant brand reputation to protect, than Transhex, the communities approach the mine sale as something of an 'end-game' scenario, which raises the stakes. Engagement options thus move from interests and rights to power-based approaches, and this explains the launch of media campaigns in October 2011, in which community members teamed up with NGOs, including CSA and the Bench Marks Foundation, which had copy-writing and other expertise and contacts in the local and international media.

This may have brought some limited success in the form of a more committed company response, demonstrated by an almost week-long visit to the area by a team of senior managers, including De Beers South Africa's corporate affairs director. A significant element of the company's response has been a media campaign, including media releases and even a video responding to an NGO-sponsored video, but company interviewees emphasized their commitment to responding positively to legitimate concerns. An important motive is clearly the company's reputation and standing in its market, with customers increasingly expecting a credible source with a clear conscience in the wake of numerous human rights controversies linked to diamond mining (Ngcobo, personal communication, 27 October 2011). Yet even if later developments in 2011 suggested a renewed commitment by the company to addressing stakeholders' concerns, and even though local community interviewees emphasized the power of De Beers, the contraction of the negotiation space brought about by the sale of the mine and the weakness of state regulation creates a difficult context for any collaborative efforts.

Acknowledgements

This chapter is indebted to all the interviewees for offering their time and perspectives on the case study, as well as to Brian Ganson, Cameron Holley, Rowena Smuts and Merle Sowman for helpful comments on an earlier draft. The research was made possible by funding from the National Research Foundation of South Africa and the UCT Vice-Chancellor's Strategic Fund.

Notes

1 'Governance' here is interpreted as 'organised efforts to manage the course of events in a social system' (Burris et al 2008, p. 3) and more specifically as the structures and processes through which commonly binding rules are established, and through which public goods or services are provided (Risse 2011). The notion of collaborative governance thus suggests that rules are established and enforced, and public goods and services are provided, not only by the state on the basis of its position of hierarchical power, but also by other actors in civil society and business, commonly premised on negotiation processes seeking consensus decisions and various actors' contribution of complementary resources.

2 Retrenchments in connection with mine downscaling commenced in 2003, and mining operations were suspended altogether in early 2010. At the time of this research in 2011, De Beers employed about 130 people in the area in environmental rehabilitation work (MacDonald, personal communication, 16 September 2011).

3 There are, of course, other controversial issues in this case, such as the proclamation of the Kleinsee town, which has been a private, De Beers-owned settlement since its establishment in the late 1920s (Carstens 2001), but they will not be discussed in any detail in this chapter.

4 This chapter is based on a review of literature and documents such as media reports, press releases and research reports, and on semi-structured interviews and more informal discussions with representatives from CSA and De Beers, as well as with local community members and officials in the area, some of whom insisted on anonymity, given the controversial nature of the issues. Face-to-face interviews or discussions were held with, among others, the individuals listed under 'Personal Communications' in the reference list at the end of this chapter. Additional communication with some interviewees took place via email. As for sampling, identifying interviewees from the NGO and company was relatively straightforward: most of the key individuals involved in the initial partnership and the more acrimonious correspondence that followed were interviewed. The sampling of local community interviewees, however, was more complex. These individuals were initially recommended by Frazee (personal communication, 24 August 2011) and also identified by company interviewees as outspoken community members

who enjoyed a significant degree of respect among the communities generally. The latter judgement was partly substantiated through casual conversations with diverse, randomly identified community residents.

5 The description of the partnership is based largely on Smuts (2010), while the discussion of more recent developments is based on interviews.

6 www.skep.org.za

7 Indeed, CSA's interest in Namaqualand Mine was triggered when an increase in livestock grazing was observed in the upland area of the Kamiesberg, conducted by retrenched workers from the mine looking for alternative livelihood options (Frazee, personal communication, 24 August 2011).

8 http://.edenproject.com

9 The collaboration in Namaqualand led to a decision to establish a global partnership between CI and De Beers, including the signing of a collaboration agreement in December 2007. This led to, among other things, the involvement of CI in the review and benchmarking of De Beers' environmental policies. Wickens (personal communication, 16 September 2011) describes De Beers' objective in this collaboration as being 'to gain the knowledge of an outside organisation' with specific reference to the company's policy commitment to no net loss of biodiversity.

10 However, Pisane (personal communication, 16 September 2011) argued that, from De Beers' perspective, limiting the involvement of community members in the charrettes was probably a good thing, because 'it is important not to raise too many expectations among community members – you need to develop feasible projects first, then go to the communities'.

11 The amended EMP was completed in early 2010 after the completion of the mine closure plan (Pisane, personal communication, 16 September 2011).

12 Wickens (personal communication, 16 September 2011) points out that De Beers has had significant experience in developing tourism projects as a means to provide post-mining employment and economic development opportunities (see, in particular, www.diamondroute.co.za and http://coastofdiamonds.co.za), and that this experience had demonstrated the need to carefully assess economic feasibility based on accessibility.

13 The selection criteria were technical, including financial capability to mine alluvial deposits (considering the associated uncertainties and difficulties), black economic empowerment criteria, and community and sustainability issues, with the latter comprising about 30–40 per cent of the total weighting of the conditions (MacDonald and Wickens, personal communication, 16 September 2011).

14 In fact, some of the commitments have been retained by De Beers because they are close to fulfilment. The mariculture project, for instance, was due to be opened before the end of 2011 (MacDonald et al personal communication, 16 September 2011).

15 A general concern has been the challenges faced by municipalities in actually preparing IDPs according to the regulatory guidelines, which emphasise community participation. Another has been their capacity to implement the activities and attain the goals consequently envisaged in the IDPs. (For a discussion on the development and implementation of IDP policy, see Harrison 2006.) This was echoed by community interviewees: 'The Kamiesberg Municipality [incorporating Hondeklip Bay and Komaggas] has an IDP but doesn't have the capacity to implement it . . . They had IDP meetings, but they're just empty promises' (Markus, personal communication, 26 September 2011). A more specific concern that has been prominent in some of this author's previous work on related issues is the lack of systematic involvement by mining companies in the public deliberations that are meant to feed into the preparation of IDPs (e.g. Hamann 2004; Hamann et al 2011).

16 Andy Pienaar is the founding director of the Komaggas Advice Centre, a local NGO focused on community education and development. The other interviewees (Thomas Cupido, Johan Cloete and Christine Cloete) were affiliated to the 'Karusab Committee', which was involved in developing the community's land claim, but emphasised that

they were participating in the interviews as community members, not having received a mandate from the committee to speak on its behalf.

17 The De Beers managers identified by community interviewees as having caused tensions in the company's relations with the community did not include any of the managers interviewed for this research.

18 Illicit diamond mining has been a prominent feature of this area since diamonds were first discovered in 1927 (Carstens 1962).

19 It is this positive outlook that underpinned the sale of the Oppenheimers' stake in De Beers to Anglo American for US$5.1 billion (Seccombe 2011).

20 At the time of writing, the amended EMP had not been made public – and indeed, this was one of the main concerns raised by CSA. De Beers argued that the amendment did not require public consultation and that it should not be published due to the commercial aspects of the sale (citing section 68 of the Promotion of Access to Information Act) (Webber Wentzel 2011). This was vigorously disputed by CSA and its lawyers (CER 2011).

Bibliography

Barry, M. (2004) 'Now another thing must happen: Richtersveld and the dilemmas of land reform in post-apartheid South Africa', *South African Journal of Human Rights*, vol 20, p. 355

Blaine, S. (2011) 'Community tries to halt De Beers mine sale', *Business Day*, 13 October 2011, http://businessday.co.za/Articles/Content.aspx?id=155933, accessed 2 July 2012

Blair, I. (2011) 'In the wake of diamond mining: A critical assessment of environmental governance and corporate social responsibility in the Namaqualand coastal region', MPhil thesis, Department of Environmental and Geographical Science, University of Cape Town

BPD (2002) *Putting Partnering to Work*, Business Partners for Development, London

Bregman, J. (2010) 'Land and Society in the Komaggas region of Namaqualand', MA thesis, Faculty of Humanities, University of Cape Town

Burris, S., Kempa, M. and Shearing, C. (2008) 'Changes in governance: A cross-disciplinary review of current scholarship', *Akron Law Review*, vol 41, no 1, pp. 1–67.

Carstens, J. (1962) *A Fortune Through My Fingers*, Howard Timmins, Cape Town

Carstens, P. (2001) *In the Company of Diamonds: De Beers, Kleinsee, and the Control of Town*. Ohio University Press, Athens, OH

CER (2011) 'Letter from Centre for Environmental Rights to Webber Wentzel', 17 November

CSA (2011) 'De Beers avoids best practice protocol in mine sale', press release, Conservation South Africa, 11 May

Covey, J. and Brown, L.D. (2001) 'Critical cooperation: An alternative form of civil society – business engagement', Report No 17(1), Institute for Development Research, Boston

Creamer, M. (2011) 'De Beers targets 5 000 non-mining Namaqua jobs', *Mining Weekly*, 27 October

Donahue, J. D. (2004) 'On collaborative governance', Corporate Social Responsibility Initiative Working Paper No. 2, John F. Kennedy School of Government, Harvard University, Cambridge, MA

Esteves, A. M. and Barclay, M. A. (2011) 'New approaches to evaluating the performance of corporate-community partnerships: A case study from the minerals sector', *Journal of Business Ethics*, vol 103, no 2, pp. 189–202

Flyvberg, B. (2006) 'Five misunderstandings about case study research', *Qualitative Inquiry*, vol 12, no 2, pp. 219–45

Gosling, M. (1992) 'An assessment of the potential socio-economic impact of the future closure of the DBNM's diamond mines in Namaqualand', master's dissertation, University of Cape Town

Hamann, R. (2004) 'Corporate social responsibility, partnerships, and institutional change: The case of mining companies in South Africa', *Natural Resources Forum*, vol 28, no 4, pp. 278–90

Hamann, R., Kapelus, P. and O'Keefe, E. (2011) 'Mining companies and governance in Africa', in J. Sagebien and N. M. Lindsay (eds) *Governance Ecosystems: CSR in the Latin American Mining Sector*, Palgrave Macmillan, London, pp. 260–76

Harrison, P. (2006) 'Integrated development plans and Third Way politics', in U. Pillay, R. Tomlinson and J. du Toit (eds) *Democracy and Delivery*, HSRC Press, Durban, pp. 186–207

Huxham, C. and Vangen, S. (2000) 'Leadership in the shaping and implementation of collaboration agendas: How things happen in a (not quite) joined-up world', *Academy of Management Journal*, vol 43, no 6, pp. 1,159–75

Jordan, B. (2011) 'De Beers disputes land claim,' *Sunday Times*, 26 March

Lytle, A. L., Brett, J. M. and Shapiro, D. L. (1999) 'The strategic use of interests, rights and power to resolve disputes', *Negotiation Journal*, vol 15, no 1, pp. 31–51

Moon, J. (2002) 'Business social responsibility and new governance', *Government and Opposition*, vol 37, no 3, pp. 385–408

Nicholson, Z. (2011) 'Two diamond miners crushed to death,' *Cape Times*, 1 August.

Ramutsindela, M. (2003) 'Land reform in South Africa's national parks: A catalyst for the human-nature nexus', *Land Use Policy*, vol 20, no 1, pp. 41–9

Risse, T. (2011) 'Governance in areas of limited statehood: Introduction and overview', in T. Risse (ed.) *Governance without a State? Policies and Politics in Areas of Limited Statehood*, Columbia University Press, New York, pp. 1–35

Seccombe, A. (2011) 'Landmark De Beers sale adds lustre to Anglo', *Business Day*, 7 November, http://businessday.co.za/Articles/Content.aspx?id=157966, accessed 2 July 2012

Smuts, R. (2010) *Are Partnerships the Key to Conserving Africa's Biodiversity? Four Partnership Case Studies between Mining Companies and Conservation NGOs*, Conservation International, Arlington, VA

Swyngedouw, E. (2005) 'Governance innovation and the citizen: The Janus face of governance-beyond-the-State', *Urban Studies*, vol 42, no 11, pp. 1,991–2,006

Van Wyk, D., Cronjé, F. and Van Wyk, J. (2009) *Corporate Social Responsibility in the Diamond Mining Industry on the West Coast of South Africa*, Bench Marks Foundation, Pretoria

Webber Wentzel (2011). Letter from Webber Wentzel to Centre for Environmental Rights, 26 October

Personal communications

Cloete, Christine: Komaggas resident (27 September 2011)

Cloete, Johan: Komaggas resident (27 September 2011)

Cupido, Thomas: Komaggas resident (27 September 2011)

Frazee, Sarah: director, Southern Africa Hotspots Programme, Conservation South Africa (24 August 2011)

Mabusela, Innocent: Northern Cape corporate affairs manager, De Beers Consolidated Mines (27 October 2011)

MacDonald, William: site manager, Namaqualand Mine, De Beers Consolidated Mines (16 September 2011)

Markus, Dawid: Hondeklip Bay resident (26 September 2011)

Ngcobo, Sakhile: corporate affairs director, De Beers Consolidated Mines (27 October 2011)

Pienaar, Andy: Komaggas resident (27 September 2011)

Pisane, Christie: Northern Cape heritage general manager, De Beers Consolidated Mines (16 September 2011, 27 October 2011)

Smith, Henk: attorney, Legal Resources Centre (email, 7 July 2012)

Smuts, Rowena: former mining engagement adviser, Conservation International (12 April 2011)

van Achterbergh, Anton: Head: Legal, Chamber of Mines (email, 10 July 2012)

Wickens, Patti: environment principal, De Beers Group (16 September 2011)

(Other interviewees insisted on anonymity.)

14 Governance, equity and sustainability in non-timber forest product value chains

Rachel Wynberg and Jaci van Niekerk

Introduction

The past two decades have seen growing attention given to the economic potential of biodiversity, spurred in part by dramatic scientific developments in biotechnology, such as the potential to transfer genes between species through genetic engineering, increased consumer demand for 'natural' products and a regulatory environment that facilitates the patenting of products and processes based on biodiversity (Dutfield 2011; Laird and Wynberg 2012). Intended products include new drugs, climate-resilient crops, industrial processing tools and novel ingredients for the food, herbal medicine and personal care industries. These products generate significant benefits for society, financial returns for the companies producing and marketing them and a range of benefits for countries that provide the biological material.

At the same time, livelihood opportunities have opened up for rural communities engaged in commercially harvesting and producing species with high value in global and local markets. Indeed, several million people in sub-Saharan Africa earn their primary cash income from the sale of non-timber forest products (NTFPs)[1] (Kaimowitz 2003), which, in addition to providing income, play important roles as natural subsidies for food, medicines, building materials, crafts and a multitude of other purposes, fulfilling the role of safety nets in times of scarcity (Shackleton and Shackleton 2004; Timko et al 2010; Shackleton et al 2011). More often than not, however, the primary producers involved in NTFP harvesting and processing are among the poorest in the world (Neumann and Hirsch 2000). This is especially the case for internationally traded products, in which primary producers are typically represented 'invisibly' in exploitative trade chains that, in keeping with the nature of extractive markets, tend to keep wages and prices for producers low, while profit-sharing rises with increased processing as the product moves closer to the consumer (Southgate et al 1996; King et al 1999; Neumann and Hirsch 2000). Such inequalities have intensified with the rise in the power of supermarkets and transnational corporations (Gibbon and Ponte 2005; Harilal et al 2006).

Increasing consumer demand for the natural products derived from NTFPs also means that sustainability questions have become central. Consumers want to know that they are purchasing a product that is ecologically sustainable, companies are under increasing pressure to demonstrate that they are making a positive environmental and social impact, and regulators want to be sure that the species is used sustainably and is not threatened by trade. Overharvesting is a common concern, especially for species that are used commercially (e.g. Ticktin 2004; Rijkers et al 2006; Ticktin and Shackleton 2011), typically leading to a shift from wild harvesting towards cultivation of resources and a 'transition from without agriculture to within, from nature to culture' (Dove 1995).

While cultivation is touted by some as arguably an environmentally benign alternative to wild harvesting (e.g. Anderson 1992; Homma 1996), its environmental impact can be significant due to land transformation, pollution and degradation (Schippman et al 2002). Moreover, small producers are likely to be at best marginalized, and at worst excluded entirely, from activities that require significant capital, land and infrastructure (Homer-Dixon et al 1993; Sunderland et al 2004).

As awareness grows of the problems of overharvesting, inequality and unfair labour practices, a string of mandatory measures have unfolded, aiming to bring greater ethical and sustainability standards to trade in biological and genetic resources.[2] This has been due in large part to the 1973 Convention on International Trade in Endangered Species of Wild Fauna and Flora (CITES), the 1992 Convention on Biological Diversity (CBD), and the related 2010 Nagoya Protocol on Access to Genetic Resources and the Fair and Equitable Sharing of Benefits Arising from their Utilization.[3] As a result of the CBD and the Nagoya Protocol, companies and researchers wishing to obtain access to biological material and associated traditional knowledge are now required to show how the providers of resources and knowledge will benefit. Moreover, such access is conditional on benefits being fair and equitable and on the prior informed consent of providers. These activities are encapsulated in the term 'access and benefit sharing', with the conclusion of benefit-sharing agreements between users and providers of resources and traditional knowledge premised on sustainable use and biodiversity conservation.

Growing consumer pressures have also led to the parallel development of a range of voluntary social and environmental certification schemes, developed largely at national and regional levels by several different industry-affiliated and independent labelling organizations to address consumer demands for fair and sustainable supply chains (Shanley et al 2008). Certification schemes relating to the sourcing of raw materials include those for fair trade and for organic and ecological or sustainable harvesting. FairWild,[4] for example, focuses on the sustainable and ethical sourcing of ingredients and incorporates the International Standard for Sustainable Wild Collection of Medicinal and Aromatic Plants.

Despite a prodigious number of international meetings and workshops on topics relating to access and benefit sharing, NTFP trade (or 'biotrade'[5] as it is commonly referred to) and rural livelihoods, and a significant investment of development aid to facilitate CBD implementation, there is little evidence that policies and laws on access and benefit sharing have achieved greater equity (see, for example, Wynberg and Laird 2007). Moreover, little attention has been given to the governance of these value chains, ways in which different actors benefit from the natural product trade, and how value chains might be modified to bring greater benefit to the poor. In addition, although several studies describe the biological characteristics of different commercially used species and the implications for their conservation and sustainable use (see Ticktin and Shackleton 2011, for a review), few have examined the interface between governance, ecological sustainability and equity.

Strong state involvement – in the commercialization of a species, for example – may well lead to a rigorous regulatory environment for conservation, but could also act against the interests of rural producers and the effective involvement of the private sector (Ribot 1995; Lele et al 2010; Novellino 2010). An exclusively market-driven and corporate approach, on the other hand, is unlikely to cater adequately for environmental considerations or for social equity. Non-governmental organizations (NGOs) may play a major role in assisting producer communities to negotiate fairer trade terms, but could also be considered an unsustainable option in the long term because of social and financial dependencies that may arise, as well as potential conflicts of interest. Producer organizations and cooperatives may be critical in facilitating more equitable trade, but their effectiveness may require strong external support and hinge upon internal robustness. Emerging ethical and fair-trade initiatives for natural products introduce an additional layer of networks and power relations to these trends.

This chapter aims to enhance understanding about the influence of these different governance arrangements in NTFP value chains on ecological sustainability and equity outcomes. We do this through the lens of two plant species endemic to southern Africa (see Figure 14.1). In the first case, we explore the commercialization of *Hoodia gordonii*, a succulent plant developed as an appetite suppressant based on traditional knowledge of the indigenous San peoples. Thereafter, we examine the progression of *Pelargonium sidoides* from a widely used traditional remedy to a successful global phytomedicine. We examine the evolving dynamics of actors involved in the governance of these value chains, changes in governance over time and implications for equity and sustainability. Southern Africa is a hub for biodiversity commercialization, and trade in these two plant species has been caught up in the turbulent world of access and benefit sharing, intellectual property rights, traditional knowledge, uncertain land tenure, unclear regulatory regimes and tension between producers, the private sector, NGOs and traditional authorities on one hand and democratically elected government structures on the other.

Hoodia and the intricacies of governing across cultures, nations and interests

Hoodia *value chains*

In a world where obesity is an escalating problem, the commercialization of species of the succulent plant *Hoodia* is particularly interesting because of the plant's traditional use to stave off hunger and thirst by the indigenous San peoples, the

Figure 14.1 Distribution ranges of *Hoodia gordonii* and *Pelargonium sidoides.*

oldest human inhabitants of Africa (Pappe 1862; White and Sloane 1937). Bringing *Hoodia* to market as an appetite suppressant involved a complex chain of events and multiple actors, including the indigenous San, Nama and Damara peoples; national and provincial government departments in several different countries sharing the resource; the South African-based Council for Scientific and Industrial Research (CSIR), a parastatal research organization; *Hoodia* growers from South Africa and Namibia; NGOs supporting the San and acting against biopiracy;[6] the British-based company Phytopharm; and large corporations, including the pharmaceutical and consumer giants Pfizer and Unilever. It also sparked a chain of regulatory and policy measures with wide ramifications.

Figure 14.2 illustrates the two *Hoodia* value chains that emerged through this process. The first value chain (Figure 14.2a) was structured around an agreement between the CSIR, Phytopharm and Unilever based on material cultivated, mostly by commercial farmers, with the San acknowledged as knowledge holders and thus beneficiaries but not having any specific role in the production process. The second value chain (Figure 14.2b) was initially based on wild-harvested material, but is now almost entirely cultivated, supplied by commercial farmers in South Africa and Namibia to herbal supplement markets, mostly in the United States. In some cases, but not all, the San received a proportion of sales income in this value chain, but again have seldom been involved in the production process.

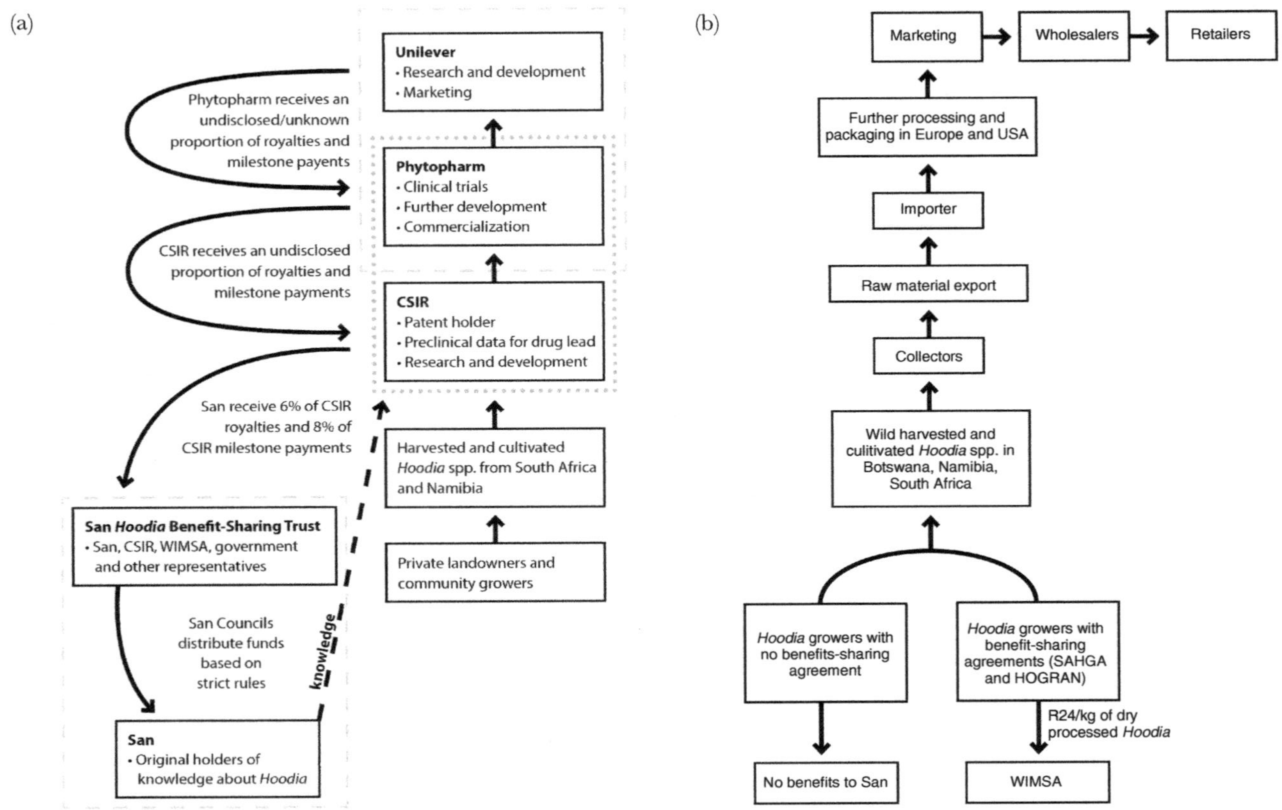

Figure 14.2 (a and b) *Hoodia* spp. value chains.

Initial commercialization of Hoodia

Comprehensive reviews of the *Hoodia* case are recounted by Wynberg (2004) and Wynberg and Chennells (2009), who describe how colonial botanical accounts of the traditional use of *Hoodia* lent the impetus for the CSIR to instigate an intensive research and development programme into the plant's properties. As Figure 14.3 illustrates, this led to the 1997 filing of a patent by the CSIR for active components of *Hoodia* species responsible for suppressing appetite, the subsequent signing of a licence agreement for its further development and commercialization with the British company Phytopharm, and the sublicensing to American corporation Pfizer for clinical development.

At the time, no legal requirements for benefit sharing were in place, although South Africa was a signatory to the CBD, and these actions took place without the consent of the San or other indigenous groups that held knowledge about *Hoodia*. Moreover, although *Hoodia* species had been listed as protected since the 1970s,[7] the absence of pressure on the resource at the time allowed the government to adopt a largely passive approach towards regulating the plant's use.

Entry of the state

Governance of the *Hoodia* chain was thus almost entirely led by the private sector prior to 2000, but this changed dramatically soon after. In 2001, a campaign spearheaded by NGOs Biowatch and Action Aid alerted the British media to the commercial development of *Hoodia* without the consent or involvement of the San, spurring the San to demand negotiations with the CSIR and the eventual development of a benefit-sharing agreement in 2003. It was partly the unfolding of these experiences and the high-profile nature of the case that gave impetus to the development of binding laws in South Africa, encapsulated in the National Environmental Management: Biodiversity Act (10 of 2004) and the later promulgation of regulations governing access and benefit sharing in 2008. These laws now require a benefit-sharing agreement to be developed between different stakeholders to ensure that holders of traditional knowledge and custodians of biodiversity are fairly compensated.

Stronger involvement of the state was also spurred by concerns about the ecological status of the resource. Publicity generated by the agreement, the marketing opportunities offered by the San use of the plant and the CSIR patent led to a frenzied interest in *Hoodia* among plant traders. A parallel market for *Hoodia* herbal and dietary supplements emerged, based on the export of dried wild-harvested *Hoodia*, escalating exponentially from just a few tons in 2002 to more than 600 tons of wet, harvested material in 2005, sold as ground powder for incorporation into non-patented dietary supplements. In South Africa and Namibia, illegal trade and harvesting of *Hoodia* resulted in a number of prosecutions and arrests; the high prices commanded for the dry product of up to USD 200 per kilogram had led to the incorporation of the plant into a global underground network of diamonds, drugs and abalone (Wynberg and Chennells 2009).

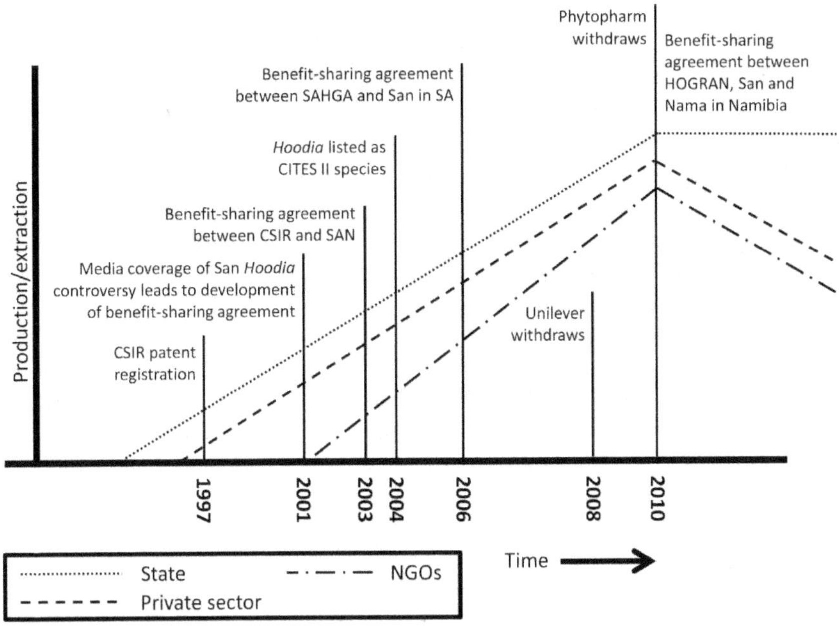

Figure 14.3 Timeline of *Hoodia* spp. commercialization.

Regional governance and the emergence of new regulatory approaches

This escalation in demand necessitated new regulatory approaches, including the listing of *Hoodia* species in Appendix II of the CITES (2004). In response, southern African governments began developing a more tightly regulated permitting system for *Hoodia* use and trade, although this was done differentially, both within and between countries. In Namibia and Botswana a moratorium was put in place on wild harvesting and trade of any *Hoodia* species. In South Africa, different sets of regulatory approaches evolved in the Northern Cape and Western Cape provinces, the areas in which most *Hoodia* species occur, with the Northern Cape using a moratorium to curb trade, while the Western Cape issued a so-called 'open permit' to several traders, arguing that a moratorium would simply drive the *Hoodia* industry underground and make the trade more difficult to track and manage (Wynberg 2010). As the only legal point of export for *Hoodia*, the Western Cape thus saw a flood of material pass through it, and the open-ended nature of the permit provided the perfect means through which illegally harvested material could be included and legitimately exported under a CITES permit (Wynberg 2010).

Increased awareness of these problems, combined with concerns about the quality and safety of material sold as *Hoodia*, and recognition of the need to ensure the sustainability of *Hoodia* supply, led to a rapid response from conservation authorities across the region. Faced with a resource crisis, Namibia, South Africa and Botswana strengthened permitting as well as export and border controls, to bring improved policy cohesion and standardization alongside greater efforts to ensure the cultivation of the species.

Regional cohesion was, however, more challenging when it came to policies of community identity and benefit sharing. South Africa, through the parastatal CSIR, had played a leading role in lodging the patent, developing the commercial partnerships for *Hoodia* and negotiating benefit-sharing agreements with the San (Wynberg and Chennells 2009). Moreover, South Africa had adopted legislation on access and benefit sharing and fully recognized the San as a community with clear rights to benefit from *Hoodia*. Namibia and Botswana, on the other hand, did not have such laws in place and considered benefits from *Hoodia* to belong to the state, rather than to traditional knowledge holders. These different approaches were not contrary to the CBD, which allows states to decide how traditional knowledge holders should be acknowledged, but caused confusion because the distribution of the San, just like that of the resource, straddles national borders.

Community governance

Such complexities were to unravel further at the community level. San leaders believed they needed to be united across the region, regardless of political boundaries. In 1996, they had established their own advocacy organization, the Working Group of Indigenous Minorities in Southern Africa (WIMSA), charged with uniting and representing San communities from Botswana, Namibia and South Africa. Through WIMSA, the South African San Council was formally mandated to represent the San of Namibia and Botswana as well as those in South Africa in all benefit-sharing negotiations. The South African San Council, supported by the South African San Institute, a San service NGO helping San-based organizations access funding and expertise, thus became an important part of the governance of the *Hoodia* commercialization process.

San involvement was strongly informed by experiences of their land claim under South Africa's post-apartheid land restitution programme, which had led to conflict between San communities (Robins 2002), as well as other agreements on access and benefit sharing that had fallen apart because of disputes as to which communities and countries should participate (e.g. Hayden 2003). In a bid to avoid similar controversies, the San decided in principle that their heritage was collectively owned, and that any benefits arising from its use were to be divided equally among countries with San populations (Wynberg et al 2009). This provided the foundation for the establishment of a trust to distribute benefits.

With the establishment of the trust, questions of internal governance became paramount, bringing together Western and traditional systems of decision-making in an uneasy marriage of convenience. The CSIR, for example, was adamant that the body should be exclusively South African and saw no place in it for Namibia and Botswana. But the San believed that 'it was our knowledge and therefore our money' (San-*Hoodia* Trustee, Upington, personal communication, 18 September

2006) and that the principle of collective ownership could not be compromised. The final outcome was a compromise, with representation of Botswana, Namibia and Angola on the trust by three South African San nominated by the three countries. A decision was also taken to allocate 75 per cent of all trust income (a total of some ZAR 600,000 or USD 60,000) equally between San councils in Namibia, Botswana and South Africa, with the remaining amount to be used for trust administration.

The need to select representative decision-makers also caused decision-making power within San ranks to shift from its relatively egalitarian distribution to a Western model. Traditional decision-making had been based on a highly participatory and consensual approach seeking to avoid conflict. Engaging with the CSIR, however, required individual representation, not recognized by traditional systems and often contested, fraught and with weak accountability. The San councils involved in deliberations were mostly no more than a few years old and had been set up to assert San rights rather than as traditional governance systems. These fragile institutional dynamics were further complicated by the need to acknowledge other groups holding traditional knowledge of *Hoodia*, such as the Nama and Damara.

Market-initiated changes

At the same time as institutional arrangements were being established to share benefits arising from the agreement between the CSIR and the San, an array of independent *Hoodia* growers and traders were emerging in South Africa and Namibia to supply the herbal and dietary supplement market. *Hoodia* was now being successfully cultivated, and growers were mindful of the need to be compliant with South Africa's laws on access and benefit sharing and the CBD. This led to the 2006 establishment of the South African *Hoodia* Growers Association (SAHGA), which subsequently developed a benefit-sharing agreement with the South African San Council that was later validated by the state and the parallel development of the *Hoodia* Growers Association of Namibia (HOGRAN), which in 2010 signed a benefit-sharing agreement with the San and Nama communities of Namibia (Figure 14.3).

As the *Hoodia* industry became more organized, it was dealt a blow by changes to the market. The pharmaceutical company Pfizer discontinued clinical development of an appetite-suppressant drug based on *Hoodia* and handed the rights back to Phytopharm in 2003. Consumer giant Unilever stepped into Pfizer's place through a joint development agreement with Phytopharm, and began investigating *Hoodia* as an ingredient for its line of Slim-Fast drinks. A massive cultivation programme was launched, involving over 300 hectares of *Hoodia* in South Africa and Namibia, clinical safety trials, manufacturing and an agreement to develop a ZAR 750 million (approximately USD 75 million)[8] extraction facility. In 2008, however, Unilever made a sudden announcement that it was abandoning plans to develop *Hoodia* as a functional food, because of safety and efficacy concerns. Although some *Hoodia* herbal products remain on the market today, based almost entirely on cultivated material, the multi-million dollar profits originally projected remain elusive.

Achieving sustainability and equity

The impacts of these initiatives on sustainability and equity outcomes are important to examine and highlight the complexities of governing a species undergoing rapid commercialization, where information about both the biology of the species and its trade is incomplete and scarce, where several nation states are involved and where multiple laws apply to regulate harvesting, trade and commercialization.

Increased awareness of problems associated with illegal and unsustainable *Hoodia* harvesting led to a rapid response from conservation authorities and the establishment of a system to incentivize cultivation. This response, best characterized as 'reactive' and 'experimental', had clear drawbacks in its lack of coherence, comprehension and foresight, but also had its benefits. Significant changes in *Hoodia* markets, availability and demand necessitated an iterative and flexible approach by government towards regulation, suggesting that reactive decision-making may well be a vital mechanism to cope with rapidly changing situations, in this case market and trade fluctuations.

In contrast to its response to resource depletion, the state played a very minimal role in securing equity outcomes, with deliberations centred instead on the San and commercial partners. Here too, however, there were concerns. The benefit-sharing agreement that was negotiated between the San and the CSIR was hailed initially as a significant breakthrough in the access and benefit-sharing impasse. Here was an example of how the CBD could work in practice to benefit both indigenous communities and those seeking to reap profit from traditional knowledge and biodiversity. The dietary control of obesity is valued at USD 3 billion per annum in the United States alone, and thus returns were expected to be lucrative.

But very soon the cracks began to show. Analysis of the agreement revealed that although the San might receive a considerable amount of money, this would be less than 0.01 per cent of the total royalties generated (Wynberg 2004). Monies received by the San would be extracted from royalties received by the CSIR, but the profits of Pfizer, Unilever and Phytopharm were to remain untouched. Was this equitable benefit sharing? Moreover, the CSIR had insisted on an exclusive agreement with the San. This had been agreed to prior to the boom in trade in *Hoodia* in 2006 and the emergence of many other potential business partners. The exclusive nature of the agreement was now reducing opportunities for the San to benefit from the use of *Hoodia*, although in practice other benefit-sharing agreements continued to be signed between the San and *Hoodia* growers and traders.

Additionally, the inflow and distribution of potentially huge sums of money to the San was very worrying because of the fragility of local San institutions. What impact would this have on the San, and how could a system be created that ensured fairness and equity across three countries? This matter was especially complex and fraught because of the wide dispersal of San across very remote parts of southern Africa. Experiences of distributing the USD 60,000 received by the

San-*Hoodia* Trust in 2006 had already highlighted the immense challenges of this task (Wynberg et al 2009).

Some five years later these issues remain unresolved, largely thanks to the near collapse of the *Hoodia* market. The withdrawal of Unilever, and later Phytopharm, from the commercial development of *Hoodia* due to worrying results from clinical trials has severely reduced demand for the plant, although a small number of products are still sold on the herbal supplement market by other players. Nonetheless, the case has yielded valuable experiences for the San, who have subsequently launched benefit claims based on their traditional knowledge of other indigenous species from the region such as *Sceletium tortuosum*, rooibos and honeybush.

A very different set of experiences emerges in the next case, that of *Pelargonium sidoides*, but with some remarkably similar outcomes. Investigating the actions of those directly involved, and the consequences of their interactions, offers a glimpse into the complexity of the governance of this NTFP, exposing the pressure and friction that can result when global ambitions such as access and benefit sharing under the CBD collide with local realities (Morris 2012).

Pelargonium sidoides: Governing a resource in times of regulatory transformation

A medicinal plant from the geranium family, *Pelargonium sidoides* (hereinafter *Pelargonium*) is endemic to South Africa and Lesotho. Long used as a traditional

Figure 14.4 Cultivated *Hoodia gordonii* (photo: Rachel Wynberg).

medicine by local communities, the plant is also incorporated into a cold-care remedy manufactured and distributed worldwide by German pharmaceutical company Schwabe. The raw material supplying the industry is largely sourced from the wild, harvested by rural communities on communal lands in the Eastern Cape Province of South Africa, and across Lesotho, where the resource is most abundant (Figure 14.1).

Community members from these areas suffer high levels of poverty, widespread dependence on state welfare grants and inadequate provision of basic services such as water, sanitation and health care – circumstances that still haunt the former homelands[9] of the Transkei and Ciskei (Westaway 2012). These conditions, which are also prevalent in Lesotho, one of the least developed countries in the region (Global Policy Network [GPN] 2006), create a situation conducive to the wide use of the plant in place of modern medicines, which are either unobtainable or in short supply, and high reliance on the income earned by the hundreds of people engaged in harvesting the plant.

With a history of commercialization spanning more than a century, the *Pelargonium* trade has seen significant changes in its governance. Once characterized by unfettered access, the industry has become subject to progressively stricter regulatory measures, which have affected a variety of actors: rural community members, who use the plant traditionally and rely on collecting the resource to supplement their livelihoods; traditional authorities responsible for land and natural resource allocation; members of the private sector bringing *Pelargonium*-based products to market; and NGOs concerned about biopiracy and inequality in the value chain.

Pelargonium commercialization

The transition from local remedy to global phytomedicine commenced around the turn of the twentieth century. In 1897, Charles Henry Stevens, an Englishman suffering from tuberculosis, travelled to southern Africa on his doctor's advice ('An English Physician' 1931). In the highlands of Lesotho, a healer treated him with a remedy derived from *Pelargonium* tubers, and upon his return to England Stevens was declared healthy.

He then set about commercializing a remedy based on the healer's preparation which he called 'Umckaloabo'. The industry gradually spread to the European continent, and once the identity of the plant material was established in 1974, renewed interest led to clinical trials which proved the efficacy of the herbal remedy for treating respiratory ailments. Since the 1980s, when Schwabe acquired the rights to sell Umckaloabo, it has been one of the company's best sellers, particularly in Germany, where the value of the local market for the product increased from €8 million in 2001 to €80 million in 2006 (Brendler and Van Wyk 2008).

For the past three decades, Schwabe has had a near-monopoly on the *Pelargonium* industry (see Figure 14.5). Until the end of 2011, the *Pelargonium* supply chain consisted of a large number of harvesters – the precise number has not been determined – resident in the Eastern Cape and Lesotho who sold the material to a handful of local buyers. In Lesotho, the main local buyer also did some

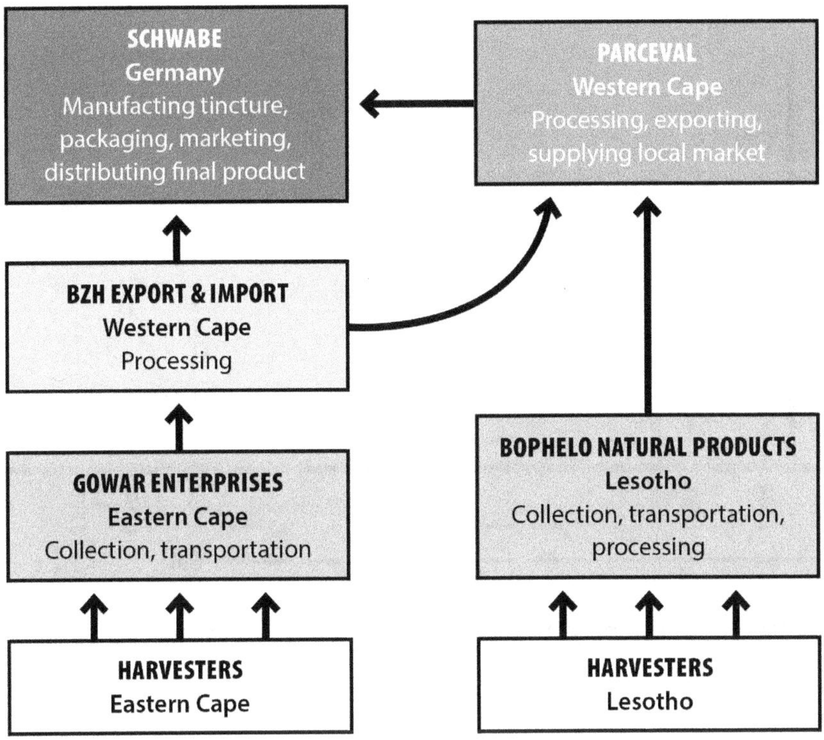

Figure 14.5 Pelargonium sidoides value chain, circa 2010.

rudimentary processing, whereas the local buyer in South Africa was responsible for transporting the raw material to an intermediary buyer in the Western Cape, where processing took place. Ultimately all material was shipped to Germany via Parceval, an erstwhile subsidiary of Schwabe based in the Western Cape. The majority of value-adding, and therefore profit, such as manufacturing the tincture, marketing and selling the final product, thus happened on German soil.

Challenging the patents

Schwabe's domination of the *Pelargonium* industry, strengthened by a number of *Pelargonium*-based patents held by the company, did not go unnoticed. Around 2007, a South African-based NGO, the African Centre for Biosafety (ACB), started drawing attention to the industry, expressing concern about the high rates of wild harvesting and the low returns received by harvesters, and accusing Schwabe of 'biopiracy' (ACB 2008). Together with representatives from a community in the Eastern Cape and the Berne Declaration, a Swiss NGO, the ACB challenged four Schwabe patents related to *Pelargonium* at the European Patent Office (EPO) in 2008 (ACB 2010a). Challenges were also launched by three competitors of

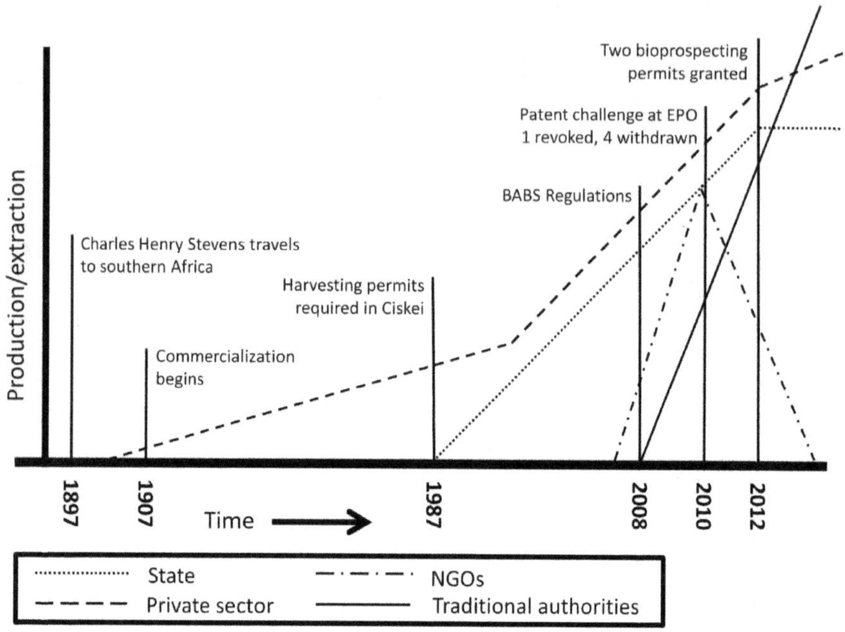

Figure 14.6 Timeline of *Pelargonium sidoides* commercialization.

Schwabe. Almost two years later, on 26 January 2010, the EPO revoked the patent for the method of extracting the substance used to make Umckaloabo on the conventional grounds of lack of inventive step[10] (ACB 2010b).

Three months later, Schwabe announced that it would voluntarily withdraw four other *Pelargonium* patents, stating that the company did not want to be drawn into the ongoing international debate around patents and biopiracy, an issue which the company ascribed to incompatibility between the CBD and the international patent system (Schwabe 2010).

Southern African states grapple with regulation

As demand for Umckaloabo rose towards the late 1990s, wild harvesting increased, raising concerns about sustainability. Provincial authorities responded by imposing limitations on harvesting permits, such as the requirement that a certain percentage of harvested material be replanted. Existing legislation (the Nature Conservation Act No. 10 of 1987, enacted in the former Ciskei, but remaining in force after the dissolution of the homelands) already required local buyers to apply for harvesting permits in the Ciskei area, although similar laws did not exist in other parts of the province. Permitting conditions were not adhered to, meaning that the majority of material was harvested illegally. Permit restrictions also had the effect of relocating the industry across the border to Lesotho, where escalating harvesting, most of it illegal, led to the listing of the species as protected in 2004 through an amendment (Legal Notice No. 93) to Lesotho's Historical Monuments, Relics, Fauna and Flora Act (No. 41 of 1967) (Newton et al 2009).

Concerns about continued illegal harvesting led authorities in the Eastern Cape to impose a temporary ban on wild harvesting in the province from 2007 to 2009. During this period, the Bioprospecting, Access and Benefit-Sharing Regulations (BABS Regulations) giving effect to the access and benefit-sharing provisions in the Biodiversity Act were promulgated in South Africa (Department of Environmental Affairs and Tourism [DEAT] 2008). Although the trade in *Pelargonium* raw material could be viewed as biotrade, or bulk trade in plant material, rather than bioprospecting, the broad definition for 'bioprospecting' in the Biodiversity Act meant that the industry would be subject to the conditions of the national access and benefit-sharing legislative framework (Van Niekerk and Wynberg 2012).

Provincial authorities were still responsible for granting harvesting permits, but only did so once users had obtained a bioprospecting permit from the national Department of Environmental Affairs (DEA), which hinged on their demonstrating sustainable use, material disclosure to stakeholders and evidence of prior informed consent and arrangements such as benefit-sharing and/or material transfer agreements concluded with providers to the satisfaction of the Minister of Environmental Affairs. With the introduction of the BABS Regulations, those already active in the industry were given time to align their activities to the regulations and submit their applications to the DEA. The ban on wild harvesting was thus partially lifted to give stakeholders with applications awaiting approval the opportunity to ply their trade (Van Niekerk and Wynberg 2012).

Figure 14.7 *Pelargonium sidoides* growing in the wild, Eastern Cape Province (photo: Jaci van Niekerk).

In South Africa, bioprospecting permits were granted in 2011 to companies that had concluded benefit-sharing agreements with various traditional leaders: Gowar Enterprises, formerly part of the Schwabe chain, and Essential Amathole, a newcomer to the industry. This was done on the understanding that harvesting permits would be issued to traditional leaders – via pre-existing development bodies or 'trusts' – on behalf of communities. Gowar Enterprises agreed to share benefits with the King Sandile Development Trust of the Rharhabe Kingdom in the Eastern Cape, while Essential Amathole concluded agreements with the Essential Amathole Community Trust, the community development arm of the Amabhele Traditional Authority. By the end of 2013, the bioprospecting permit applied for by Schwabe, in conjunction with its local partner, Parceval, after agreeing to share benefits with chiefs of the Rharhabe Kingdom in 2008, had not been approved – marking an end to the domination by the Schwabe supply chain.

Sustainability and equity in the **Pelargonium** *industry*

While these developments may seem to represent an important step towards greater sustainability and equity, questions remain about what they mean in practice. As a result of the guarded nature of the industry, the lack of monitoring and repeated shifts in wild harvesting between South Africa and Lesotho, calculating volumes of wild-harvested *Pelargonium* has been difficult, if not impossible (Newton et al 2009; Van Niekerk and Wynberg 2012). This scenario is set to change, however, as bioprospecting permits issued in terms of the BABS Regulations require that reports be submitted to the issuing authorities within one year of the permit being issued and annually thereafter. Once these reports are received, information on trade volumes, trade pathways and resource provenance will become known (Republic of South Africa [RSA] 2013). The DEA intends for sustainable wild harvesting to be attained by rotational harvesting. This means that each village under a certain chief's control will be given a chance to collect material in a ten-year cycle in order to give the tubers time to regenerate. Critics suggest this is unlikely to happen, as the lack of resources allows little monitoring and enforcement on the ground (A.P. Dold, curator, Selmar Schonland Herbarium, Grahamstown, personal communication, 2012).

Equity outcomes are equally ambiguous. In many ways the *Pelargonium* industry typifies the historically unequal relationship between Southern providers of NTFPs and Northern users, which persists despite decades of attention to this inequality via instruments such as the CBD. Examination of actors in the *Pelargonium* value chain reveals that rural harvesters are reaping limited benefits from their engagement with the industry. On the one hand the structure of the industry, in particular the near-monopoly of the Schwabe supply chain, has meant that harvesters have been unable to negotiate better returns for their labour, earning between ZAR 2 and ZAR 4 (USD 0.2 and USD 0.4) per kilogram of tubers – a price set by the industry. In addition, harvesters have limited access to markets, credit and technology, and unclear rights of access to the resource, and thus remain unable to establish enterprise ownership and improve their prospects (Van Niekerk and Wynberg 2012).

Driven by the desire to alleviate rural poverty and address inequalities, the DEA has turned to the implementation of access and benefit sharing as a solution. Regrettably, early attempts at doing so, as witnessed in the *Pelargonium* industry, have not accomplished these goals. In fact, the need for benefit-sharing agreements has stimulated the interest of the ruling elite in rural communities, which does not bode well for harvesters without close ties to traditional leaders. This form of elite capture may well have been driven by the industry and government themselves, albeit unintentionally, since both sets of actors are in favour of dealing with traditional leaders or established groups because they see it as 'easier to work with an overarching structure rather than individual communities' (industry representative, personal communication, 2009), and believe that 'chiefs will take care of their communities' (DEA 2010).

Prior to the introduction of access and benefit sharing, harvesting permits were obtained by local buyers who dealt directly with harvesters, but the permitting process is now routed via the chiefdoms. While this might not affect the price harvesters receive, it does mean that any residents of the former homelands who do not accept the authority of traditional leaders are forced to defer to those leaders in order to participate in the industry (Morris 2012; Van Niekerk and Wynberg 2012). Indeed, pressing questions remain about the validity of traditional authorities who are not universally accepted and have been criticized for executing their roles in an undemocratic, unaccountable and inequitable manner, often placing the interests of individuals before community interests (Ntsebeza 2002; Logan 2009).[11]

Like *Hoodia*, the *Pelargonium* case presents valuable lessons for the unfolding process of access and benefit sharing, especially with respect to the complexities of regulating both for sustainability and for equity. Unlike *Hoodia*, though, providers in this case have not been empowered. In fact, the opposite is true, as local harvesters have been disregarded in terms of participation and benefit sharing, with ruling elites prominent. Issues of identity and leadership are not easily resolved by access and benefit sharing, which is an approach designed for commercial agreements. Moreover, the contested nature of traditional leadership and concomitant land tenure questions in South Africa are much larger issues that require debate in many forums before any kind of satisfactory solutions can be determined.

Discussion

An examination of the trade chains of these two commercially valuable southern African resources yields valuable insights into the way in which the state, the private sector, NGOs, rural communities and other relevant institutions engage in and respond to the development of a natural product and changes in its supply and demand. There are a number of differences between the *Hoodia* and *Pelargonium* industries in their governance arrangements, levels of equity and approaches to sustainability (Table 14.1), yet their congruencies allow for useful cross-analyses.

Table 14.1 Comparative table of the two case studies

	Hoodia gordonii	*Pelargonium sidoides*
Biological characteristics	Spiky stems are utilized.	Underground tubers are utilized; can possibly be sustainably harvested, but tubers take eight to nine years to reach harvestable size.
Harvesting	Initially wild-harvested, but since 2007 cultivated by commercial farmers.	Mostly harvested from the wild by rural poor, but cultivation under way by commercial farmers in the Free State Province and Schwabe-funded operations in Kenya and Mexico.
Sustainability	Listed in CITES Appendix II in 2004. The ease of cultivating *Hoodia* and a decline in market interest have reduced threats from wild harvesting.	Little cultivation in southern Africa. Wild harvesting raises concerns about sustainability: listed as threatened in Lesotho in 2004, South African resource assessment declares it not threatened, with localized overharvesting in 2010.
Traditional knowledge	Acknowledged as the foundation of product development, but only after NGO intervention. Benefit sharing is based on use of traditional knowledge rather than provision of resources.	Traditional knowledge exists widely and was indeed used by Stevens in early commercialization, but is not explicitly linked to products and is not given recognition in the benefit-sharing agreements concluded thus far.
Market	Peaked in 2002 and 2003, but fake products, inconclusive clinical trials and a lack of interest from business partners have caused market stagnation.	Market has steadily grown since the 1990s. Schwabe has plans to roll out to other countries, but the company's near-monopoly has diminished since early 2012, when new market entrants were awarded bioprospecting permits.
Patents	The CSIR and Phytopharm hold a number of patents related to *Hoodia's* appetite-suppressant properties.	Schwabe held a number of patents, one revoked in 2010. Subsequently four were withdrawn, but others remain.
Equity	Benefit-sharing agreements developed between the San and the CSIR, and South African and Namibian *Hoodia* growers. Whether or not these are equitable is open to debate, but the slump in trade makes this a largely moot point.	It is likely that two thirds of monies collected will be retained by the King Sandile Development Trust, with each chiefdom allocated a percentage of remaining funds. However, there have been claims of a lack of transparency in earlier cases, raising questions of fairness.

The *Hoodia* case reveals interesting changes to the cycles of governance that occur with the commercialization of a high-value species associated with traditional knowledge. State intervention was initially triggered when an NGO lobbied for the San to be acknowledged for their traditional knowledge. As the value of the plant increased in the early stages of commercialization and overharvesting became a problem, state intervention strengthened, in response to peaked commercial interest and declines in the availability of the resource. As cultivation emerged as a viable option, the role of the state in regulating use tapered off and the industry became more self-regulated.

In contrast, while the state was initially not interested in questions of equity and benefit sharing, its agency in these issues became increasingly prominent – ironically at a time when the benefits were dwindling. NGOs were initially responsible for catalysing changes towards greater equity in the commercialization process, and have remained involved in supporting the San with capacity development and negotiation. The approach of industry has also changed, from one of aloofness from equity considerations, to one based on a perception that benefit-sharing agreements are a vital part of securing regulatory approval and niche markets.

Although *Pelargonium* has a much longer history of commercial use than *Hoodia*, rapid escalation in demand for the raw material prompted a similar response from the state. Initially reacting to fears of overexploitation, environmental regulators in South Africa tightened permitting conditions for harvesting, and, with the advent of legislation on access and benefit sharing, became increasingly heavy-handed. In Lesotho, wild harvesting increased as a direct result of stricter regulation in South Africa, leading to tighter controls. In 2009, a requirement for harvesting permits to provide evidence of benefit sharing precipitated increased involvement of the state in equity deliberations. Governance of the *Pelargonium* industry has thus become increasingly state-centric, despite the long history of commercialization.

Unease about inequality in the supply chain also troubled NGOs, as did notions that the patents held by Schwabe had misappropriated traditional knowledge. Whether or not this was the case was never fully explored, but the successful challenge of the Umckaloabo process patent nevertheless served to highlight the power imbalances inherent in the value chains of commercially successful NTFPs. Following the patent challenge, and with benefit sharing a prerequisite for trading *Pelargonium*, the structure of the industry altered dramatically. Schwabe's 30-year near-monopoly, sustained by both patents and control over the supply chain, came to an end when bioprospecting permits were allocated to entities outside the remaining partners in the chain.

Despite attempts to bring about equity and secure wider benefits through the sustainable use of *Pelargonium*, the lot of the rural community members who engage in the trade has remained largely unchanged. In fact, for those who do not acknowledge traditional leadership, the opportunity to collect *Pelargonium* to supplement their livelihoods is compromised, as both government and industry have identified traditional authorities as the legitimate 'resource and access providers' envisioned by the CBD. The agency of traditional authorities has thus strengthened alongside increased state intervention, but not necessarily with more equitable outcomes for harvesters. Further confusing the picture in these rural areas is the effect of plural, overlapping and, in some cases, contradictory governance regimes between customary and statutory systems.

These findings provide pointers as to the way in which the governance of natural product value chains is unfolding, in a context that increasingly incorporates sustainable use and equity as integral parts of the policy landscape. The private sector, for example, is well known for its pioneering nature and ability to identify market opportunities, its lead in the initial commercialization and later cultivation of a product and the way in which it thrives and declines in parallel with supply and demand trends. As specific industries evolve and mature, the private sector typically becomes more able to influence government policy and more likely to regulate itself. While this approach may work for cultivated and managed resources, questions remain as to how far the private sector can be assigned to self-regulate wild-harvested resources, or to determine equity along the chain.

The crucial role played by NGOs in the governance of natural product value chains has been well demonstrated and is likely to become stronger in the face of dwindling government capacities and in a policy space that is increasingly unclear, uncoordinated and confusing. NGO involvement typically lags behind that of the private sector and initially corrects for market failures by bringing in neglected social and environmental dimensions and lobbying for change. As the *Hoodia* case demonstrates, this can range from actively campaigning against biopiracy to supporting communities in negotiating conditions and contracts and catalysing market and policy interest as activities become more established. An increased global emphasis on equity and benefit sharing is likely to enhance this role, especially among public interest organizations that provide legal support to communities.

The role of the state, commonly recognized as a key factor influencing natural product development (e.g. Dove 1995; Schreckenberg 2003; Laird et al 2009) is similarly changing with greater scrutiny of resource sustainability and equity. The state has typically been reactive and interventionist in the early stages of commercialization, initially responding to peaked commercial interest and declines in resource availability through policy measures that regulate resource access. This role has often tapered off as the resource has become better managed, generally through increased cultivation (Wynberg 2006).

The introduction of regulations on access and benefit sharing, however, is set to change this pattern, signalling a new era of increased state intervention in determining who gets what benefits, and how this occurs. This in turn raises the question: does access and benefit sharing promote or hinder equity and sustainability? The requirement for benefit-sharing agreements in South Africa's Biodiversity Act, for example, has generated its own suite of complications. While the inclusion of prior informed consent and benefit sharing in this law represents a major step forward in redressing past imbalances in the way in which biodiversity and traditional knowledge have been exploited, implementation presents major challenges (Crouch et al 2008; Taylor and Wynberg 2008). Because the Biodiversity Act fails to vest ownership of genetic resources in the state, an approach has been pursued that looks to bilateral arrangements between specific communities or individuals and companies as the *modus operandi* for securing benefit sharing (see Taylor and Wynberg 2008). As the two case studies explain, singling out 'access providers' or 'knowledge providers' with whom to conclude benefit-sharing agreements is highly contentious and problematic, in current practice encouraging elite capture rather than establishing equity.

The examination of the trade chains of two high-value NTFPs from southern Africa – each industry having its own nuances and determinants, each species and production system its own ecological characteristics, each product its own particularities and each actor its own political and cultural configurations and dynamics – reveals the realities that result when global ideals such as the equity and sustainable use envisioned by the CBD play out on the ground. Interventions initiated by the state, transformations in the private sector and mediations offered by NGOs offer great promises of benefit distribution, sustainable resource use and poverty alleviation, yet these initiatives may well get caught up in a policy snare that is fundamentally self-serving, and ultimately detrimental to rural producers and traditional knowledge holders.

With demand for natural products set to rise in future, it is vital that decision-makers take heed of the need for tailored solutions that are better informed, reflect the heterogeneity of the actors and are rooted in realities on the ground that seek to reduce inequality and achieve ecological sustainability.

Acknowledgements

We warmly thank the resource and knowledge providers who contributed to this research, as well as key informants who provided insights. This work is based on research supported by the South African Research Chairs Initiative of the Department of Science and Technology and National Research Foundation of South Africa. Any opinions, findings, conclusions or recommendations set out in this material are those of the authors and not necessarily those of the NRF.

Notes

1 Meaning 'all biological materials other than timber which are extracted from forests' (De Beer and McDermott 1989).
2 'Genetic resources' is defined by the CBD as 'genetic material of actual or potential value', and 'biological resources' includes 'genetic resources, organisms or parts thereof, populations, or other biotic components of ecosystems with actual or potential use or value for humanity' (United Nations Environment Programme [UNEP] 1992).
3 The CBD is a leading international treaty on biodiversity. It came into force in 1993, is legally binding and has 193 parties. Its three objectives are: the conservation of biological diversity, the sustainable use of its components, and the fair and equitable sharing of benefits arising out of the utilization of genetic resources. The Nagoya Protocol is a supplementary agreement to the CBD providing a legal framework for the implementation of the CBD's third objective. It will enter into force 90 days after its fiftieth ratification, whereupon obligations in terms of access to resources and traditional knowledge, benefit sharing and compliance will become a legal requirement for parties.
4 http://fairwild.org
5 'Biotrade' refers to the commercial collection, processing and sale of products derived from biodiversity, usually in large volumes.
6 'Biopiracy' refers to the misappropriation of genetic resources or traditional knowledge through the patent system, and/or the unauthorized collection for commercial ends of genetic resources or traditional knowledge (Dutfield 2002).
7 Provincial legislation in South Africa – the Environmental Conservation Ordinance No. 19 of 1974 in the Northern Cape Province and Nature Conservation Ordinance 19 of 1974 in the Western Cape Province – lists *Hoodia* species as protected, requiring a permit for certain activities related to the plants.
8 Exchange rate as at June 2013: USD 1 = ZAR 9.99

9 The apartheid state demarcated certain predominantly rural areas, such as the Transkei and Ciskei, as 'homelands', or nominal states where the black majority were concentrated and to which they were given obligatory citizenship.

10 According to patent law, for an invention to be patentable, it must be novel, non-obvious or involve an inventive step, and it must have useful application.

11 When the apartheid government, which was in power from 1948 until 1994, established the homelands, traditional leaders were given land administration roles and had uncontested powers in rural areas. The legitimacy of these leaders remains a fiercely debated issue. In the Ciskei, for example, where no formally recognized tribal authorities existed at the time of its declaration as a homeland, chiefdoms such as the Rharhabe were 'resuscitated' by apartheid rulers (Peires 1989).

Bibliography

ACB (2008) 'Knowledge not for sale: Umckaloabo and the Pelargonium patent challenges', briefing paper, African Centre for Biosafety, Johannesburg

ACB (2010a) 'Biopiracy under fire: The Pelargonium patent hearing', briefing paper, January, African Centre for Biosafety, Johannesburg

ACB (2010b) 'Joy as Pelargonium patent revoked', press release, African Centre for Biosafety, Johannesburg, 26 January, http://acbio.org.za/index.php/media/64-media-releases/260-joy-as-pelargonium-patent-revoked

Anderson, A. B. (1992) 'Land-use strategies for successful extractive economies in Amazonia', in D. C. Nepstad and S. Schwartzman (eds) *Advances in Economic Botany 9: Non-Timber Product Extraction from Tropical Forests: Evaluation of a Conservation and Development Strategy*, New York Botanical Garden, pp. 67–77

Brendler, T. and van Wyk, B. E. (2008) 'A historical, scientific and commercial perspective on the medicinal use of *Pelargonium sidoides* (Geraniaceae)', *Journal of Ethnopharmacology*, vol 119, pp. 420–33

CITES (2004) Amendments to Appendices I and II of CITES, proposal to the 13th meeting of the Conference of the Parties, Bangkok, Thailand, 2–14 October

Cronkleton, P., Taylor, P. L., Barry, D., Stone-Jovicich, S. and Schmink, M. (2008) 'Environmental governance and the emergence of forest-based social movements', Occasional Paper No. 49, Center for International Forestry Research, Bogor, Indonesia

Crouch, N., Douwesa, E., Wolfson, M., Smith, G. F. and Edwards, T. J. (2008) 'South Africa's bioprospecting, access and benefit-sharing legislation: Current realities, future complications, and a proposed alternative', *South African Journal of Science*, vol 104, no 9/10, pp. 355–66

DEA (2010) Biotraders' workshop on the Bioprospecting, Access and Benefit Sharing Regulations, 5 August, Department of Environmental Affairs, Cape Town

DEAT (2008) 'National Environmental Management: Biodiversity Act – Regulations on Bio-prospecting, Access and Benefit Sharing', Government Notice R138, *Government Gazette* 30739, Department of Environmental Affairs and Tourism, Pretoria

De Beer, J. H. and McDermott, M. J. (1989) *The Economic Value of Non-Timber Forest Products in Southeast Asia with Emphasis on Indonesia, Malaysia and Thailand*, Netherlands Committee for IUCN, Amsterdam

Dove, M. R. (1995) 'Political versus techno-economic factors in the development of non-timber forest products: Lessons from a comparison of natural and cultivated rubbers in Southeast Asia (and South America)', *Society and Natural Resources*, vol 8, pp193–208

Dutfield, G. (2002) 'Sharing the benefits of biodiversity: Is there a role for the patent system?', *Journal of World Intellectual Property*, vol 5, no 6, pp. 899–932

Dutfield, G. (2011) *Intellectual Property Tools for Products Based on Biocultural Heritage: A Legal Review of Geographical Indications, Trademarks and Protection from Unfair Competition*, International Institute for Environment and Development, London

'An English Physician' (1931) *Tuberculosis: Its Treatment and Cure with the Help of Umckaloabo (Stevens)*, B Fraser & Co, London

Gibbon, P. and Ponte, S. (2005) *Trading Down: Africa, Value Chains and the Global Economy*, Temple University Press, Philadelphia, PA

GPN (2006) *Highlights of Current Labor Market Conditions in Lesotho*, Global Policy Network

Harilal, K. N., Kanji, N., Jeyaranjan, J., Eapen, M. and Swaminathan, P. (2006) 'Power in global value chains: Implications for employment and livelihoods in the cashew nut industry in India', International Institute for Environment and Development, London, http://dlc.dlib.indiana.edu/dlc/handle/10535/6180

Hayden, C. (2003) 'From market to market: Bioprospecting's idioms of inclusion', *American Ethnologist*, vol 30, no 3, pp. 359–71

Homer-Dixon, T., Boutwell, J. and Rathjens, G. (1993) 'Environmental change and violent conflict', *Scientific American*, vol 268, pp. 38–45

Homma, A. K. O. (1992) 'The dynamics of extraction in Amazonia: A historical perspective', in D. C. Nepstad and S. Schwartzman (eds) *Advances in Economic Botany 9: Non-Timber Product Extraction from Tropical Forests: Evaluation of a Conservation and Development Strategy*, New York Botanical Garden, pp. 23–31

Homma, A.K.O. (1996) 'Modernisation and technological dualism in the extractive economy of Amazonia', in M. Ruiz-Pérez and J. E. M. Arnold (eds) *Current Issues in Non-Timber Forest Products Research*, Centre for International Forestry Research, Bogor, Indonesia, pp. 59–82

Kaimowitz, D. (2003) 'Not by bread alone . . . Forests and rural livelihoods in sub-Saharan Africa', in T. Oksanen, B. Pajari, T. Tuomasjukka (eds) *Forestry in Poverty Reduction Strategies: Capturing the Potential*, EFI Proceedings No. 47, European Forest Institute, Joensuu, Finland, pp. 45–64

King, S. R., Meza, E. N., Carlson, T. J. S., Chinnock, J. A., Moran, K. and Borges, J. R. (1999) 'Issues in the commercialisation of medicinal plants', *Herbalgram*, vol 47, pp. 46–51

Koelble, T. A. and LiPuma, E. (2011) 'Traditional leaders and the culture of governance in South Africa', *Governance*, vol 24, no 1, pp. 5–29

Laird, S. A. and Wynberg R. (2012) 'Diversity and change in the commercial use of genetic resources: Implications for access and benefit-sharing policy', special issue: Socio-economics and management of bioprospecting, *International Journal of Ecological Economics and Statistics*, vol 26, no 3

Laird, S. A., Wynberg R. and McLain, R. J. (2009) 'Wild product governance: Laws and policies for sustainable and equitable non-timber forest product use', policy brief, a collaboration of United Nations University; Centre for International Forestry Research; People and Plants International; Environmental Evaluation Unit, University of Cape Town; and the Institute for Culture and Ecology

Lele, S., Pattanaik, M. and Rai, D. N. (2010) 'NTFPs in India: Rhetoric and reality', in S. Laird, R. J. McLain and R. P. Wynberg (eds) *Wild Product Governance: Finding Policies that Work for Non-Timber Forest Products*, Earthscan, London

Logan, C. (2009) 'Selected chiefs, elected councillors and hybrid democrats: Popular perspectives on the co-existence of democracy and traditional authority', *Journal of Modern African Studies*, vol 47, no 1, pp. 101–28

Morris, C. (2012) 'Pharmaceutical bioprospecting and the law: The case of Umckaloabo in a former apartheid homeland of South Africa', *Anthropology News*, 3 December,

American Anthropological Association, http://www.anthropology-news.org/index.php/2012/12/03/pharmaceutical-bioprospecting-and-the-law/

Neumann, R. and Hirsch, E. (2000) *Commercialisation of Non-Timber Forest Products: Review and Analysis of Research*, Center for International Forestry Research, Bogor, Indonesia

Newton, D., Letsela, T., Lijane, T., Mafatle, N., Manyama, P., Naha, S., Ntloko, B., Ntoshi, R., Paetzold, B., Pires, A., Polaki, M., Raimondo, D., Rouget, M., T'sele T., Wistebaar, N. and Zimudzi, C. (2009) 'A non-detriment finding for *Pelargonium sidoides* (DC) in the Kingdom of Lesotho', draft document prepared as a contribution to a Regional BMP (biodiversity management plan) for *Pelargonium sidoides*

Novellino, D. (2010) 'From indigenous customary practices to policy interventions: The ecological and socio-cultural underpinnings of the non-timber forest trade on Palawan Island, the Philippines', in S. Laird, R. J. McLain and R. P. Wynberg (eds) *Wild Product Governance: Finding Policies That Work for Non-Timber Forest Products*, Earthscan, London

Ntsebeza, L. (2002) 'Decentralisation and natural resource management in rural South Africa: Problems and prospects', paper presented to the 9th Conference of the International Association for the Study of Common Property, 17–21 June, Victoria Falls, Zimbabwe

Pappe, L. (1862) *Silva Capensis: A Description of South African Forest Trees and Arborescent Shrubs Used for Technical and Economical Purposes*, 2nd ed., Ward & Co, London

Peires, J. B. (1989) 'Ethnicity and pseudo-ethnicity in the Ciskei', in L. Vail (ed.) *The Creation of Tribalism in Southern Africa*, James Currey, London, and University of California Press, Berkeley and Los Angeles, CA, pp. 395–413

Ribot, J. (1995) 'From exclusion to participation: Turning Senegal's forest policy around?', *World Development*, vol 23, no 9, pp. 1,587–99

Rijkers, T., Ogbazghi, W., Wessel, M. and Bongers, F. (2006) 'The effect of tapping for frankincense on sexual reproduction in *Boswellia papyrifera*, *Journal of Applied Ecology*, vol 43, no 6, pp. 1,188–95

Robins, S. (2002) 'NGOs, 'bushmen', and double vision: The Khomani San land claim and the cultural politics of "community" and "development" in the Kalahari', in T.A. Benjaminsen, B. Cousins, and L. Thompson (eds) *Contested Resources: Challenges to the Governance of Natural Resources in South Africa*, Programme for Land and Agrarian Studies, School of Government, University of the Western Cape, Cape Town, pp. 208–27

RSA (2013) 'National Environmental Management: Biodiversity Act (10/2004): Biodiversity Management Plan for *Pelargonium sidoides* in South Africa 2011–2020', *Government Gazette* No. 36411, Pretoria, 26 April

Schippmann, U., Leaman, D. J. and Cunningham, A. B. (2002) 'Impact of cultivation and gathering of medicinal plants on biodiversity: Global trends and issues', in *Biodiversity and the Ecosystem Approach in Agriculture, Forestry and Fisheries*, satellite event on the occasion of the ninth regular session of the Commission on Genetic Resources for Food and Agriculture, Rome, 12–13 October

Schreckenberg, K. (2003) 'Appropriate ownership models for natural product-based small and medium enterprises in Namibia', prepared for the Trade and Investment Development Programme of the Ministry of Trade and Industry, Namibia

Schwabe (2010) 'Umckaloabo in discussion: Current position and sustainability perspectives', media release, Schwabe Pharmaceuticals, 26 April

Shackleton, C. and Shackleton, S. (2004) 'The importance of non-timber forest products in rural livelihood security and as safety nets: A review of evidence from South Africa', *South African Journal of Science*, vol 100, no 11/12, pp. 658–64

Shackleton, S., Delang, C. O. and Anglesen, A. (2011) 'From subsistence to safety nets and cash income: Exploring the diverse values of non-timber forest products for livelihoods and poverty alleviation', in S. Shackleton, C. Shackleton and P. Shanley (eds) *Non-Timber Forest Products in the Global Context*, Springer-Verlag, Berlin, pp. 55–82

Shanley, P., Pierce, A., Laird, S. and Robinson, D. (2008) *Beyond Timber: Certification of Non-Timber Forest Products*, Center for International Forestry Research, Bogor, Indonesia

Southgate, D., Coles-Ritchie, M. and Salazar-Canelos, P. (1996) 'Can tropical forests be saved by harvesting non-timber products? A case study for Ecuador' in W. L. Adamowicz (ed.) *Forestry, Economics and the Environment*, CAB International, Wallingford, UK, pp. 68–80

Sunderland, T. C. H., Harrison, S. T., and Ndoye, O. (2004) 'Commercialisation of non-timber forest products in Africa: History, context and prospects' in T. Sunderland and O. Ndoye (eds) *Forest Products, Livelihoods and Conservation: Case Studies of Non-Timber Forest Product Systems – Volume 2: Africa*, Center for International Forestry Research, Bogor, Indonesia, pp. 1–24

Taylor, M. and Wynberg, R. (2008) 'Regulating access to South Africa's biodiversity and ensuring the fair sharing of benefits from its use', *South African Journal of Environmental Science and Policy*, vol 15, no 2, pp. 217–43

Ticktin, T. (2004) 'The ecological consequences of harvesting non-timber forest products', *Journal of Applied Ecology*, vol 41, pp. 11–21

Ticktin, T. and Shackleton, C. (2011) 'Harvesting non-timber forest products sustainably: Opportunities and challenges', in S. Shackleton, C. Shackleton and P. Shanley (eds) *Non-Timber Forest Products in the Global Context*, Springer-Verlag, Berlin, pp. 149–69

Timko, J. A., Waeber, P. O. and Kozak, R. A. (2010) 'The socio-economic contribution of non-timber forest products to rural livelihoods in Sub-Saharan Africa: Knowledge gaps and new directions', *International Forestry Review*, vol 12, no 30, pp. 284–94

UNEP (1992) *Convention on Biological Diversity: Text and Annexes*, United Nations Environment Programme, Chatelaine, Switzerland

Van Niekerk, J. and Wynberg, P. (2012) 'The trade in *Pelargonium sidoides*: Rural livelihood relief or bounty for the "bio-buccaneers"?', *Development Southern Africa*, vol 29, no 4, pp. 530–47

Westaway, A. (2012) 'Rural poverty in the Eastern Cape Province: Legacy of apartheid or consequence of contemporary segregationism?', *Development Southern Africa*, vol 29, no 1, pp. 115–25

White, A. and Sloane, B. L. (1937) *The Stapelieae III*, 2nd ed., Abbey San Encino Press, Pasadena, CA

Wynberg, R. (2004) 'Rhetoric, realism and benefit-sharing: Use of traditional knowledge of *Hoodia* species in the development of an appetite suppressant', *World Journal of Intellectual Property*, vol 6, no 7, pp. 851–76

Wynberg, R. (2006) 'Identifying pro-poor, best practice models of commercialisation of southern African non-timber forest products', PhD thesis, University of Strathclyde, Glasgow

Wynberg, R. (2010) 'Navigating a way through regulatory frameworks for *Hoodia* use, conservation, trade and benefit sharing', in S. Laird, R. McLain and R. Wynberg (eds) *Wild Product Governance: Finding Policies That Work for Non-timber Forest Products*, Earthscan, London, pp. 309–26

Wynberg, R. and Chennells, R. (2009) 'Green diamonds of the South: An overview of the San- *Hoodia* case', in R. Wynberg, D. Schroeder and R. Chennells (eds) *Indigenous Peoples, Consent and Benefit Sharing: Lessons from the San-Hoodia Case*, Springer, London, pp. 89–126

Wynberg, R. P. and Laird, S. L. (2007) 'Bioprospecting: Tracking the policy debate', *Environment*, vol 49, no 10, pp. 20–32

Wynberg, R., Schroeder, D., Williams, S. and Vermeylen, S. (2009) 'Sharing benefits fairly: Decision-making and governance', in R. Wynberg, D. Schroeder and R. Chennells (eds) *Indigenous Peoples, Consent and Benefit Sharing: Lessons from the San-Hoodia Case*, Springer, London, pp. 231–60

15 Governing the ungovernable?

Climate change and social justice in southern Africa

Penny Urquhart

Introduction

The threats posed by global climate change are severe, and are already impacting on natural resources, food security, human health, the environment, economic activity and physical infrastructure. The quest for social justice is central to international human rights, development and environmental conventions and the global issues they aim to govern. This is nowhere more so than for climate change: massive questions of distributional justice between countries now, and of equity intergenerationally, result from what has been termed the biggest and widest-ranging market failure ever seen (Stern 2006).

While the concepts of 'equity' and 'common but differentiated responsibilities' in the response to climate change do lie at the heart of the United Nations Framework Convention on Climate Change (UNFCCC), resolving these dilemmas may seem impossible in practice, particularly given the stagnation at the level of international negotiations. The problem is so complex, so interlinked, so wide-ranging, so global and yet so locally specific, that we may wonder whether we are attempting to govern what may, in fact, be ungovernable.

Our failure to respond adequately to climate change is foreclosing options for future generations: as we hurtle towards dangerous climate change with annually increasing emissions of greenhouse gases, changes in weather patterns are accelerating and the relative climate stability we have enjoyed for centuries is now a thing of the past (Solomon et al 2007; Anderson and Bows 2008). There can be nothing just about condemning future generations to climatic mayhem.

Sobering as the question of treating future generations equitably is, it should not deflect us from examining current climate-justice-development nexuses, which include human rights questions, responsibility for past emissions, links between poverty and climate change and responsibility for financing adaptation to climate change (Huq et al 2006; Gardiner 2010; Arora-Jonsson 2011; Moellendorf 2012). It is where existing vulnerabilities and a high degree of climate risk come together, such as in the marginalized dryland areas of southern Africa, that we may most need to interrogate the fairness of evolving responses to the climate challenge. What are the key equity issues associated with climate change in southern Africa, and how will natural resource management and governance systems need to be modified to address these?

This chapter does not attempt to set out a comprehensive framework for evolving effective and equitable governance systems for managing climate change, but rather aims to provide an initial exploration of these questions. Firstly, the context of climate change risks and multiple livelihood stresses that act upon existing high levels of vulnerability in southern Africa is set. This leads to a discussion of key equity issues in climate change and some reflections on existing governance regimes in the region. The focus then shifts to practical and policy issues for the two main areas of adaptation and mitigation, in which some of the emerging hazards on the route to a fair and just response are sketched out. In conclusion, some ways forward are explored, which include new approaches and orientations in the way we go about the multi-level processes of adaptation and mitigation.

Southern Africa's vulnerability to climate change

Southern Africa is extremely vulnerable[1] to the temperature and precipitation changes linked to climate change, according to the Intergovernmental Panel on Climate Change (IPCC). Nearly all climate models used in the IPCC Fourth Assessment Report show overall drying for southern Africa, with increased rainfall variability. Subsequent multi-model analysis for the region projects a delay in onset of the rainy season, an early cessation in many parts, a decrease in mean summer rainfall in some areas and an increase in rainfall intensity in others (Conway 2011).

Analysis for southern Africa carried out by the Climate Systems Analysis Group at the University of Cape Town found that the median (average) of 21 model simulations[2] comparing the period 1980–1999 with the period 2080–99 projected the following changes:

- a 3.4 °C increase in annual temperature (up to 3.7 °C in spring)
- a 23 per cent decrease in winter rainfall
- a 13 per cent decrease in spring rainfall.

However, these changes will not be experienced uniformly across the region: for instance, the western parts of South Africa are projected to become drier while the eastern parts become wetter, and central Botswana is likely to warm up much faster than surrounding areas. The warming over Africa is very likely to be larger than globally, and the observed temperature changes for southern Africa are higher than the increases reported for other parts of the world (Solomon et al 2007). In most parts of eastern and southern African, potential evaporation is almost twice as high as rainfall totals; thus less than 15 per cent of rainfall contributes to rivers and groundwater (Mwendera 2010).

There is growing evidence of the interactions between current climate variability and extreme weather events and long-term climate change (Field et al 2012). In southern Africa, despite the paucity of observational data (Tadross et al 2009), there is medium confidence that droughts will intensify in the twenty-first century, due to reduced precipitation and/or increased evapotranspiration (Field et al 2012).

Recent analyses indicate that two thirds of the countries most vulnerable to climate variability and change are in Africa. Southern Africa's vulnerability is considered to be high due to the large number of people who make their living off the land, as well as to underlying causes such as structural poverty, combined with high climate risks (Darkoh 2009; Midgley et al 2011). A recent exercise to map current and future climate-related vulnerability in southern Africa found a current band of high exposure lying between 12° S and 25° S, which will, it is projected, extend south to the 30° S latitude and into the north-western parts of the region by 2050 (Midgley et al 2011). This analysis found that vulnerability to climate impacts would intensify in the following areas: eastern and northern Angola, parts of the Democratic Republic of the Congo, southern Malawi, the Highveld of South Africa, parts of Madagascar and southern and western Zambia (Midgley et al 2011). A key factor in this water-stressed region is the impact of climate change on water availability and access: as Schulze (2007) points out, water poverty is already acute in many mesoscale catchments and will in all likelihood be exacerbated by climate change.

Climate change both introduces new sources of risk and acts as a risk multiplier, for example by exacerbating existing food insecurity or placing additional stress upon already degraded ecosystems. Key risks relate to increased average temperature, heightened intensity and/or the frequency of extreme events such as floods and droughts, more severe heat and water stress, rising sea levels and landslides and soil erosion. The sectors upon which the livelihoods and food security of rural people depend are the very ones most vulnerable to climate change: agriculture, forestry (including non-timber forest products) and fisheries, including water-related aspects of these sectors.

There is a growing understanding of urban vulnerability to climate change in the region too, key factors being the rapid urban growth over the past three decades, resulting in large proportions of inhabitants living in informal settlements with high levels of vulnerability to extreme events such as flooding, and low levels of services. The movement of people from the rural areas has not necessarily reduced climate vulnerability: as Chapman and Sasman (2012: iv) note, 'Urbanisation has largely replaced one type of climate vulnerability experienced in rural areas with another.'

Climate change threatens development gains, makes sustainable natural resource management increasingly difficult, and is likely to further entrench poverty, in the absence of appropriate and sufficient responses (Parry et al 2007; Shackleton and Shackleton 2012). While climate change impacts are locally specific, the effects where many livelihoods are directly dependent on rainfed agriculture and natural resources will include an increased likelihood of crop failure; an increase in diseases and mortality and forced sales of livestock and other assets; increased livelihood insecurity, indebtedness, migration and dependency on food aid; and a downward spiral in human development indicators such as health and education (Parry et al 2007; Midgley et al 2011).

Negotiations under the UNFCCC have centred on keeping the average global temperature to no more than 2 °C above pre-industrial levels, although it is now widely held that a safer goal is 1.5 °C, equivalent to 350 parts per million (ppm) of atmospheric carbon dioxide (Hansen et al 2008; Anderson and Bows 2008; Rockström et al 2009). If we reach the two-degree boundary, between 350 million and 600 million Africans will suffer increased water scarcity (Boko et al 2007). If the global temperature increases by 3 °C to 4 °C, changed run-off patterns and glacial melt could force an additional 1.8 billion people to live in a water-scarce environment by 2080 (United Nations Development Programme [UNDP] 2008a). James Hansen, a pre-eminent climate scientist formerly attached to the USA's National Aeronautics and Space Administration, has stated that stabilizing emissions at 450 ppm will not, in fact, limit the increase in temperatures to two degrees above pre-industrial levels – still the goal of the UNFCCC negotiating process – but will instead result in an increase of six degrees this century (Hansen et al 2008). This takes us into uncharted territory for projections of climate impacts.

A particularly severe climate impact on humanity in the near future is likely to be an increase in hunger (Oxfam 2009; Ericksen et al 2011). By 2050, under what we must now consider to be optimistic scenarios, the number of people at risk of hunger is expected to increase by 10 to 20 per cent more than would be expected without climate change, with at least 24 million additional malnourished children, and the worst impacts would be in sub-Saharan Africa (Parry et al 2007). This is of grave concern in southern Africa, with its widespread food insecurity, high burden of disease and rapidly increasing population densities in areas of high agricultural productivity, which correspond with areas of high vulnerability to current and future climatic factors (Osbahr et al 2010; Midgley et al 2011).

Notwithstanding progress made since the 1990s, figures from the 2004–6 period reveal that approximately 95 million people in southern Africa, or 40 per cent of the population, are undernourished (De Wit and Midgley 2012). Given that food production in the region is already greatly influenced by high degrees of climatic unpredictability and extreme events like droughts and floods, there are serious concerns about the impacts of climate change, which will exacerbate these variables.

Of course, food production is just one element of food availability, which is one of four dimensions of food security (Food and Agriculture Organization [FAO] 2008). Nevertheless, the projected decreases in food production and changes in the nutritional value of crops as a result of climate change are likely to worsen food security in the region, as attaining food security relies on fulfilling all four dimensions simultaneously (Misselhorn 2005; De Wit and Midgley 2012). In southern Africa, food insecurity has numerous social and political dimensions, which include diminished social capital linked to poverty, conflict and HIV/AIDS. These are also fundamental constraints to food production (Misselhorn 2005).

A 2010 Stanford University report notes that by 2050, under the two-degree warming scenario, the growing season *average* temperature will be hotter than any year in historical experience for nine years out of ten, which could result in a 32 per cent production loss for key staple crops like maize and sorghum. This would be bad enough if the overall global temperature increase was indeed going to be kept to two degrees or less above pre-industrial levels. However, current pledges for emissions reductions are setting a course towards an average global temperature increase of at least four degrees (see, for example, Anderson and Bows 2008), which implies a higher increase in Africa. Moreover, emissions are now beyond even the most extreme of the IPCC's emissions scenarios (Anderson and Bows 2008).

Climate change will have grave impacts on biodiversity and ecosystems in southern Africa (Biggs et al 2004; Parry et al 2007), with implications for the livelihood opportunities and food security of communities that eke out their existence from these sources. Such impacts could affect agro-ecological functions, such as crop pollination, which could be severely affected by the loss of large numbers of insect species in what is being referred to as the 'sixth great extinction'. This highlights the importance of achieving synergy between biodiversity conservation and actions to address climate change and ensure that the 'dynamic and complex linkages and feedbacks between human vulnerability and ecosystem vulnerability' (Shackleton and Shackleton 2012) are taken into account in responses to climate change.

The convergence of these climate risks and other threats would constitute an extremely serious situation for the livelihoods of poor people and the ecosystems upon which they depend, as well as for national economies in southern Africa and elsewhere, even if the processes and institutions for governance were well developed and functioned effectively. However, this is not the case.

Climate governance and justice

A broad definition of governance is the 'system of rules that shape the actions of social actors' (in Treib et al 2007, p. 3). While recognizing a range of actors and stakeholders, the concept of governance implies the centrality of institutions at different levels. As Sowman and Wynberg point out in their introduction to this volume, governance for justice and sustainability is concerned not only with state regulation and enforcement, but also with the 'political, institutional, and cultural frameworks through which diverse interests in natural and cultural resources are coordinated and controlled' (Cronkleton et al 2008, p. 1). Consequently, the interaction between private and public actors in natural resource management is an important element of governance for justice and sustainability.

In Africa, multiple uncertainties that affect the lives and livelihoods of poor people relate not only to the changing climate, but to the numerous environmental, social, economic, institutional and political stressors acting at different levels and across scales, sectors and stakeholder groups (Boko et al 2007; Ziervogel and Taylor 2008). And in southern Africa, as in many developing countries, climate-sensitive sectors such as water, agriculture, forestry and fisheries are not well adapted to current climatic risks and variability (Midgley et al 2011; Smith et al 2011), and the region as a whole has low levels of adaptive capacity, defined by the IPCC in 2007 as 'the ability of a system to adjust to climate change (including climate variability and extremes) to moderate potential damages, to take advantage of opportunities, or to cope with the consequences' (Parry et al 2007; Woolfrey 2012). While this definition emphasizes systems, it is relevant to consider the adaptive capacity of individuals and institutions as well.

The low levels of adaptive capacity in southern Africa are linked to many shortcomings in the environmental and climate governance system, including fragmented legislative and policy frameworks, inadequate institutional coordination, a failure to recognize and integrate customary arrangements and limited participation by civil society organizations and communities. This has been well described in relation to the governance of natural resources elsewhere in this volume, and more specifically to climate issues by, for example, Madzwamuse (2010a). Adaptive capacity is directly related to governance at different levels. At the local level, adaptive capacity is influenced by the institutional environment, access to information, finances and technology, political power and also by governance resources and processes at the regional level (Smit and Wandel 2006). Regional, national and global processes, including governance regimes and economic forces such as globalization, constitute the enabling environment in which communities and individuals must try to adapt (Smit and Wandel 2006; Jones et al 2010; Ludi et al 2012).

Legislative and policy frameworks for climate change and adaptation in southern Africa remain fragmented (Chandani 2011). This is the case for Africa as a whole: adaptation policy approaches fail to consider realities in the political and institutional spheres (Naess et al 2011; Lockwood 2012). While African countries have initiated comprehensive planning processes by developing national adaptation programmes of action or climate change response strategies, in general these have not been facilitated, integrated responses, as they often focus on biophysical vulnerabilities, adopt sectoral and project approaches to adaptation, and do not incorporate adaptation needs at the micro level (Madzwamuse 2010a). Current responses in the region do not recognize or address the interdependent factors causing vulnerability, nor do the governance frameworks yet create an enabling environment that reflects and could support the variety of interlinked components that comprise adaptive capacity.

Given the extremely wide-ranging and cross-cutting nature of climate change, and of the determinants of adaptive capacity, the situation calls for transdisciplinary research and action which recognizes that natural and social systems (society and ecology) are interlinked. Apart from overarching planning frameworks that incorporate these elements, critical areas to address for southern Africa include water and human and food security. Governance institutions at the regional, national and subnational levels will need to develop and promote adaptive management systems to deal with the rapidly changing prognoses.

Yet the Southern African Development Community (SADC) as a regional entity has been weak on translating policy statements and declarations into concrete implementation plans (Richards, 2008) and still lacks a clear agenda on climate change (Ruppel and Ruppel-Schlichting 2012). The SADC region's high levels of vulnerability to global developments such as the financial crisis, oil and food price increases, climate change and the regional energy crisis indicate the need to build institutional capacity for conflict management, mediation and, in a broad sense, forecasting (Giuffrida and Müller-Glodde 2009).

Taking one example, adaptation and development would require the management of transboundary water resources in the region to ensure equitable sharing, particularly for people in the most vulnerable sub-basins (Midgley et al 2011). While SADC does have institutions and processes designed to promote this, successful implementation and conflict management will require sustained political commitment. At the national level, climate policy institutional frameworks are not yet able to coordinate existing adaptation projects and programmes, a shortcoming that results in a donor-driven, ad-hoc and project-level approach, which is not optimal for promoting local or national ownership.

There is now broad consensus that questions of equity, justice and ethics are central to managing climate change (Eakin and Luers 2006; Biermann et al 2012). Climate justice dimensions operate at various levels and across different timescales. They include justice between generations – there is little doubt that climate change is foreclosing options for future generations, which is unjust – and social justice in the present. This is particularly pertinent for Africa, as people in poor countries, especially in Africa, are already far more likely to die as a result of climatic changes that occurred up to 2000 – and these regions and populations are the least responsible for the increase in greenhouse gases that is driving climate change (Patz et al 2007).

The literature on climate change, equity and social justice has several themes: climate change and human rights, responsibility for past emissions and the polluter-pays principle, the setting of mitigation targets, questions concerning entitlements for future emissions and responsibility for financing adaptation (including technology transfer) (Paavola and Adger 2002; Brown et al 2007; Gardiner 2010; Jamieson 2010; Winkler et al 2011; Moellendorf 2012). These themes include both procedural and distributive justice matters (Paavola and Adger 2002), and are clearly set out in the Rock Ethics Institute's White Paper on the Ethical Dimensions of Climate Change (Brown et al 2007). This chapter does not discuss all of the ethical themes of relevance to climate change, but instead focuses on components of a number of different themes as they relate to the key developmental and governance issues discussed here.

In the global negotiations arena, the concepts of 'equity' and 'common but differentiated responsibilities' lie at the heart of the UNFCCC, the centrepiece of the evolving international climate change regime (International Council on Human Rights Policy [ICHRP] 2008). The UNFCCC sets down principles for distributive justice to address the fact that countries and communities that are not responsible for creating the problem, in that their contribution to greenhouse gas emissions has been negligible, are often the most vulnerable to climate change, as they generally have lower adaptive capacity (Stern 2006; Fankhauser and Schmidt-Traub 2010). In addition to equity issues linked to causal responsibility, there are justice issues related to the response to climate change.

Thus article 3 of the UNFCCC provides that parties to the convention should respond to climate change 'on the basis of equity and in accordance with their common but differentiated responsibilities and respective capacities', specifying that developed countries should take the lead in combating climate change, and that responses should take into account the development needs of developing countries. In this way, the UNFCCC is relevant to the defining characteristics of our world: global inequality and extreme poverty (Moellendorf 2012). Emerging economies and developing countries, including those in southern Africa, have stated that equitable access to sustainable development and other issues of equity should be the foundation of the proposed global climate change agreement (Winkler et al 2011).

The recent IPCC special report on extreme events has highlighted how inequalities among people influence local coping and adaptive capacity. These inequalities, which also constitute significant challenges to disaster risk management and adaptation from local to national levels (Field et al 2012), comprise social and power inequities based on gender, class, caste, race, age, education, religion and health. An equitable response to climate change necessitates governance across national boundaries too. In southern Africa, 'hotspot' regions (in terms of current and future climate-related vulnerability) correspond with climate and land use systems that cut across political boundaries, but national governance systems and socio-economic factors influence the risks (Midgley et al 2011).

While climate change is a global problem that does indeed require global political solutions, the context-specific nature of vulnerability and adaptive capacity highlights the importance of localized responses. However, globalization and other external forces may heighten susceptibility to the effects of climate change for the most vulnerable countries at different levels, for example where increasing scarcity and/or contestation for resources exacerbates inequities between and within communities, regions and nations. As Gomera et al (2012, p. 301) note, climate change is 'expanding the suite of global claimants on local resources'.

The discussion above sets a broad context for the examination of some specific equity issues that impinge on the two core policy areas of climate change: adaptation and mitigation, which some have characterized as a false dichotomy (Simon 2011). The following discussion is not exhaustive; it merely highlights some key areas in which equity and social justice considerations may be particularly relevant to natural resource-based livelihoods in southern Africa, and begins with a consideration of the interlinked concepts of adaptation, vulnerability and adaptive capacity.

Adaptation

'Adaptation' has been defined as the adjustment in natural or human systems in response to actual or expected climatic stimuli or their effects, which moderates harm or exploits beneficial opportunities (Parry et al 2007). Different types of adaptation can be distinguished according to timing, intent and form, with common types being anticipatory, autonomous and planned adaptation (Smit and Wandel 2006; Parry et al 2007; Collier et al 2008). Adaptation responses may occur at different levels. As Paavola and Adger (2002) note, 'Adaptation is comprised of inaction and proactive and reactive responses at the international, national, local and individual levels'.

There is growing acknowledgement that adaptation responses must include increased resilience in both social and ecological systems (Girot et al 2012). The focus on resilience has developed largely in response to the many uncertainties about future carbon emissions and their impacts, as well as the multi-stressor context in which people live and pursue their livelihoods (Conway 2011).

The other side of building resilience is reducing vulnerability, usually considered in terms of a number of different components: inherent vulnerability within a system, which may be termed sensitivity; the magnitude of change imposed upon the system; and exposure to that change (Eakin and Luers 2006; Smit and Wandel 2006). There are, however, diverse definitions of vulnerability that draw from natural and social sciences and focus, respectively, on the climate impacts and on the social, political or economic context (Adger et al 2004; O'Brien et al 2007; Kazmierczak and Handley 2011). Vulnerability is linked to adaptive capacity, which we can consider as the ability of individuals, institutions and systems to adjust to climate change. For effective adaptation, vulnerability (which encompasses exposure, as used here) needs to be reduced and adaptive capacity built.

Paavola and Adger's (2002) framework for analysing justice issues in adaptation has two main components: distributive justice, which includes how the beneficial and adverse effects of climate impacts and adaptation to them are distributed across groups of people and time, and procedural justice, which concerns how and by whom decisions on adaptive responses are made.

According to this framework, distributive justice issues centre on the concepts of equity and fairness. Vulnerability to climate change is highly differentiated, socially and spatially (Adger 2008), with clear implications for equity. A decade ago, Burton et al (2002) raised the issue of the differential distribution of adaptation benefits and costs. While this is now receiving greater attention in policy and practice (see, for example, Hesse 2011, concerning drylands development and adaptation), little attention, in general, has been paid to equity and social justice issues in adaptation initiatives (Thomas and Twyman 2005). There is no doubt that adaptation costs are highly distorted at the global scale (Adger 2001); this applies as well at the local scale, where some groups and individuals who may least be able to respond, for various reasons, are more vulnerable than others to the effects of climate change. Significant thematic areas for considering equity at the local level include gender and the involvement of vulnerable or marginalized groups such as children and the elderly.

Low levels of gender equality in southern Africa have implications for adaptation to climate change in the region. While it is recognized that not all women are poor (Arora-Jonsson 2011), poor women often have limited access to resources, restricted rights, restricted mobility and little say in shaping decisions, all of which makes them highly vulnerable to climate change. Gender inequalities in the natural resource arena include inequitable access to resources, a lack of empowerment and participation in decision-making and the burden of unpaid household work (Raworth 2008; Arora-Jonsson 2011). Despite significant efforts over the past few decades, the gender equality bar remains depressingly low: in 2012 it was estimated that 34–77 per cent of the global population fell below the social foundation for gender equality (Raworth 2012).

Sub-Saharan Africa scores lowest among regions on the gender inequality index (UNDP, 2011), with most SADC countries scoring low on gender empowerment measures (UNDP and SADC 2000). Gender inequalities intensify climate risks and vulnerabilities, and climate change threatens to exacerbate existing inequalities and jeopardize gains made in this area (UNDP 2008a; Raworth 2008; Patt et al 2009; UNFPA 2009). Actions to deal with the depletion of natural resources linked to climatic changes in Africa increase women's workload (UNDP 2010; Romero González et al 2011), and can reverse educational achievements, if girls are kept out of school to assist with heavier household duties (Raworth 2008).

Secure tenure is considered essential to enhancing the adaptive capacity of poor and vulnerable people (African Development Forum 2010). Differential security of tenure over land and resource access between men and women, often linked to cultural norms, further reduces the ability of women to adapt to climate change, and places them at greater risk of losing customary informal tenure arrangements in the face of 'land grabs'. Such loss of land, which affects men too, is increasingly linked to mitigation initiatives and the evolution of the global carbon market, discussed below.

Children and the elderly are disproportionately vulnerable to the impacts of climate change (Drimie and Casale 2009; Seballos and Tanner 2011). The multidimensional nature of children's vulnerability is shaped by the context in which they live and the physical, social, and emotional changes that occur during the childhood years, which are exacerbated by their lack of agency and voice (United Nations Children's Fund South Africa [UNICEF] 2011). Children and youth form the majority of Africa's population, but are largely excluded from national policy development, and their needs are not incorporated into climate change negotiations (African Development Forum 2010). Studies point to serious climate impacts aggravating child malnutrition and hunger. The International Food Policy Research Institute estimates that by 2050, 20 per cent more children will be malnourished than would have been the case in a world without climate change (Nelson et al 2009).

These differential aspects of vulnerability, which demand an analysis from the perspectives of equity and justice, are very much present in southern Africa, where women and unemployed young people are considered particularly at risk from climate change (Midgley 2011). Moreover, the increase in extreme events projected for the region may worsen vulnerability levels for those most at risk. Research in disaster management has highlighted the fact that women and other marginalized social groups suffer more during disasters and find it harder to bounce back afterwards (Adger et al 2007), although Arora-Jonsson (2011) has questioned the evidence for this assertion. If it is true, then the increased intensity and frequency of disasters linked to climate change could exacerbate existing social inequalities. In southern Africa, this relates to the increased incidence and severity of drought and floods, as well as cyclones along the eastern coast.

As we have seen, entrenched cycles of vulnerability characterize southern Africa's food-insecure communities, caused largely by poverty, environmental stressors and conflict. Multiple stresses related to HIV and AIDS, land degradation, trends in economic globalization and exposure to violent conflict aggravate exposure to climate risks (Adger et al 2007) and affect the capacity of poor and vulnerable people in southern Africa to adapt to climate change (Drimie 2010; Midgley et al 2011). The incorporation of considerations of both climate change and vulnerability to climate change into development activities is essential to preserving and increasing development gains and to supporting the kind of adaptation that reduces both vulnerability and poverty: that is, *sustainable adaptation measures* (Eriksen et al 2007).

There is a range of climate adaptation programmes in southern Africa, including both inter-country and civil-society-driven programmes (Chishakwe 2010). Most of these are donor-funded and range in implementation approach from community-based interventions to government disaster management activities. While it is beyond the scope of this chapter, an analysis of these programmes aimed at understanding the degree to which equity and justice considerations are incorporated would be instructive.

Many programmes in the region have included vulnerability assessments, for example under the Assessments of Impacts and Adaptation to Climate Change programme[3] and the more recent Regional Climate Change Programme for Southern Africa.[4] Vulnerability indices and vulnerability assessment methodology are being refined to better elucidate the underlying causes of vulnerability that relate to the inequity issues discussed above, such as a lack of voice or agency, and inequitable access to land. At the international level, the Climate and Regional Economics of Development integrated assessment[5] (or climate-economics) model is designed to emphasize issues of international equity while using the most up-to-date estimates of physical climate impacts and emissions abatement costs.

The greater emphasis now being placed on understanding the social dimensions of climate change has both analytical and normative components (SDCC Task Team 2011). This focus both highlights the fundamentally social aspects of climate processes and elaborates equity and social justice principles, particularly for the most vulnerable people. The social dimensions of climate change include social, institutional, economic, knowledge, informational and behavioural aspects, emphasizing the socio-economic dimension of vulnerability (Bauer and Scholz 2010; SDCC Task Team 2011; African Development Forum 2010; Chambwera and Anderson 2011). Taking account of this broader context, there is a growing focus on the social, institutional, policy, knowledge and informational approaches to responding to climate change. This evolving view allows for enhanced linkages between the diverse range of adaptation options and the multiple stressors at livelihood level – or, in the words of Tschakert and Dietrich (2010), the 'multi-faceted livelihood-vulnerability risks' faced by many people in Africa.

Successful adaptation strategies will need to embrace social and behavioural change. Recognition of this is evident in the emerging trend of transforming adaptation projects from *technical* interventions into *social or process* interventions (Urquhart 2009). Political aspects, in addition to social dimensions, have often been downplayed in the early emphasis on technological and infrastructural adaptation approaches (Brown et al 2007). However, power dynamics, access to information and services and control over resources are important determinants of people's capacity to adapt to climate change, as indeed they are for any development process. Institutions (both formal, such as government bodies, and informal, such as kinship networks), social processes and political context influence how people decide on adaptation actions, reflecting the centrality of these preoccupations of governance to understanding the adaptation process.

Work carried out by the Africa Climate Change Resilience Alliance (ACCRA) in Uganda, Mozambique and Ethiopia highlights the pivotal role of social processes and institutions in influencing people's response to changing external pressures (Jones et al 2011). In the context of climate variability and change, formal and informal institutions hold sway over individual and collective decision-making and may often control entitlements to key resources in lean times. In addition to rules for collective ownership rights to natural resources, informal social institutions include religious and ethnic belief systems and household gender roles and responsibilities (Jones et al 2011). Adaptation also requires a long-term approach: thus it necessitates institutions and governance structures capable of managing evolving climate change risks and adaptation priorities over time (UNDP 2008b).

Equity and human rights issues are finding new focus in the growing literature on environmental migrants and climate refugees. It is likely that climate change will exacerbate the existing movement of people towards areas of actual or perceived higher resilience, so it will require careful management to reduce inherent conflicts and promote sustainable livelihoods for host communities and new residents. However, SADC currently lacks the necessary political engagement, institutions, policy and management frameworks for reducing the potentially serious impacts of climate change on regional security (Midgley et al 2011; Ruppel and Ruppel-Schlichting 2012).

Despite a range of adaptation initiatives in the region (International Development Research Centre [IDRC] 2010; Climate Change Adaptation in Africa Program [CCAA] 2011; Chambwera and Anderson 2011; Jones et al 2011; Global Environment Facility [GEF] 2011; Koelle and Anneke n.d.) and the prospects of additional international financing, limits to the ability of poor households to adapt (in particular autonomously) are becoming evident (Adger et al 2009; Ludi et al 2012). Related factors include reduced access to assets and resources, linked to environmental degradation or elite capture; less functional social networks and reciprocity related to migration and displacement; and limited options for livelihoods diversification. This emphasizes the importance of a focus on enhancing the adaptive capacity of communities dependent on natural resources, so that they can adjust management and use patterns in response to climate change (UNDP 2008b).

Mitigation

Mitigation consists of actions taken to reduce greenhouse gas emissions and improve sinks for carbon. The aim is to reduce the concentration of greenhouse gases in the atmosphere and limit the extent of dangerous climate change. While effective mitigation requires action by major greenhouse gas emitters globally, including far-ranging moves in the industrial, transport and energy generation sectors (Klein et al 2007), the diverse portfolio of mitigatory initiatives includes some that can be coupled with adaptation actions to benefit poor rural people (Eriksen et al 2007; Urquhart 2010). An example is systems that reward smallholders and natural resource users for conserving environments such as forests, woodlands and peatlands, known as 'payments for ecosystem services'. Given new understandings of the urgency of the climate situation – briefly stated, a global mean temperature increase will reduce the biosphere's ability to store carbon emissions (Anderson and Bows 2008) – it is important to integrate adaptation and mitigation, especially through activities that can reduce carbon cycle feedbacks.

Activities that can serve both adaptation and mitigation goals simultaneously are agroforestry, conservation agriculture – including conservation tillage and appropriate cropping systems – and the rehabilitation of degraded environments such as grasslands (Verchot et al 2006; Syampungani et al 2010; World Bank 2011). Afforestation and reforestation have similar potential, if the negative ecological effects associated with monocrop plantations can be avoided (International Fund for Agricultural Development [IFAD] 2009). Through these and other activities, such as maintaining forests and managing rangelands, small-scale farmers and grassroots natural resource managers provide a wide range of environmental services that support carbon sequestration and limit the emissions of other greenhouse gases.

However, equity considerations are important here: such people account for only a small part of agricultural emissions, and are very often the most highly vulnerable to and severely affected by climate change (IFAD 2009). Similarly, it is widely accepted that indigenous peoples are among those most harmed by climate change, despite having made negligible contributions to the creation of the climate challenge. The global mitigation imperative has a valuable role for these indigenous and local communities, a role which should be recognized and rewarded. Payments for mitigation services do come from a number of sources, but are currently limited in scale and benefits.

Among the substantial external threats to the natural resource-based livelihoods of poor communities are deforestation (including the conversion of forests to monocultures), impacts from the growing number of dams for hydropower and fast-paced land privatization for plantations and mining (Brown et al 2007). The impacts of climate change are likely to increase the risks to the poor which are already evident in external processes such as globalization, particularly where they involve powerful actors like agricultural industries making new claims over scarce land and water (Brown, Slaymaker et al 2007). Heightened competition for natural resources and a worsening asset situation both put the welfare of the poor at risk. The vulnerability of indigenous and other traditional local communities to external interests that are keen to take over control of their traditional resources (Swiderska et al 2009; Gomera et al 2012) is exacerbated by environmental degradation, which may in turn be exacerbated by climate change.

The global carbon market is a driver behind some of these threats, as transnational interests scramble for land in Africa for mitigation-related activities such as the cultivation of crops for biofuels (Gomera et al 2012). Problematic areas related to the expansion of biofuels include negative effects on biodiversity and ecosystems, competition for land and water between fuel and food crops, risks to land tenure security and reduced livelihood opportunities for women, pastoralists and migrant farmers related to reduced access to land (German et al 2011; Unruh 2008).

The UN's Reducing Emissions from Deforestation and Forest Degradation initiative, known as REDD+, embodies approaches that can provide financial and non-financial incentives for the protection and restoration of forests and degraded land, and, if so designed, can also offer tools to leverage more secure access or tenure. Despite this potential, caution should be exercised, as REDD+ and similar approaches could result in the loss of access to land if collective and individual rights and customary laws are not recognized, and in negative impacts on biodiversity (Phelps et al 2012). Such approaches may also not deliver ecological and mitigation benefits, as there are doubts about whether deforestation can be prevented by carbon markets.[6] Stronger provisions will be needed than just trading the carbon stored in forests and hoping that the magic of the markets will keep the profits from carbon trading higher than the profits from palm oil trading (Lang 2010).

In an analysis of the threats to community resource governance arising from the global carbon market, Gomera et al (2012) discuss issues of devolved versus centralized governance, and highlight the vested interests that operate at every level. Discussing REDD+, they note that forest governance is critical to the operational effectiveness of this regime. They point out the paradox inherent in needing strong local institutions and resource rights to capture the benefits of REDD+, in a context in which the greater the benefits, the more contested the rights are likely to be. This is strongly reminiscent of the contested landscape for community-based natural resource management (CBNRM), emphasizing the desirability of drawing on lessons learned from the many years of CBNRM in southern Africa in designing and implementing carbon governance regimes.

Security of tenure and/or usufruct is a key basic requirement for ensuring that local communities are able to benefit from initiatives like REDD+ and other carbon market possibilities. It is also, of course, a basic human right, as well as an important step for supporting local and indigenous knowledge that may itself be important for responding to climate change. For example, Swiderska et al (2009) show how maintaining indigenous knowledge is critically linked to ensuring ongoing access to traditional lands.

Not only should equitable benefit-sharing mechanisms be adopted for implementing REDD+ at the local level, but such mechanisms should also diversify the asset base for all groups within communities, and include steps to ensure synergies with adaptation (Graham 2011). Equitable benefit-sharing mechanisms necessitate effective participation from all groups in their design, as well as the targeting of benefits to particular groups in order to attain distributional equity (Graham 2011).

Discussion and ways forward

Promoting more accountable and effective decision-making and governance mechanisms is an integral part of building adaptive capacity (Jones et al 2010; Ludi et al 2012). We have seen that equity and social justice are central considerations of governance and responding to climate change. This chapter has highlighted significant and unresolved equity issues relating to climate change and natural resource management in southern Africa, and the inadequacy of existing governance systems to address existing constraints. Given the very serious threats posed by climate change in the near future, what are some of the ways forward towards developing more responsive and adaptive governance systems that can help to build adaptive capacity, mediate the response to climate change and promote climate justice? Five key approaches are proposed.

Developing a new orientation for adaptation

This chapter has highlighted the importance of a focus on equity and justice when designing and implementing adaptation and mitigation measures. As Brooks et al (2005) note, adaptive capacity is strongly associated with governance, civil and political rights and literacy. The discussion above has pointed to the move to see adaptation as an ongoing social process, rather than consisting of discrete steps to adapt to one or more specific impacts, within a multi-stressor context (see, for example, Eriksen et al 2011).

Sustainable or structural adaptation aims to address poverty and the underlying causes of vulnerability through the adaptation process, and as such is critical for a focus on redressing inequity. This kind of pro-poor adaptation incorporates a 'soft path' approach in which participatory action learning is emerging as a key mechanism to drive an iterative adaptation process that is cognizant of climate uncertainties and the dynamic multiple-stressor livelihood contexts in which people live. Research carried out under the CCAA programme has highlighted that adaptation actions are most likely to also help reduce poverty when they are the result of participatory processes that promote local ownership (CCAA 2011). At the same time, critical reflections on how 'participatory approaches can miss social dynamics of power at the community level and in the broader context' (Roncoli et al 2011) must be acknowledged.

Other criteria for pro-poor adaptation indicated by CCAA research are institutional strengthening to sustain community-based adaptation, also involving policy actors in order to develop supportive policy frameworks. However, while adaptation must take place in the context of development, it does not merely demand doing more of the same. Speaking at a meeting of the Overseas Development Institute on climate change and development, Nigel Arnell, director of the Walker Institute for Climate Systems Research at the University of Reading, argued that the extra dimension in adaptation is not the techniques used but the degree of uncertainty, demanding new criteria of robustness and flexibility (Overseas Development Institute [ODI] 2009).

Valuing the role of indigenous and local knowledge

Attention is increasingly being paid to the role of indigenous knowledge systems, and to the potential that the blending of indigenous and scientific knowledge holds for more robust and locally specific adaptation responses or integrated adaptation-mitigation measures (Tschakert and Dietrich 2010; Ziervogel and Opere 2010; Newsham and Thomas 2011). Interventions that recognize and recover local knowledge and culture, and highlight the value of indigenous environmental knowledge, are likely to be critical in speeding up adaptation responses.

However, reports from different parts of the world increasingly indicate that as a result of the pace of climate change, indigenous knowledge about natural resource use and agriculture, passed down from generation to generation, is now becoming obsolete or unable to support adaptive responses (Roncoli et al 2009). While many indigenous technologies in southern Africa embody ecologically sustainable development and promote adaptation to climate change, others may have negative adaptation effects, particularly given changing socio-political contexts. For example, slash-and-burn cultivation is no longer sustainable in many areas, due to population pressure and limited land access resulting in shorter fallow periods, and is also believed to aggravate soil degradation, as well as dryness in an already semi-arid area (Syampungani et al 2010; Culas 2012).

Despite these reservations, the multiple livelihood strategies of people living in marginal lands in southern Africa remain underpinned by indigenous knowledge systems, developed through cumulative experience of the interactions between land use management systems and changing environmental conditions, as Dube and Sekhwela (2007) show for Botswana. Moving forward on both adaptation and mitigation in such areas will require the constraints upon the utility of indigenous knowledge to be understood, and addressed where possible.

Promoting equity through human-rights-based and empowerment approaches

Human rights-based approaches to development are designed to integrate international human rights principles into development planning and practice. Specifically, human rights-based approaches promote understanding and integration of equity issues. Using these approaches means asking the critical questions: 'What, why, who and what capacities?' (UNDP 2012). For example, adopting a rights-based approach to understand climate-related disaster risk would call for a risk analysis that probed the root causes of the vulnerabilities that heighten the disaster risk of certain groups. Analysis would include determining who has the duty to reduce these disaster risks, and what capacities are needed to address disaster risk, both for rights holders and for duty bearers.

Former United Nations secretary-general Kofi Annan set it out this way:

> [A] rights-based approach to development describes situations not simply in terms of human needs, or of development requirements, but in terms of

> society's obligations to respond to the inalienable rights of individuals, empowers people to demand justice as a right, not a charity, and gives communities a moral basis from which to claim international assistance when needed.
>
> (UN 1998)

Few dispute that climate change is already destabilizing the realization of many internationally protected human rights: rights to health and to life; rights to food, water, shelter and property; and rights associated with livelihood and culture, with migration and resettlement and with personal security in the event of conflict (International Council on Human Rights [ICHRP] 2008).

The agency of individuals and communities, or their 'ability to act independently and to make their own choices' (Jones et al 2011, p. 37; Ludi et al 2012) is a critical component of adaptive capacity, with some suggesting that agency and adaptive capacity are synonymous (Jones et al 2011). Agency relates strongly to adaptation's procedural justice concerns about how and by whom decisions on adaptive responses are made – concerns that centre on concepts of recognition, participation and legitimacy (Paavola and Adger 2002). Empowerment, which can be seen as a mechanism to achieve agency, is linked to innovation: interventions should seek to empower rural poor people to develop their own adaptive strategies in response to new challenges and opportunities resulting from trade liberalization, the dismantling of state welfare programmes, climate change, conflict and pest outbreaks, rather than mutely accept these as they descend from above (IFAD 2007).

Agency is an important element in unlocking the potential contribution to changing power relationships of a rights-based approach, just as mainstreaming gender is in climate change initiatives. The differential impacts, as discussed, are real, but there has been little research aimed at understanding different adaptive strategies of benefit for women and men. Moreover, tenure security over land and resources is frequently not sufficiently extended to women in adaptation responses (Madzwamuse 2010b).

While the UNFCCC's principles of 'equity' and 'common but differentiated responsibilities' may open the way to a consideration of human rights claims across borders relating to climate change impacts, this potential has not yet been realized in practice (ICHRP 2008), although reparations for climate change damage constitute a growing field of enquiry, closely linked to the likely costs of adaptation. The evolving system of monitoring, reporting and verification is one of the key areas where the focus on equity and human rights issues can be entrenched, both internationally and within countries.

Southern African countries are beginning to move on this component of the governance system, by pushing for the inclusion of key equity aspects in international negotiations, and by setting in place the required institutions nationally. For example, South Africa's Department of Environmental Affairs has set up an expert working group on the monitoring, reporting and verification of adaptation actions. The role of these experts (not more than 15 people, including civil society representatives) will be to provide advice and guidance for the development of the national monitoring, reporting and verification system. For countries likely to suffer loss or damage from climate change, in particular vulnerable developing countries, redress options outside the UNFCCC, such as international litigation, are being considered (Hyvarinen 2012).

Pursuing integrated approaches to adaptation

Given the cross-sectoral nature of the climate challenge, and the context of vulnerability to multiple stressors within which many people in southern Africa pursue their livelihoods, integrated approaches to adaptation are required to provide a sufficiently broad and realistic intervention capacity. Such approaches would need to consider the range of institutional, environmental, political, social and economic realities that impinge on people's lives, livelihoods and agency. The failure of developmental strategies to adequately address current climate risks reinforces the need for strong linkages between adaptation and development (Bauer and Scholz 2010). Low-regrets adaptation strategies[7] that produce developmental co-benefits are called for, as well as integrated ways to address environmental and climate change issues together, and to integrate adaptation and mitigation in a way that benefits poor people.

Drawing on past regional experience, it is clear that there are possibilities inherent in Community-Based Natural Resource Management (CBNRM) for improving links between ecosystem services and poverty reduction (Shackleton et al 2010), which will be an important component of sustainable adaptation approaches. This is reflected in recent debate on the integration of community-based adaptation and ecosystem-based adaptation approaches (Girot et al 2012). Given that growing populations in southern Africa are likely to remain highly vulnerable, or to become increasingly so, as climate change impacts accelerate, there is a real risk they will be unable to maintain any existing sustainable natural resource management and land use practices, as they increasingly rely on natural resources for food and livelihoods (Midgley et al 2011). Enhancing natural resource management now, and planning to maintain this in the future, is essential. Chishakwe et al (2012) point out the synergies between CBNRM and adaptation at the community level, based on an analysis of lessons learned in Botswana, Malawi, Mozambique, Namibia, Tanzania, Zambia and Zimbabwe.

Formulating adaptive and flexible governance frameworks

Perhaps most importantly, all of this needs to occur within adaptive and flexible governance and management frameworks capable of dealing with rapid and uncertain rates of change. The adaptive capacity of societies relies on their ability to act collectively, through institutions that govern social relations between individuals and social groups, for natural resource management at multiple scales (Adger et al 2004). These institutions need to be flexible and responsive: work carried out under the Africa Climate Change Resilience Alliance (ACCRA) programme points to the importance of what the alliance terms 'flexible, forward-looking governance' (Jones et al 2011, p. 30; Ludi et al 2012), and the need for a shift from 'optimal' decisions (the best ones in current conditions) to 'robust' decisions (reliable ones over the widest range of likely conditions) (Jones et al 2011, p. 35).

Institutions that facilitate cross-sectoral and cross-scale adaptation planning in order to create multiple benefits are required at all levels (Bizikova et al 2010; Brown et al 2010; Madzwamuse 2010a; Bronen 2011). Participatory monitoring and evaluation involving local users, which feeds back into an action learning approach at different levels, is essential for the kind of rapid and proactive responses required.

Conclusion

These new approaches and orientations are important ingredients for climate governance systems that promote equity, social justice and environmental sustainability:

- *Developing a new orientation for adaptation* that is pro-poor and emphasizes social dimensions and processes will also serve to address the root causes of vulnerability.
- *Valuing the role of indigenous and local knowledge* provides a basis for engaging with and supporting autonomous and community-based adaptation.
- *Promoting equity through human rights-based and empowerment approaches* will develop the agency that is seen as critical for enhancing adaptive capacity.
- *Pursuing integrated approaches to adaptation* more accurately reflects the context of vulnerability to multiple stressors in which many people in southern Africa have to pursue their livelihoods.
- *Formulating adaptive and flexible governance frameworks*, including frameworks for sustainable natural resource management, is a critical precursor to sound and equitable adaptation responses.

There are interconnections and areas of overlap among these five themes: for example, a human-rights-based approach highlights the need to focus adaptation efforts on vulnerable groups and to address the root causes of vulnerability, which is consistent with a pro-poor and sustainable orientation to adaptation; and pursuing integrated approaches to adaptation necessitates institutional frameworks that can effectively coordinate the range of adaptation initiatives being implemented, across sectors and between scales, which is a component of an adaptive governance framework.

These five approaches do not make up a comprehensive framework for evolving governance systems, but they are vital ingredients of this. They constitute entry points for tackling current constraints and for evolving the kinds of adaptive, multi-level, integrated and responsive governance systems that can deal with the complex socio-ecological conditions of the region, subject to a changing climate. Making progress on each of the five themes will mean that we can begin to govern what at times seems ungovernable – at least from equity and ecological sustainability perspectives – and ensure that the response to climate change in the southern African region is just, as well as effective.

Notes

1 'Vulnerability' is here defined as the degree to which a system is susceptible to, and unable to cope with, adverse effects of climate change, including climate variability and extremes. Vulnerability is a function of the character, magnitude and rate of climate change and variation to which a system is exposed, its sensitivity and its adaptive capacity (Parry et al 2007).

2 This analysis was conducted for the land area falling within the coordinates 12–35°S and 10–52° E, and for the emissions scenario named 'A1B' (see http://media.csag.uct.ac.za/faq/qa_3impacts.html). This area encompasses South Africa, Lesotho, Swaziland, Namibia, Botswana, Zimbabwe, Mozambique, Madagascar, Zambia, most of Malawi and the southern half of Angola.

3 http://aiaccproject.org

4 http://rccp.org.za

5 http://sei-international.org/projects?prid=1778

6 Persson and Azar (2009) found that putting a price on carbon is unlikely to prevent forests being cleared for oil palm plantations. This is partly because a higher carbon price drives up demand for biofuels as an alternative to expensive fossil fuels, which in turn increases both the price of biofuels and the likelihood that forests are converted to oil palm plantations. http://www.redd-monitor.org/2010/02/17/why-a-price-on-carbon-will-not-stop-deforestation/?utm_source=feedburner&utm_medium=feed&utm_campaign=Feed%3A+Redd-monitor+%28REDD-Monitor%29

7 Low-regrets adaptation measures provide benefits under current climate and a range of future climate change scenarios. Potential low-regrets measures include early warning systems, sustainable land management, ecosystem management and restoration, improvements to health surveillance, water supply and sanitation, changes to building codes and better education and awareness (Field et al 2012).

Bibliography

Adger, N. (2008) 'Climate change and insecurity', inaugural lecture, University of East Anglia, UK, 26 February, www.tyndall.ac.uk/sites/default/files/neil_adger_full_speech08_0.pdf

Adger, W. N. (2001) 'Scales of governance and environmental justice for adaptation and mitigation of climate change', *Journal of International Development*, vol 13, pp. 921–31

Adger, W. N., Brooks, N., Bentham, G., Agnew, M. and Eriksen, S. (2004) 'New indicators of vulnerability and adaptive capacity', Technical Report 7, Tyndall Centre for Climate Research, Norwich, UK

Adger, W. N., Agrawala, S., Mirza, M. M. Q., Conde, C., O'Brien, K., Pulhin, J. Pulwarty, R., Smit, B. and Takahashi, K. (2007) 'Assessment of adaptation practices, options, constraints and capacity', in M. L. Parry, O. F. Canziani, J. P. Palutikof, P. J. van der Linden and C. E. Hanson (eds) *Climate Change 2007: Impacts, Adaptation and Vulnerability – Contribution of Working Group II to the Fourth Assessment Report of the Intergovernmental Panel on Climate Change*, Cambridge University Press, Cambridge, UK, pp. 717–43

Adger, W. N., Dessai, S., Goulden, M., Hulme, M., Lorenzoni, I., Nelson, D. R., Naess, L. O., Wolf, J. and Wreford, A. (2009) 'Are there social limits to adaptation to climate change?', *Climatic Change*, vol 93, pp. 335–54

African Development Forum (2010) 'Climate change and human development', Issues Paper No. 3 in *Acting on Climate Change for Sustainable Development in Africa*, report of the proceedings of the Seventh African Development Forum, 10–15 October, Addis Ababa, Ethiopia

Anderson, K. and Bows, A. (2008) 'Reframing the climate change challenge in light of post-2000 emission trends', *Philosophical Transactions of the Royal Society A*, vol 366: pp. 3863–3882, http://rsta.royalsocietypublishing.org/content/366/1882/3863.full.pdf

Arora-Jonsson, S. (2011) 'Virtue and vulnerability: Discourses on women, gender and climate change', *Global Environmental Change*, vol 21, no 2, pp. 744–51

Bauer, S. and Scholz, I. (2010) 'Adaptation to climate change in southern Africa: New boundaries for sustainable development?', *Climate and Development*, vol 2, pp. 83–93

Biermann, F., Abbott, K., Andresen, S., Bäckstrand, K., Bernstein, S., Betsill, M. M., Bulkeley, H., Cashore, B., Clapp, J., Folke, C., Gupta, A., Gupta, J., Haas, P. M., A. Jordan, N. Kanie, T. Kluvánková-Oravská, L. Lebel, D. Liverman, J. Meadowcroft, R.B. Mitchell, P. Newell, S. Oberthür, L. Olsson, P. Pattberg, Sánchez-Rodríguez, R., Schroeder, H., Underdal, A., Camargo Vieira, S., Vogel, C., Young, O. R., Brock, A. and Zondervan, R. (2012) 'Navigating the Anthropocene: Improving earth system governance, *Science*, vol 335, no 6,074, pp. 1,306–7

Biggs, R., Bohensky, E., Desanker, P. V., Fabricius, C., Lynam, T., Misselhorn, A. A., Musvoto, C., Mutale, M., Reyers, B., Scholes, R. J., Shikongo, S. and van Jaarsveld, A. S. (2004) *Nature Supporting People: The Southern African Millennium Ecosystem Assessment, Integrated Report*, Council for Scientific and Industrial Research, Pretoria, South Africa

Bizikova, L., Boardley, S. and Mead, S. (2010) 'Participatory scenario development (PSD) approaches for identifying pro-poor adaptation options', Development and Climate Change Discussion Paper Number 18, December, World Bank, Washington, DC

Boko, M., Niang, I., Nyong, A., Vogel, C., Githeko, A., Medany, M., Osman-Elasha, B., Tabo, R. and Yanda, P. (2007) 'Africa', in M. L. Parry, O. F. Canziani, J. P. Palutikof, P. J. van der Linden and C. E. Hanson (eds) *Climate Change 2007: Impacts, Adaptation and Vulnerability – Contribution of Working Group II to the Fourth Assessment Report of the Intergovernmental Panel on Climate Change*, Cambridge University Press, Cambridge UK, pp. 433–67

Bronen, R. (2011) 'Climate-induced community relocations: Creating an adaptive governance framework based in human rights doctrine', *New York University Review of Law & Social Change*, vol 35, pp. 356–406

Brooks, N., Adger, W. N. and Kelly, P. M. (2005) 'The determinants of vulnerability and adaptive capacity at the national level and the implications for adaptation', *Global Environmental Change*, vol 15, no 2, pp. 151–63

Brown, D., Slaymaker, T. and Kaur Mann, N. (2007) 'Access to assets: Implications of climate change for land and water policies and management', draft paper, Overseas Development Institute, London

Brown, D., Tuana, N., Averill, M., Baer, P., Born, R., Brandao, C., Frodeman, R., Hogenhuis, C., Heyd, T., Lemons, J., McKinstry, R., Lutes, M., Müller, B., Gonzalez Miguez, J. D., Munasinghe, M., Muylaert de Araujo, M. S., Nobre, C., Ott, K., Paavola, J., Pires de Campos, C., Pinguelli Rosa, L., Rosales, J., Rose, A., Wells, E., Westra, L. (2007) *White Paper on the Ethical Dimensions of Climate Change*, Rock Island Ethics Institute, Pennsylvania State University, University Park, PA, www.ndsciencehumanitiespolicy.org/resources/climate_change_white_paper.pdf

Brown C., Nkem, J., Sonwa, D. and Bele, Y. (2010) 'Institutional adaptive capacity and climate change response in the Congo Basin forests in Cameroon', *Mitigation and Adaptation Strategies for Global Change*, vol 15, no 3, pp. 263–82

Burton I., Huq, S., Lim, B., Pilifosova, O. and Schipper, E. L. (2002) 'From impact assessment to adaptation priorities: The shaping of adaptation policy', *Climate Policy*, vol 2, pp. 145–9

CCAA (2011) 'Climate Change Adaptation in Africa Program: 2010–11 annual report', International Development Research Centre, Ottawa, http://idrc.ca/EN/Resources/Publications/Pages/ArticleDetails.aspx?PublicationID=1034

Chambwera, M. and Anderson, S. (2011) 'Integrating climate change into agricultural research for development in Africa', briefing paper, September, International Institute for Environment and Development, London, http://pubs.iied.org/pdfs/17099IIED.pdf?

Chandani, A. (2011) 'Improving leadership on climate change', *Reflect & Act*, briefing paper, March, International Institute for Environment and Development, London, http://pubs.iied.org/pdfs/G03110.pdf?

Chapman, R. A. and Sasman, N. (2012) *Cities and Climate Change: Urban Vulnerability and Resilience in Southern Africa*, synthesis report for the Regional Climate Change Programme for Southern Africa and UK Department for International Development, OneWorld Sustainable Investments, Cape Town

Chishakwe, N. (2010) *Southern Africa Sub-Regional Framework on Climate Change Programmes Repor,*, first draft (Working Document), unpublished document, Southern African Development Community and United Nations Environment Programme

Chishakwe, N., Murray, L. and Chambwera M. (2012) *Building Climate Change Adaptation on Community Experiences: Lessons from Community-Based Natural Resource Management in Southern Africa*, International Institute for Environment and Development, London, http://pubs.iied.org/pdfs/10030IIED.pdf?

Collier, P., Conway G. and Venables, T. (2008) 'Climate change and Africa', *Oxford Review of Economic Policy*, vol 24, no 2, 337–53

Conway, D. (2011) 'Adapting climate research for development in Africa', *Wiley Interdisciplinary Reviews: Climate Change*, vol 2, no 3, pp. 428–50

Cronkleton, P., Taylor, P. L., Barry, D., Stone-Jovicich, S. and Schmink, M. (2008) 'Environmental governance and the emergence of forest-based social movements', Occasional Paper No. 49, Center for International Forestry Research, Bogor, Indonesia

Culas, R. J. (2012) 'Chapter 8: Technological change and productivity growth for food security: The Case of Shifting Cultivation and the REDD Policy' In Jones, M. A. and Hernandez, F. E. (eds) *Food Security: Quality Management, Issues and Economic Implications*, Nova Science Publishers, Inc., Hauppauge, NY

Darkoh, M. (2009) 'An overview of environmental issues in Southern Africa', *African Journal of Ecology*, vol 47, no 1, pp. 93–8

De Wit, M. P. and Midgley, S. J.E. (2012) *Hunger and Climate Change: An Analysis of Key Variables in Southern Africa*, for the Regional Climate Change Programme for Southern Africa (RCCP), UK Department for International Development, OneWorld Sustainable Investments, Cape Town

Drimie, S. (2010) 'Adaptation to climate change in southern Africa: Factoring in AIDS', *Environmental Science and Policy*, vol 13, no 8, pp. 778–84

Drimie, S. and Casale, M. (2009) 'Multiple stressors in southern Africa: The link between HIV/AIDS, food security, poverty and children's vulnerability now and in the future', *AIDS Care: Psychological and Socio-Medical Aspects of AIDS/HIV*, vol 21, no 1, pp. 28–33

Dube, O. P. and Sekhwela, M. B. M. (2007) 'Community coping strategies in semiarid Limpopo basin part of Botswana: Enhancing adaptation capacity to climate change', Working Paper No. 47, April, Assessments of Impacts and Adaptations to Climate Change,

Washington, DC, http://www.aiaccproject.org/working_papers/Working%20Papers/AIACC_WP47_Dube.pdf

Eakin, H., and Luers, A. L. (2006) 'Assessing the vulnerability of social-environmental systems', *Annual Review of Environment and Resources*, vol 31, no 1, pp. 365–94

Ericksen, P., Thornton, P., Notenbaert, A., Cramer, L., Jones, P., Herrero, M. (2011) 'Mapping hotspots of climate change and food insecurity in the global tropics', CCAFS Report No. 5, CGIAR Research Program on Climate Change, Agriculture and Food Security, Copenhagen, Denmark, http://cgspace.cgiar.org/bitstream/handle/10568/3826/ccafsreport5-climate_hotspots_final.pdf?sequence=13

Eriksen, S. E. H., Klein, T. J., Ulusd, K., Næss, L. O. and O'Brien, K. L. (2007) 'Climate change adaptation and poverty reduction: Key interactions and critical measures', report prepared for the Norwegian Agency for Development Cooperation, University of Oslo, Norway

Eriksen, S., Aldunce, P., Bahinipati, C. S., d'Almeida Martins, R., Molefe, J. I., Nhemachena, C., O'Brien, K., Olorunfemi, F., Park, J., Sygna, L. and Ulsrud, K. (2011) 'When not every response to climate change is a good one: Identifying principles for sustainable adaptation', *Climate and Development*, vol 3, no 1, pp. 7–20

Fankhauser, S. and Schmidt-Traub, G. (2010) 'From adaptation to climate-resilient development: The costs of climate-proofing the Millennium Development Goals in Africa', policy paper, Centre for Climate Change Economics and Policy and Grantham Research Institute on Climate Change and the Environment, February

FAO (2008) 'An introduction to the basic concepts of food security', *Food Security Information for Action*, Practical Guides, Food and Agricultural Organization, Rome

Field, C., Barros, V., Stocker, T. F., Qin, D., Dokken, D. J., Ebi, K. L., Mastrandrea, M. D., Mach, K. J., Plattner, G. K., Allen, S. K., Tignor, M., and Midgley, P. M. (eds) (2012) *Managing the Risks of Extreme Events and Disasters to Advance Climate Change Adaptation: A Special Report of Working Groups I and II of the Intergovernmental Panel on Climate Change*, Cambridge University Press, New York, NY

Gardiner, S. M. (2010) 'Ethics and climate change: An introduction', *Wiley Interdisciplinary Reviews: Climate Change*, vol 1, no 1, pp. 54–66

GEF (2011) 'Evaluation of the Special Climate Change Fund', October, unedited version of final report, Global Environment Facility Evaluation Office, Washington, DC, http://www.thegef.org/gef/SCCF%20Evaluation

German, L., Schoneveld, G. C. and Gumbo, D. (2011) 'The local social and environmental impacts of smallholder-based biofuel investments in Zambia', *Ecology and Society*, vol 16, no 4, pp. 12–27

Girot, P., Ehrhart, C. and Oglethorpe, J. (2012) 'Integrating community and ecosystem-based approaches in climate change adaptation responses', briefing paper, Ecosystem and Livelihoods Adaptation Network, http://elanadapt.net/sites/default/files/siteimages/elan_integratedapproach_15_04_12.pdf

Giuffrida, L. and Müller-Glodde, H. (2009) 'Chapter 6: Strengthening SADC institutional structures: Capacity development is the key to the SADC secretariat's effectiveness', in A. Bösl, W. Breytenbach, T. Hartzenberg, C. McCarthy, K. Schade (eds) *Monitoring Regional Integration in Southern Africa Yearbook 2008*, Trade Law Centre for Southern Africa, Stellenbosch, South Africa, http://kas.de/upload/auslandshomepages/namibia/MRI2008/MRI2008_06_Giuffrida.pdf

Gomera, M., Rihoy, L. and Nelson, F. (2012) 'A changing climate for community resource governance: Threats and opportunities from climate change and the emerging carbon market', in F. Nelson (ed.) *Community Rights, Conservation and Contested Land: The Politics of Natural Resource Governance in Africa*, Routledge, pp. 293–309

Graham, K. (2011) 'REDD+ and adaptation: Will REDD+ contribute to adaptive capacity at the local level?', paper, REDD-net, http://redd-net.org/files/REDD%20ADAPTION%20LONG%20-%20MASTER%20final.pdf

Hansen, J., Sato, M., Kharecha, P., Beerling, D., Berner, R., Masson-Delmotte, V., Pagani, M., Raymo, M., Royer, D. L. and Zachos, J. C. (2008) 'Target atmospheric CO_2: Where should humanity aim?', *Open Atmospheric Science Journal*, vol 2, pp. 217–31, http://benthamscience.com/open/toascj/articles/V002/217TOASCJ.pdf

Hesse, C. (2011) 'Ecology, equity and economics: Reframing dryland policy', IIED Opinion: Lessons from Adaptation in Practice, November, International Institute for Environment and Development, London, http://pubs.iied.org/pdfs/17106IIED.pdf

Huq, S., Reid, H. and Murray, L. A. (2006) 'Climate change and development links', IIED Gatekeeper Series 123, International Institute for Environment and Development, London, http://pubs.iied.org/pdfs/14516IIED.pdf

Hyvarinen, J. (2012) 'Loss and damage caused by climate change: Legal strategies for vulnerable countries', paper, Foundation for International Environmental Law and Development, London, www.field.org.uk/files/field_loss__damage_legal_strategies_oct_12.pdf

ICHRP (2008) *Climate Change and Human Rights: A Rough Guide*, International Council on Human Rights Policy, Versoix, Switzerland

IDRC (2010) 'Improving accessibility and usability of seasonal forecasts for food security in Africa: Lessons from CCAA participatory action research projects', background paper, International Development Research Centre, Ottawa, www.africa-adapt.net/media/resources/327/Background%20Paper%2001March.pdf

IFAD (2007) 'Innovation strategy', International Fund for Agricultural Development, Rome, www.ifad.org/pub/policy/innovation/e.pdf

IFAD (2009) 'Research and innovation for smallholder farmers in the context of climate change', Discussion Paper for Round Table 3, Governing Council of the International Fund for Agricultural Development, 18–19 February, Rome, www.ifad.org/events/gc/32/roundtables/3.pdf

Jamieson, D. (2010) 'Climate, nature and ethics: An editorial essay', *Wiley Interdisciplinary Reviews: Climate Change*, vol 1, no 5, 621–3

Jones, L., Ludi, E. and Levine, S. (2010) 'Towards a characterisation of adaptive capacity: A framework for analysing adaptive capacity at the local level', background note, December, Overseas Development Institute, London, www.odi.org.uk/resources/docs/6353.pdf

Jones, L., Ayorekire, F., Barihaihi, M., Kagoro, A. and Ruta, D. (2011) 'Preparing for the future in Uganda: Understanding the influence of development interventions on adaptive capacity at the local level', Africa Climate Change Resilience Alliance (ACCRA) Uganda Synthesis Report, http://community.eldis.org/.5a35bbfb/Uganda_Synthesis_Report.pdf

Kazmierczak, A. and Handley, J. (2011) 'The vulnerability concept: Use within GraBS', briefing paper, School of Environment and Development, University of Manchester, May

Klein, R. J. T., Huq, S., Denton, F., Downing, T. E., Richels, R. G., Robinson, J. B., Toth, F. L. (2007) 'Inter-relationships between adaptation and mitigation', in M. L. Parry, O. F. Canziani, J. P. Palutikof, P. J. van der Linden and C. E. Hanson (eds) *Climate Change 2007: Impacts, Adaptation and Vulnerability – Contribution of Working Group II to the Fourth Assessment Report of the Intergovernmental Panel on Climate Change*, Cambridge University Press, Cambridge, UK, pp. 745–77

Koelle, B. and Anneke, W. (n.d.) 'Community-based climate change adaptation (CBA)',

Adaptation and Beyond No. 1: Lessons Learned from the Community Based Adaptation in Africa (CBAA) Programme, Indigo Development & Change, Nieuwoudtville, South Africa

Lang, C. (2010) 'Why a price on carbon will not stop deforestation', *Redd-Monitor*, 17 February, www.redd-monitor.org/2010/02/17/why-a-price-on-carbon-will-not-stop-deforestation

Lockwood, M. (2012) 'Adaptation policy, governance and politics in sub-Saharan Africa', Institute of Development Studies paper for an international symposium on the Governance of Adaptation, Amsterdam, 22–23 March

Ludi, E., Jones, L. and Levine, S. (2012) 'Changing focus? How to take adaptive capacity seriously', ODI Briefing Paper 71, January, Overseas Development Institute, London. http://odi.org.uk/resources/docs/7532.pdf

Madzwamuse, M. (2010a) 'Climate governance in Africa: Adaptation strategies and institutions', synthesis report, Heinrich Böll Stiftung, Cape Town

Madzwamuse, M. (2010b) 'Drowning voices: The climate discourse in South Africa', Policy Brief 05/2010, Heinrich Böll Stiftung, Cape Town

Midgley, S. (2011) 'Gender- and age-responsive adaptation in southern Africa', Research Brief 11, Regional Climate Change Programme for Southern Africa, Department for International Development, London, OneWorld Sustainable Investments, Cape Town

Midgley, S. J. E., Davies, R.A.G. and Chesterman, S. (2011) 'Climate risk and vulnerability mapping in southern Africa: Status quo (2008) and future (2050)', Technical Reports 1, Regional Climate Change Programme for Southern Africa, Department for International Development, London, OneWorld Sustainable Investments, Cape Town

Misselhorn, A. (2005) 'What drives food insecurity in southern Africa? A meta-analysis of household economy studies', *Global Environmental Change*, vol 15, no 1, pp. 33–43

Moellendorf, D. (2012) 'Climate change and global justice', *Wiley Interdisciplinary Reviews: Climate Change*, vol 3, no 2, pp. 131–43

Mwendera, M. (2010) 'Situation analysis for water and wetlands sector in eastern and southern Africa', International Union for the Conservation of Nature, Nairobi

Naess, L. O., Polack, E. and Chinsinga, B. (2011) 'Bridging research and policy processes for climate change adaptation', *IDS Bulletin*, vol 42, no 3, May

Nelson, G. C., Rosegrant, M. W., Koo, J., Robertson, R., Sulser, T., Zhu, T., Ringler, C., Msangi, S., Palazzo, A., Batka, M., Magalhaes, M., Valmonte-Santos, R., Ewing, M. and Lee, D. (2009) 'Climate change: Impact on agriculture and costs of adaptation', Food Policy Report, International Food Policy Research Institute, Washington, DC, www.ifpri.org/sites/default/files/publications/pr21.pdf

Newsham, A. J., and Thomas, D. S. G. (2011) 'Knowing, farming and climate change adaptation in north-central Namibia', *Global Environmental Change*, vol 21, no 2, pp. 761–70

O'Brien, K., Eriksen, S., Nygaard, L. P. and Schjolden, A. (2007) 'Why different interpretations of vulnerability matter in climate change discourses', *Climate Policy*, vol 7, no 1, pp. 73–88

ODI (2009) 'Climate change and water: Understanding impacts, formulating responses', report on presentations at an event in the Overseas Development Institute series on climate change and development, 30 March, www.odi.org.uk/events/431-climate-change-water-understanding-impacts-formulating-responses

Osbahr, H., Twyman, C., Adger, W. N. and Thomas, D. S. G. (2010) 'Evaluating successful livelihood adaptation to climate variability and change in southern Africa', *Ecology and Society*, vol 15, no 2, www.ecologyandsociety.org/vol15/iss2/art27/

Oxfam (2009) 'Suffering the science: Climate change, people and poverty', Briefing Paper 130, Oxfam International

Paavola, J. and Adger, W. N. (2002) 'Justice and adaptation to climate change', Working Paper 23, Tyndall Centre for Climate Change Research, School of Environmental

Sciences, University of East Anglia, Norwich, UK, http://academia.edu/639039/Justice_and_adaptation_to_climate_change

Parry, M. L., Canziani, O. F., Palutikof, J. P., van der Linden, P. J. and Hanson, C. E. (eds) (2007) *Climate Change 2007: Impacts, Adaptation and Vulnerability – Contribution of Working Group II to the Fourth Assessment Report of the Intergovernmental Panel on Climate Change*, Cambridge University Press, Cambridge, UK

Patt, A., Dazé, A. and Suarez, P. (2009) 'Gender and climate change vulnerability: What's the problem, what's the solution?', in M. Ruth and M. E. Ibarraran (eds) *Distributional Impacts of Climate Change and Disasters: Concepts and Cases*, Edward Elgar, Cheltenham, UK

Patz, J. A., Gibbs, H. K., Foley, J. A., Rogers, J. V. and Smith, K. R. (2007) 'Climate change and global health: Quantifying a growing ethical crisis', *EcoHealth*, vol 4, pp. 397–405

Persson, U. M. and Azar, C. (2009) 'Preserving the World's Tropical Forests – A Price on Carbon May Not Do', *Environment, Science and Technology*, vol 44, no 1, pp. 210–15

Phelps, J., Friess, D. A. and Webb, E. L. (2012) 'Win-win REDD+ approaches belie carbon–biodiversity trade-offs', *Biological Conservation*, vol 154, pp. 53–60

Raworth, K. (2008) 'Gender and climate adaptation', Briefing Paper, February, Oxfam GB, Oxford, UK

Raworth, K. (2012) 'A safe and just space for humanity: Can we live within the doughnut? Oxfam Discussion Papers, February, Oxfam International, Oxford, UK

Richards, R. (2008) 'Assessing progress on climate change policy and practice: Observations from South Africa and SADC', *Policy: Issues & Actors*, vol 21, no 1, Centre for Policy Studies, Johannesburg, http://cps.org.za/cps%20pdf/pia21_1.pdf

Rockström, J., Steffen, W., Noone, K., Persson, Å., F. S. Chapin, Lambin, E., Lenton, T. M., Scheffer, M., Folke, C., Schellnhuber, H., Nykvist, B., de Wit, C. A., Hughes, T., van der Leeuw, S., Rodhe, H., Sörlin, S., Snyder, P. K., Costanza, R., Svedin, U., Falkenmark, M., Karlberg, L., Corell, R. W., Fabry, V. J., Hansen, J., Walker, B., Liverman, D., Richardson, K., Crutzen, P. and Foley, J. (2009) 'Planetary boundaries: Exploring the safe operating space for humanity', *Ecology and Society*, vol 14, no 2, article 32

Romero González, A., Belemvire, A. and Saulière, S. (2011) 'Women farmers in Burkina Faso: Impact and adaptation policies and practices', Oxfam Research Reports, July, Oxfam International, Oxford, UK

Roncoli, C., Crane, T. and Orlove, B. (2009) 'Fielding climate change in cultural anthropology', in S. A. Crate and M. Nuttarll (eds) *Anthropology and Climate Change: From Encounters to Actions*, Left Coast Press, Walnut Creek, CA, pp. 87–115

Roncoli, C., Orlove, B. S., Kabugo, M. R. and Waiswa, M. M. (2011) 'Cultural styles of participation in farmers' discussions of seasonal climate forecasts in Uganda', *Agriculture and Human Values*, vol 28, no 1, pp. 123–38

Ruppel, O. C. and Ruppel-Schlichting, K. (2012) 'Chapter 2: Climate change and human security: Relevant for regional integration in SADC?', in T. Harzenberg, G. Erasmus and André du Pisanie (eds) *Monitoring Regional Integration in Southern Africa Yearbook 2011*, Trade Law Centre for Southern Africa, Stellenbosch, South Africa, pp. 32–71

Schulze, R.E. (2007) 'Some foci of integrated water resources management in the "South" which are oft-forgotten by the "North": A perspective from southern Africa', *Water Resources Management*, vol 21, pp. 269–94

SDCC Task Team (2011) *The Social Dimensions of Climate Change: Discussion Draft*. United Nations Task Team on Social Dimensions of Climate Change

Seballos, F. and Tanner, T. (2011) 'Enabling child-centred agency in disaster risk reduction', Global Assessment Report on Disaster Risk Reduction 2011, UN International Strategy for Disaster Risk Reduction, Geneva

Shackleton, C. M., Shackleton, S. E., Gambiza, J., Nel, E., Rowntree, K., Urquhart, P., Fabricius, C. and Ainslie, A. (2010) *Livelihoods and Vulnerability in the Arid and Semi-Arid Lands of Southern Africa: Exploring the Links Between Ecosystem Services and Poverty Alleviation*, NOVA Science Publishers, New York

Shackleton, S. E. and Shackleton, C. M. (2012) 'Linking poverty, HIV/AIDS and climate change to human and ecosystem vulnerability in southern Africa: Consequences for livelihoods and sustainable ecosystem management', *International Journal of Sustainable Development and World Ecology*, vol 19, no 3, pp. 275–86

Simon, D. (2011) 'Reconciling development with the challenges of climate change: Business as usual or a new paradigm?, in D. J. Kjosavik and P. Vedeld (eds) *The Political Economy of Environment, Development and Conflict in a Globalised World: Exploring the Frontiers, Essays in honour of Nadarajah Shanmugaratnam*, Tapir Academic, Trondheim, Norway, pp. 195–217

Smit, B. and Wandel, J. (2006) 'Adaptation, adaptive capacity and vulnerability', *Global Environmental Change*, vol 16, pp. 282–92

Smith, J. B., Dickinson, T., Donahue, J. D. B., Burton, I., Haites, E., Klein, R. J. T. and Patwardhan, A. (2011) 'Development and climate change adaptation funding: Coordination and integration', *Climate Policy*, vol 11, no 3, pp. 987–1000

Solomon, S., Qin, D., Manning, M., Chen, Z., Marquis, M., Averyt, K. B., Tignor, M. and Miller, H. L. (eds) (2007) '*Climate Change 2007: The Physical Science Basis – Contribution of Working Group I to the Fourth Assessment Report of the Intergovernmental Panel on Climate Change*, Cambridge University Press, Cambridge UK and New York, NY

Stern, N. (2006) *Stern Review on the Economics of Climate Change*, independent review commissioned by HM Treasury, United Kingdom, http://webarchive.nationalarchives.gov.uk/+/http://www.hm-treasury.gov.uk/independent_reviews/stern_review_economics_climate_change/stern_review_report.cfm

Swiderska, K., Argumedo, A., Pant, R., Vedavathy, S., Munyi, P., Mutta, D., Herrera, H., Song, Y. and Li, J. (2009) 'Protecting traditional knowledge from the grassroots up', IIED Briefing Paper, June, International Institute for Environment and Development, London, http://pubs.iied.org/pdfs/17067IIED.pdf

Syampungani, S., Chirwa, P. W., Akinnifesi, F. K. and Ajayi, O. C. (2010) 'The potential of using agroforestry as a win-win solution to climate change mitigation and adaptation and meeting food security challenges in southern Africa', *Agricultural Journal*, vol 5, no 2, pp. 80–8

Tadross, M., Suarez, P., Lotsch, A., Hachigonta, S., Mdoka, M., Unganai, L., Lucio, F., Kamdonyo, D. and Muchinda, M. (2009) 'Growing-season rainfall and scenarios of future change in southeast Africa: Implications for cultivating maize', *Climate Research*, vol 40, pp. 147–61

Thomas, D. S. G. and Twyman, C. (2005) 'Equity and justice in climate change adaptation amongst natural-resource-dependent societies', *Global Environmental Change*, vol 15, pp. 115–24

Treib, O., Bähr, H. and Falkner, G. (2007) 'Modes of governance: Towards a conceptual clarification', *Journal of European Public Policy*, vol 14, no 1, pp. 1–20

Tschakert, P. and Dietrich, K. A. (2010) 'Anticipatory learning for climate change adaptation and resilience', *Ecology and Society*, vol 15, no 2, p. 11

UN (1998) *Annual Report on the Work of the Organization*, United Nations, New York

UNDP and SADC (2000) *SADC Regional Human Development Report 2000: Challenges and Opportunities for Regional Integration*, Southern African Regional Institute for Policy Studies of the Southern African Political Economy Series Trust, Harare, http://hdr.undp.org/en/reports/regional/africa/sadc_2000_en.pdf

UNDP (2008a) *Human Development Report 2007/2008 – Fighting Climate Change: Human Solidarity in a Divided World*, United Nations Development Programme, New York

UNDP (2008b) *Operational Guidance for Climate Change Adaptation: Technical Papers in Six Priority Thematic Areas*, United Nations Development Programme, New York

UNDP (2010) 'Africa Adaptation Programme experiences: Gender and climate change: Advancing development through an integrated gender perspective', Discussion Paper Series, vol 1, March, United Nations Development Programme, New York

UNDP (2011) *Human Development Report 2011*, United Nations Development Programme, New York

UNDP (2012) 'Mainstreaming human rights in development policies and programming: UNDP experiences', issue brief, March, United Nations Development Programme, New York, http://hrbaportal.org/wp-content/files/Mainstreaming-HR-in-Development-Policies.pdf

UNFPA (2009) *State of World Population 2009 – Facing a Changing World: Women, Population and Climate*, United Nations Population Fund, New York

UNICEF (2011) *Exploring the Impact of Climate Change on Children in South Africa*, United Nations Children's Fund South Africa, Pretoria

Unruh, J. D. (2008) 'Carbon sequestration in Africa: The land tenure problem', *Global Environmental Change*, vol 18, no 4, pp. 700–7

Urquhart, P. (2009) 'IFAD's response to climate change through support to adaptation and related actions', comprehensive report and policy summary, International Fund for Agricultural Development, Rome

Urquhart, P. (2010) 'WFP and climate change: A review of ongoing experience and recommendations for action', WFP Occasional Paper No. 23, December, United Nations World Food Programme, Rome

Verchot, L. V., van Noordwijk, M., Kandji, S., Tomich, T., Ong, C., Albrecht, A., Mackensen, J., Bantilan, C., Anupama, K. V. and Palm C. (2006) 'Climate change: Linking adaptation and mitigation through agroforestry', *Mitigation and Adaptation Strategies for Global Change*, vol 12, pp. 901–18

Winkler, H., Jayaraman, T., Pan, J., Santhiago de Oliveira, A., Zhang, Y., Sant, G., Miguez, J. D. G., Letete, T., Marquard, A. and Raubenheimer, S. (2011) 'Equitable access to sustainable development: Contribution to the body of scientific knowledge', a paper by experts from BASIC countries, BASIC expert group, Beijing, Brasilia, Cape Town and Mumbai

Woolfrey, S. (2012) 'Chapter 1: Climate change and trade in southern Africa: an examination of the potential impacts of climate change and climate change response measures on the ability of SADC countries to engage in international trade', in T. Harzenberg, G. Erasmus and André du Pisanie (eds) *Monitoring Regional Integration in Southern Africa Yearbook 2011*, Trade Law Centre for Southern Africa, Stellenbosch, South Africa, pp. 6–31

World Bank (2011) *Climate-Smart Agriculture, Policy Brief*, World Bank, Washington, DC http://climatechange.worldbank.org/sites/default/files/documents/CSA_Policy_Brief_web.pdf

Ziervogel, G. and Opere, A. (eds) (2010) 'Integrating meteorological and indigenous knowledge-based seasonal climate forecasts for the agricultural sector: Lessons from participatory action research in sub-Saharan Africa', Climate Change Adaptation in Africa learning paper series, International Development Research Centre, Ottawa

Ziervogel, G. and Taylor, A. (2008) 'Feeling stressed: Integrating climate adaptation with other priorities in South Africa', *Environment*, vol 50, no 2, pp. 32–41

16 Towards robust governance for justice and environmental sustainability

Lessons from natural resource sectors in sub-Saharan Africa

Merle Sowman and Rachel Wynberg

Introduction

The overall aim of this book has been to enhance understanding of the nature of governance in relation to social justice and sustainability across a range of natural resource sectors in sub-Saharan Africa. The case studies presented have explored and critiqued various governances processes and systems, providing insight into the values and principles that shape governance in different contexts, the institutional arrangements that characterize different modes of governance and the drivers that determine governance interactions, approaches and outcomes in different contexts. Also reviewed are the local conditions that determine how governance plays out in different settings.

This final chapter attempts to synthesize the key findings and outcomes of these different governance modes, and the normative approaches adopted, including their implementation in practice. We also aim to identify the factors that enable or constrain the achievement of socially just and environmentally sustainable outcomes. The chapter concludes with recommendations for addressing challenges and strengthening enabling conditions for more robust forms of natural resource governance. These recommendations recognize, however, that natural resource governance is inextricably linked to broader governance systems and that structural, political and economic forces are major impediments to transforming governance across natural resource sectors in sub-Saharan Africa.

Many scholars have highlighted the lack of empirical research informing the natural resource governance field (Adger and Jordan 2009; Nelson 2010) and the need to improve understanding of governance in all its complexity and diversity. This volume, with its chapters on different aspects of natural resource governance in sub-Saharan Africa, goes some way towards contributing to filling that information deficit. The cases discussed do not by any means represent the full range of natural resource sectors and modes of governance encountered in sub-Saharan Africa, so the findings presented and lessons learned cannot be generalized. However, they do provide insights into the principles and norms that inform governance policies and rhetoric, the drivers that shape how governance plays out in practice and the modes and specific features of governance encountered in the natural resource arena in sub-Saharan Africa.

Histories and context

Following several decades of colonial rule, the process of transforming natural resource governance in most countries in sub-Saharan Africa was informed very much by the need for redress: improving the lives of poor and marginalized people, securing their rights to natural resources and helping them gain equitable access to these resources. Reforms in the natural resource governance arena were also shaped by global debates about the limitations of conventional technocratic, centralized and highly regulatory natural resource management systems and the need to move towards more holistic, participatory and integrated approaches (Berkes et al 2003; Garcia et al 2008; Sowman 2011; Symes 2006). There was also increasing evidence of the important role played by local collective resource institutions and customary systems in sustainable resource governance (Ostrom 1990; Ribot 2004). Growing concerns about the declining capacity of state governing institutions in Africa to execute a range of governance functions (Bratton and Rothchild 1992) and increasing calls for countries in Africa to embrace good governance principles (United Nations Development Programme [UNDP] 1997; World Bank 2002) also contributed to a shift in approaches to natural resource governance.

The transition to multiparty elections and democracy led to a major overhaul of the legal systems and the restructuring of governance institutions in many sub-Saharan countries. In the natural resource governance arena, reforms at policy, programme and institutional level were largely focused on decentralizing and devolving powers and responsibilities to local resource-dependent communities and promoting the participation of resource users in planning, management and decision-making processes.

These reforms expanded rapidly during the 1990s with the emergence of community-based natural resource management (CBNRM) programmes such as the Communal Areas Management Programme for Indigenous Resources (CAMPFIRE) in Zimbabwe and similar projects later in Botswana, the community-based conservancy programme in Namibia, and the Administrative Management Design for Game Management Areas (ADMADE) programme in Zambia.

However, despite initial optimism, research undertaken in southern Africa by several scholars across a wide range of disciplines is increasingly showing that many of these initiatives have failed to achieve their aims of providing tangible benefits to local communities, enhancing social equity and alleviating poverty. Instead, decisions regarding who gains access to land and resources, especially for high-value resources such as land, wildlife, forests and fisheries, remain vested in the state (Murombedzi 2001; Van Sittert 2002; Ribot 2004; Murphree 2005; Isaacs 2006; Anstey and Rihoy 2009; Jones 2010; Murombedzi 2010; Nelson 2010). A key obstacle to implementing these decentralization and devolution programmes has been the reluctance of central governments to relinquish power over valuable natural resources (Keeley and Scoones 2003; Nelson 2010; German et al 2010).

Evolving governance approaches to natural resource management in sub-Saharan Africa: What do the chapters tell us?

An overview

The narratives examined in this volume provide ample evidence of these trends. Table 16.1 offers an overview of all the case studies, describing the sectors and countries involved, the modes of governance adopted, the drivers of different modes of governance, the stated policy and outcomes for each case and factors that enable or constrain governance towards social justice and environmental sustainability. As the table demonstrates, the case studies cover a wide range of natural resource sectors – forestry, fisheries, non-timber forest products, land, wildlife, agriculture and freshwater and lake systems – across Botswana, Cameroon, Lesotho, Madagascar, Mozambique, Namibia, Sierra Leone, South Africa, Zambia and Zimbabwe. The governance modes examined in these cases fall mainly into three of the five categories discussed in the introductory chapter, namely, state-centric (or hierarchical), co-governance (networks) and self-governance, which includes customary governance.

The cases highlight the diverse and changing nature of natural resource governance in sub-Saharan Africa, but provide little evidence that governance reforms have enhanced social justice in a meaningful way, secured better access to resources and land or produced improved benefits for the poor. There is also little indication that governance reforms have led to improvements in ecosystem health and resource sustainability.

While there have been significant policy and institutional reforms in all of the countries covered in this volume, the overwhelming trend is that these have been ineffectual at best, and detrimental to livelihoods and resource sustainability at worst. New governance rhetoric that embraces so-called 'good governance' principles such as respect for the rule of law and for human rights, participation in decision-making, free informed prior consent, transparent and accountable institutions, subsidiarity, benefit sharing and sustainable resource use and development, has not been translated into day-to-day management and decision-making. Governance praxis has been and continues to be significantly influenced by various external drivers including globalization, the desire to accumulate capital, direct foreign investment and an expanding conservation agenda. Country-specific drivers and factors such as weak governance institutions and mechanisms for meaningful participation, limited capacity, political and economic interests and power asymmetries at all levels of governance, as well as the plurality of normative and legal systems operating in natural resource management, play a significant role in shaping governance practice.

While there is widespread evidence of governance reform, as well as experimentation with more decentralized, participatory and community-orientated forms of governance – in particular the roll-out of CBNRM programmes in Botswana, Zimbabwe, Zambia and Namibia – practices encountered on the ground are not

Table 16.1 Overview of case studies

Case study	*Sectors*	*Modes of governance*	*Drivers*	*Policy and praxis*	*Enabling and constraining factors*
Governing ancestral land: Madagascar	Forestry Conservation Mining Agribusiness	State-centric Some degree of co-management in conservation areas, typically through non-governmental organizations (NGOs)	Land and resource grabbing by foreign investors Global conservation agenda	Policies support ancestral land titling and community-based approaches. In practice, however, participation is limited, rights are poorly protected, accountability is low and benefits are skewed. Biodiversity agenda through support from conservation NGOs takes preference over local community needs.	**Constraining** Fragmentation of governance through sector-specific policies Weak governing institutions and implementation mechanisms, high rate of institutional change, conflicts between 'modern' and traditional values Power asymmetries Local interests not represented Weak civil society
Localizing global environmental governance norms: South Africa	Conservation Tourism	State-centric Some co-management agreements with land reform beneficiaries	Global conservation agendas Tourism as a vehicle for socio-economic development	South African Constitution requires co-operative governance and participation, but this is limited in practice and the community has a weak voice and powers. Access to natural resources is restricted and benefit sharing limited. Conflict exists due to tensions between normative global conservation norms and local realities.	**Constraining** Poor coordination between many institutions involved in governance and unclear roles and responsibilities Traditional authorities powers not clear Lack of recognition of customary governance systems Inadequate communication

CBNRM Chobe enclave: Botswana	Wildlife Conservation	State-centric CBNRM	Political pressure to address community needs Conservation imperative	Despite existence of CBNRM policy, central government retains decision-making power over use and management of wildlife resources and benefit distribution.	Restricted use and management rights for communities Lack of community expertise and leadership to operate tourism enterprises Few linkages between tourism enterprises and local economic development Elite capture of funds State grants reducing incentive for community involvement in conservation
Community-based conservation and protected areas: Namibia	Conservation Wildlife	CBNRM and state-centric management	Land restitution National imperative to protect wildlife and promote biodiversity Economic incentive for communities	Strong policy support of CBNRM is reflected in greater involvement of communities in self-organizing and negotiating their own boundaries. There are some local benefits, but concerns about community exclusion from tourism benefits in wider area.	**Enabling** Good support from NGOs External funding support **Constraining** Poor financial management, accountability and transparency in decision-making
CBNRM and micro-governance: Zimbabwe, Zambia and Botswana	Wildlife	CBNRM with power recentralized in some cases	Financial rewards from wildlife Controlling poaching	Policies are generally supportive of greater participation. However, different outcomes are achieved depending on whether or not representational or participatory governance is adopted.	**Enabling** Participatory governance leading to better financial flows and management and greater equity Changing forms of governance that can address equity **Constraining** Technical and administrative challenges at local level that constrain participatory democracy

Table 16.1 Continued

Case study	*Sectors*	*Modes of governance*	*Drivers*	*Policy and praxis*	*Enabling and constraining factors*
Power, poverty and CBNRM: Sierra Leone and Cameroon	Mining Forestry	State-driven community governance	Global diamond and timber markets International calls for devolution Pressure to give citizens a voice National forest laws calling for community involvement	Sierra Leonean policies require local benefit sharing from mining. While large amounts of revenue are distributed to local communities, there is significant leakage and limited benefits to the poor. Forestry law in Cameroon is progressive, but implementation is slow and state retains decision-making powers. Some benefit distribution occurs at local level.	**Constraining** Investment pressures undermining state's autonomy for managing natural resources Confusion due to multitude of legal frameworks Revenue leakage and limited local benefits due to powerful links between companies, government and local chiefs and 'invisible' and hidden power differentials Unequal power relations within communities Lack of transparency Indigenous rights to land not recognized beyond access for subsistence Lack of capacity and information
Co-management in the Okavango Delta: Botswana	Fisheries	State-centric but moving towards co-management	Conflict between different fishery sectors Recognition of customary systems and co-management as preferred approaches for small-scale fisheries	Policy support led to strong capacity development in all fishery sectors, improved communication and enhanced understanding of sustainable resource use.	**Enabling** Adoption of inclusive participatory approach **Constraining** Power differentials between fisher groups Cultural differences between fisher groups

					Formation of a local co-management structure involving government and traditional authorities Use of neutral facilitators and consensus-building approach for decision-making	Maintaining interest of participants
					Enabling	**Constraining**
A new governance framework for small-scale fisheries: South Africa	Fisheries	State-centric governance but policy shift to co-governance	Historic policies and practices and neglect of marginalized fishing communities Political pressures to address rights and needs of poor Commitments to various international instruments Legal action by fishers	Existing policies are technocratic, regulatory and natural science-based, ignoring customary practices, local knowledge and role of women. There has been limited participation in management and decisions. High levels of conflict and distrust exist among fishers and between fishers and government, state- imposed rules are transgressed and certain stocks are declining.	New policy providing framework for change Legal action and subsequent court ruling requiring inclusion of marginalized fishers	Limited local capacity Centralized allocation and management system mitigating against locally appropriate governance systems Current institutional failings minimizing the efficient roll-out of policy

Table 16.1 Continued

Case study	*Sectors*	*Modes of governance*	*Drivers*	*Policy and praxis*	*Enabling and constraining factors*	
					Enabling	**Constraining**
Legal pluralism and the governance of freshwater resources: Zimbabwe, Zambia, South Africa	Freshwater	State-centric and customary systems existing in parallel Reversion to customary systems with uncertainty and political change	Multiplicity of legal systems and erosion of customary systems as a result of colonial and apartheid systems Western norms that continue to shape policy while statutory rules ignored in favour of local norms	Principle of equitable access and sustainability has guided water reform. In some countries, customary systems are recognized in the law but difficult to implement in practice. Customary systems are flexible, adaptive and more congruent with local world view and social, religious and ecological systems. Dual systems of governance lead to confusion and tensions. No clear rules, procedures and management systems and transgressions of state rules.	Existence of customary governance systems which communities support	Lack of practical recognition of customary systems by the state Resource users ignore state-imposed rules, sometimes leading to resource degradation
					Constraining	
Diamond mining in Namaqualand: South Africa	Mining	Collaborative governance between private sector and NGO	International pressure for mining companies to implement corporate social responsibility programmes	Collaborative arrangements between parties to exchange information and support community participation unravelled, with broken promises leading to distrust, low participation and limited capacity and support from government.	Breakdown in trust Lack of government regulatory framework to ensure that environmental and social commitments of mining companies are implemented	

			National legal requirements to address historical injustices and environmental rehabilitation Pending mine closure Limited state capacity		
Use of customary systems for baobab governance: Zimbabwe	Non-timber forest products	Dual governance system – state-centric and customary	Multiplicity of legal systems arising from colonial histories, alongside political, social and economic developments State interest fuelled by commercial value of baobab	State laws recognize customary systems, but state still has decision-making power. Both systems have policies for sustainable resource use, but in practice there is increasing disregard for customary rules and mistrust of government's intentions. This has led to resource degradation and the inequitable allocation of benefits. Resource users engage in 'forum shopping' to address their interests.	**Constraining** Unclear roles and powers of traditional authorities versus state functionaries, mistrust of both Confusion caused by conflicting administrative and jurisdictional boundaries Lack of state capacity to implement by-laws and facilitate working relations among actors

Table 16.1 Continued

Case study	*Sectors*	*Modes of governance*	*Drivers*	*Policy and praxis*	*Enabling and constraining factors*	
					Enabling	**Constraining**
Governance, equity and sustainability in non-timber forest product chains: South Africa and Lesotho	Non-timber forest products	Combination of state-centric, private sector-driven and collaborative governance	Global markets for two high-value species (*Hoodia* and *Pelargonium*)	ABS laws require benefit sharing, but in practice this has had either no effect or a negative impact on communities.	Prevention of private-sector monopolies with state help	Private sector monopolies
			Resource overexploitation Biopiracy concerns	Tighter regulations have stimulated cultivation. This has improved resource sustainability, but with negative social consequences.	Important NGO role in challenging unjust systems	Reduction in benefits owing to lack of markets
			International and national laws requiring benefit sharing	Participation is supported in policy, but representational governance has limited real dialogue.	Legal support to NGOs, communities and knowledge holders	Limited capacity of state to engage in and monitor benefit sharing initiatives
						Limited capacity of harvesters and traditional knowledge holders to engage with private sector and state
						Challenges of identifying community

generally delivering on intended policy and programme objectives (Gupta; Child et al; Ferguson et al; Hansen et al; Buchy and Maconachie, this volume). Few contributors (Hoole; Child et al, this volume) have reported on positive outcomes associated with new forms of governance that theorists anticipate, such as tangible benefits for local communities, well-functioning and empowered local level institutions, improved participation and representation, especially of marginalized groups and individuals, improved mechanisms of accountability, sustainable resource use and biodiversity conservation.

Instead, the majority of chapters provide a harsh critique of the shortcomings of natural resource governance systems in sub-Saharan Africa, highlighting progressive policy and governance reforms on paper, but little tangible improvement to the lives of local communities or resource sustainability.

State remains the key governing actor

A key finding is that despite the rhetoric of new and reformed modes of governance, the de facto situation is that the state remains the key governing actor and continues to hold the power to make decisions. While various case studies report that community-based management or co-management is the intended mode of governance, they highlight – aside from the Namibian case study, which presents a positive and more promising picture of CBNRM – the central role that the state plays in decision-making with regard to access to and the ownership, use and management of resources.

In the case of CBNRM in Botswana, Gupta (this volume) suggests that local communities have restricted-use rights over wildlife, which is regarded as the property of the state. Community-level decisions are limited to those concerning the distribution of revenues generated through leasing wildlife resources to tourism and wildlife operators. Critical decisions about the allocation and use of resources and decisions about management interventions that may be required to address problems, such as culling troublesome elephants on communal land, are still within the purview of central government. Regarding another Botswanan case, Child et al (this volume) suggest conditions under which CBNRM can be more empowering and deliver benefits to households, but conclude that in many cases there is a move towards the recentralization of resource decisions.

Although the community-based management of forests in Cameroon and Madagascar (Buchy and Maconachie; Ferguson et al, this volume), is delivering some benefits to local communities, these projects typically comprise a small percentage of the forestry sector in these countries by comparison to projects under state and private control. In the case of Madagascar, the strong presence of conservation-oriented non-governmental organizations (NGOs) has resulted in a focus on biodiversity at the expense of social justice (Ferguson et al, this volume). In several of the cases (Madagascar, Cameroon, Sierra Leone, South Africa), political and economic pressures and interests, which may in turn be influenced by global processes, have significantly undermined efforts to transition to more participatory and locally appropriate forms of governance.

Disjunctures between policy and praxis

The new governance ideas and language found in the plethora of multilateral agreements, national policies, programmes and initiatives all indicate a definite shift over the past four decades. However, most of the cases suggest that there is a significant disjuncture between policy rhetoric and implementation praxis. On the one hand, the policies and programmes incorporate principles such as rights-based, community-based and participatory approaches, as well as calls for equity, securing property rights and accountable and transparent forms of governance. Yet, in practice, the cases report shortcomings in policy implementation: restricted user rights (Botswana, South Africa, Madagascar, Cameroon, Sierra Leone), limited participation in planning and decision-making (all countries), a failure to recognize and accommodate local customary systems (South Africa, Madagascar, Cameroon, Zimbabwe, Botswana), limited distribution of tangible benefits at the local level (Botswana, Zambia, South Africa, Cameroon, Sierre Leone) and generally weak governance, especially with regard to accountability.

However, in the case of South African and Botswanan fisheries, new policies and programmes now require the implementation of co-management. This approach is largely supported by politicians, resource managers and other governance actors who acknowledge that conventional top-down and technocratic approaches have failed both communities and the environment. Both countries stress the importance of the enabling and empowering role that the state needs to play to effect this shift to a more people-centred, developmental and participatory form of governance.

Paradoxical role of the state

Other cases highlight the important role of the state in regulating activities to ensure biodiversity protection and social justice. This points to a paradox regarding the role of the state: on the one hand it is censured for retaining powers, including management and decision-making authority, but on the other hand it is expected to regulate activities to protect natural resources and community interests.

In their study of the commercialization of non-timber forest products (NTFPs) in South Africa, Wynberg and van Niekerk (this volume) highlight the important role of the state in regulating the overexploitation of biodiversity products and introducing regulations to protect biodiversity and the rights of traditional knowledge holders. However, this has sometimes been at the cost of community benefits, owing to the inappropriate translation of laws at the local level and a lack of understanding of the local context on the part of the state.

In the case of a partnership between a large mining company and an NGO working on behalf of local communities in Namaqualand in South Africa, Hamann (this volume) concludes that the state has a critical regulatory role to play in safeguarding the rights of communities and ensuring that environmental commitments by the private sector are honoured. The government, this would suggest, needs to be attuned to the needs of its citizens, responsive to particular situations, flexible in executing its mandate and ready to provide guidance, leadership, regulatory control or management intervention as appropriate. This is a lot to expect.

Multiple legal systems

Another central finding is the fact of multiple legal systems operating in the natural resources arena in sub-Saharan Africa. Several chapters report on the prevalence of customary practices and institutions, with these systems being the de facto mode of governance guiding the activities and lives of rural dwellers, albeit in an eroded state. Yet these customary systems are seldom recognized and accommodated in formal state-centric systems.

Several case study authors (Pollard and Cousins; Kozanayi et al; Ferguson et al; Hansen et al, this volume) highlight the tensions between the two systems of governance, including the lack of clarity regarding the roles and powers of traditional leaders, the conflicts between 'modern' and 'traditional' values and the co-option of traditional leaders to serve the interests of powerful political players. These chapters illustrate how customary systems prevail, especially in situations of political change or uncertainty or where communities are remotely situated. They argue that customary governance systems are more congruent with local resource users' world views and cultural and religious practices and beliefs, are more flexible than Western approaches and thus better serve local interests.

These findings concur with research conducted in other parts of the world which highlights the resilient nature of customary management systems in the face of socio-economic and political changes (Jentoft et al 2003; Bernard and Kumalo 2004; Cinner and Aswani 2007). However, Kozanayi et al and Pollard et al (this volume) highlight how state-imposed laws have weakened customary governance systems, thereby undermining practices and norms that contribute to resource sustainability and social justice. The existence of parallel legal systems also creates confusion and uncertainty for both local resource users and state-appointed functionaries, resulting in 'forum shopping' (Meinzen-Dick and Pradhan 2002) as resource users select the rules and regulations that best suit their particular needs and circumstances. While customary systems are more suited to natural resource management because they are more flexible and responsive to change, such systems can entrench power relations and thus threaten the already precarious position of the most vulnerable in society.

Yet for resource users on the ground, the de facto situation is that authority derives from both customary systems and state laws. Given the dire socio-economic circumstances of many resource-dependent communities in sub-Saharan Africa and the increasing pressure on natural resources, some authors in this volume (Kozanayi et al; Ferguson et al; Pollard and Cousins) argue that both state and customary forms of governance have an important role to play in regulating resource use. Thus the challenge is to recognize and respect the existence of a plurality of legal systems and embrace a kind of 'cooperative pluralism' (New Zealand Law Commission 2006, p. 44), that can 'draw strength and inspiration' from both legal systems (p. 23). The feasibility of such a legal pluralist regime would depend on many factors, not least of them the political will and a desire among people to find a compromise (Benda-Beckman and Benda-Beckman 2006).

Participation and representational failures

Much has been written about the need to enhance participation in natural resource management and decision-making (e.g. Agrawal and Gibson 1999; Cleaver 1999; Berkes et al 2001; Pinkerton 2009) affirmed by bodies as diverse as the United Nations, the World Bank, national governments and human rights organizations (Eyben 2003; Greiber et al 2009). Furthermore, this principle is advocated in a number of international policy instruments adopted by governments worldwide (e.g. Principle 10 of the Rio Declaration 1992 [United Nations Environment Programme [UNEP 1992b]; the Convention on Biological Diversity [UNEP 1992a]; World Summit on Sustainable Development [WSSD], Johannesburg Plan of Implementation [WSSD 2002]; the Aarhus Convention [United Nations Economic Commission for Europe [UNECE] 1998]).

Given this breadth of uptake it is perhaps not surprising that experiences of what participation means in practice remain extremely divergent – from lip-service 'stakeholder consultation', which is more easily achieved, to challenges of securing genuine participation in decision-making and, in some cases, co-management (Borrini-Feyerabend et al 2000; Pomeroy and Douvere 2008). Indeed, standing out as a common theme across most chapters in this book is the failure of governance arrangements to adequately ensure effective and inclusive participation.

In their case study on Madagascar, Ferguson et al (this volume) describe how a lack of inclusivity in environmental authorization processes for mining has led to community disempowerment, leaving the communities with few opportunities for meaningful engagement and no influence over the outcome of the process. Similarly, in the conservation sector, while participatory governance has been promoted as the basis for park management, this has not always been adequate, with several protected areas failing to recognize local needs and effectively integrate customary land tenure. Thus, while policy reform has advocated participation, flaws in conception, design and implementation have precluded the attainment of social justice. Local communities often lack the capacity, literacy and legal awareness to make informed and meaningful contributions to the process.

Experiences from the iSimangaliso Wetland Park in the northern part of KwaZulu-Natal offer a similar narrative. While the South African Constitution emphasizes 'cooperative government' and 'participatory democracy', and post-apartheid policies require the inclusive participation of citizens in planning and decision-making, Hansen et al (this volume) report that the depth of participation needed for engagement in planning and development activities was lacking, with the result that community members had no voice in decision-making.

Several authors point to the institutional challenges of participation, more especially in the context of representational democracy. Gupta (this volume), for example, explains how representatives of the Chobe Enclave Community Trust, like those of many other trusts, were not downwardly accountable to the villages represented, leading to elite capture of benefits from wildlife and minimal distribution of those benefits.

Through six CBNRM case studies across Zimbabwe, Zambia and Botswana, Child et al (this volume) introduce the term 'micro-governance', concluding that 'weaknesses in benefit sharing and participation reflect the challenge of scaling down and the fact that human interactions are better managed at smaller scales'. They strongly advocate 'face-to-face' participatory democracy, where the whole community meets together to discuss issues, to make decisions, and to instruct the committee on matters of implementation. Complex systems, they say, must be 'constructed from the bottom up to serve the bottom'. These ideas resonate with scholars who argue that participatory democracy provides an opportunity for more direct forms of citizen participation in governance and is better able to address the diverse needs and interests of citizens (Holmes and Scoones 2000; Petts 2001; Manor 2004).

This is, however, more easily said than done. Buchy and Maconachie (this volume) describe the centrality of power in determining the efficacy of participation, pointing out that participatory initiatives, for example in the distribution of benefits from natural resources or minerals, are often 'invited' spaces within a country, designed and managed by a state bureaucracy. While these can lead to some benefits for local communities, more often than not they fail to address underlying, and often 'hidden' or 'invisible', power issues. As Hamann (this volume) and several other authors (e.g. Wynberg and van Niekerk; Kozanayi et al; Ferguson et al; Buchy and Maconachie, this volume) explain, this becomes all the more challenging when several other voices such as local government, state-supported traditional authorities or NGOs act on behalf of communities.

Mosepele et al and Sowman et al (this volume) suggest that when significant effort is devoted to facilitating an inclusive participatory process, the outcome has broader support and is more likely to succeed. In the first case, the establishment of a fisheries co-management committee helped to enhance local participation in the management of the Okavango Delta in Botswana and to reduce conflict and facilitate understanding about sustainable resource use. In the case of South Africa, participation of civil society in policy formulation ensured that the voices of previously marginalized fishers were heard and integrated into the final policy. Even so, there are still ongoing challenges in including the voices of those less organized and in sustaining the interest of different stakeholders. In common with many of the other case studies in this book, these authors point to the ongoing challenges of capacity development, leadership and financial support.

Lack of attention to resource sustainability

One of the more surprising findings of the case studies presented in this volume is the lack of detailed attention given by authors to questions of resource sustainability. It is an open question whether this is due to the breadth of topic, the social science discipline of many of the authors, or the underlying assumption that if governance systems perform well in delivering equitable and socially just benefits, then desirable ecological outcomes will follow (Folke 2006). As Tacconi et al (2006) note, it is imperative to interrogate whether or not this 'romantic' view of conservation following benefits is necessarily true. In a detailed analysis, they further argue that the complex linkages existing among decentralization, forest management and livelihoods imply that it is not possible to assert that the democratic decentralization of forest management leads to sustainable forest management and increased livelihood benefits (Tacconi et al 2006). This finding is reinforced by various studies on forest governance in Africa (German, Colfer et al 2010) which suggest that community-based governance can result in either sustainable forest management or deforestation.

What is clear is that the complex social-ecological systems introduced in different cases present a diverse array of biological and physical characteristics that interplay with a plurality of legal systems, institutional contexts, livelihood possibilities and political histories in ways that are bound to be indeterminate and non-linear (Ribot 2010). In eastern Zimbabwe, for example, Kozanayi et al (this volume) show how customary systems to regulate the use of the baobab tree have become increasingly entangled with and undermined by statutory systems, leading to a confusing governance landscape that has served only to further degrade the resource. Pollard and Cousins (this volume) similarly imply that overlapping customary and statutory regimes, combined with the existence of multiple laws, have had negative impacts on freshwater resources. Wynberg and Niekerk (this volume), describing regulatory frameworks for two high-value NTFPs in South Africa, illustrate how overharvesting concerns led to state intervention to regulate resource use, but with negative social and economic outcomes for local people. None of these cases, however, sketches a context where socially equitable benefits have indeed been secured, suggesting a need for further detailed, empirical research to explore the interface between equity and ecological sustainability.

Key drivers of change

It is pertinent to ask at this point why state-centric governance persists, and what factors inhibit a transition to more participatory and locally appropriate forms of governance. Table 16.1 provides a synthesis of the drivers that influence the nature of governance in the cases examined as well as the factors that constrain or enable governance for justice and sustainability outcomes. Here we focus on three main drivers of these trends: capital, profit and foreign investment; the political and economic underpinnings of natural resource governance; and international multilateral agreements and related processes.

Capital, profit and foreign investment

In many respects, the central role of African states is being backed up by foreign governments concerned about future food supplies and seeking land for food exports, agribusiness companies acquiring land for biofuels, mining companies pursuing gas, oil and minerals and a host of other investors and speculators interested in Africa's hitherto ignored natural resources (Nelson 2010; African Progress Panel 2012). Such trends are amplified by the fact that some 60 per cent of the world's uncultivated arable land is believed to be in Africa (Roxburgh et al 2010).

Investments in resource-rich African countries are having profound influences on governance in sub-Saharan Africa. Emerging economies, and China in particular, are at the forefront of setting in place a new South-South 'global architecture' for trade and investment (Cotula et al 2009; Karsenty 2010). African countries are being offered direct, condition-free financial assistance in exchange for guaranteed access to strategic resources, with the result that African states are enjoying greater autonomy, free from the conditionals imposed by international financial institutions such as the World Bank (Karsenty 2010).

Not all of the chapters focus on the driving force of investments, profit and capital, but those that do share similar threads. Investments in Madagsacar, for example, have led to numerous mining ventures and agricultural land acquisitions. As Ferguson et al (this volume) explain, this growth of interest in Madagascar from the extractive sector, from those seeking land for biofuels and from the global conservation movement, has led to cases of social injustice against the rural Malagasy people, whose customary and ancestral lands have been caught up in a 'modern natural resource rush'. Although rights to ancestral lands have been legislated, the process of formalizing these rights by providing title to local communities has suffered lengthy bureaucratic delays and, in the process, local rights holders have lost land due to foreign investors and political deals.

Often the impacts of such investments permeate to a much deeper level. In Sierra Leone and Cameroon, for example, Buchy and Maconachie (this volume) describe how the links between government departments that oversee mining and forestry, high-level politicians and the private sector filter down to ties with corrupt village elites, thus depriving the wider community of benefits. The case studies of *Pelargonium* and *Hoodia*, both high-value NTFPs, illustrate some of the challenges of introducing investment, capital and global value chains into a rural setting where the organizational infrastructure is weak and the possibilities of deepening inequality high (Wynberg and Van Niekerk, this volume). In the *Pelargonium* case, government interest was piqued by increased demand, prompting regulatory measures with negative consequences for harvesters and a requirement for benefit sharing that did little for horizontal equity and improving the lot of the poor.

Left to its own devices, unfettered capitalism – when its power and resources are combined with unaccountable governments – is almost certain to produce inequality and environmental devastation (Newell 2011). New governance configurations mean that environmental and social standards are falling by the wayside, with Asian companies often preferring rather 'to pay the heavy surcharges associated with breaches of law than the costs of compliance' (Karsenty 2010, p. 83). As resources become more scarce, and thus more valuable, it is likely that national African governments with high-value and strategic resources will strengthen their efforts to retain ownership and stewardship, rather than attempt to meet demands for local control. Urquhart (this volume) notes that the introduction of the United

Nations Reducing Emissions from Deforestation and Forest Degradation initiative, known as REDD+, may offer tools to use global carbon markets and investment to leverage more secure local access, but joins others in cautioning that it could have the reverse effect too (Naughton-Treves and Day 2012).

Political and economic drivers

The strong links between the state and private sector in the acquisition and control of natural resources is evident throughout Africa where the state facilitates and supports the accumulation of wealth by those in positions of power (Van Sittert 2002; Karsenty 2010; Nelson 2010). This powerful alliance, entrenched during the colonial period, has effectively limited democratization and the envisaged benefits associated with devolving powers to local resource users.

Notwithstanding the rhetoric of democratization and devolution, the state has retained ownership and control over land and resources, and power over decisions to assign use and management rights. It has also controlled the pace of legal reforms, illustrated by the formalization of customary title to land in Madagascar (Ferguson et al, this volume) and the implementation of community forestry laws and fisheries co-management in Cameroon and South Africa respectively (Buchy and Maconachie; Sowman et al, this volume). The lack of awareness of people's rights, the lack of capacity of institutions and individuals to make claims or challenge decisions and the absence of social movements that can monitor and put pressure on corrupt state activities or of NGOs that can support communities to demand their rights, fuels the ongoing accumulation of wealth among corrupt politicians, state functionaries and local elites, as well as the private-state alliances that serve their interests. These factors represent major constraints to promoting governance towards justice and environmental sustainability.

Several scholars have questioned the underlying assumption that the state and its political economy will serve the socio-economic interests of the poor (Bromley 2009; Davis and Ruddle 2012). They argue that neoliberal economics is market-driven and does not take account of the socio-cultural and historical characteristics of specific community contexts or the power differentials that exist. Furthermore, they argue that the neoliberal project serves the wealthy sectors of society and that alternative forms of governance that promote community involvement and co-management all work to serve the interests of the wealthy (Bromley 2009; Davis and Ruddle 2012). The lack of evidence, in most of the cases, of structural and institutional changes to enable more inclusive participation, empowerment and equitable benefit sharing suggests that powerful political and economic processes play a significant role in influencing the praxis of governance.

The issue of power and politics at different levels of governance is a key theme emerging throughout the cases, particularly at the local level, where powerful local actors capture benefits through political patronage or corruption (Ferguson et al; Buchy and Maconachie; Gupta; Kozanayi et al; Wynberg and Van Niekerk, this volume) or through relationships among actors across different governance scales – for example, the co-opting of traditional leaders by the state

(Ferguson et al; Kozanayi et al, this volume). Systematic ignorance of power relations and their influence on social relations and structures was identified by Buchy and Maconachie (this volume) as a key factor undermining progress towards social justice and poverty reduction in resource-rich areas. These authors also refer to the importance of recognizing 'invisible power', such as social values, witchcraft, gender and ethnicity, which are frequently overlooked by policy-makers and governance-implementers.

International multilateral agreements and related processes

An important driver in several of the case studies is the role of multilateral agreements, such as the Convention on Biological Diversity and the United Nations Framework Convention on Climate Change (United Nations Framework Convention for Climate Change [UNCED] 1992), in exerting influence on national and local governance processes (Hansen et al; Sowman et al; Wynberg and Van Niekerk; Urquhart, this volume).

Growing realization of the precarious balance between socio-economic development and resource limitations has prompted calls for policy and institutional reforms, and this has resulted in a plethora of bilateral and multilateral agreements. The aim of these agreements is not only to protect natural resources and ensure their long-term sustainable use, but equally to protect the rights and socio-economic needs of indigenous and local resource-dependent communities. Most countries in sub-Saharan Africa have committed to several of these conventions and agreements[1] in an effort to halt environmental degradation and protect Africa's natural resources and its peoples. However, while desirable in terms of protecting resources and natural systems, the increasing appropriation of land and marine areas for conservation purposes, referred to by some scholars as 'green and blue grabbing' (Benjaminsen and Bryceson 2012; Fairhead et al 2012) is resulting in increased restrictions on access to land and resources for resource-dependent communities, and in some cases in dispossession, with concomitant livelihood and socio-cultural impacts.

Balancing the tensions across these various instruments and national imperatives, which are often in conflict, has proven a challenging task (Brown 2009). Resource-dependent communities are being squeezed, on the one hand, by global and national economic pressures for land and development, and, on the other, by conservation expansion policies and programmes.

A number of authors in this book raise concerns about the uncritical application of these normative environmental sustainability concepts at the local level. Hansen et al (this volume) describe how the World Heritage status of iSimangaliso in KwaZulu-Natal, South Africa, raises it above the level of the local territory and often beyond the decision-making authority of local people, also placing extra responsibilities on park managers. Gupta (this volume) explains the introduction of the CBNRM concept in Botswana by the United States Agency for International Development as a tool to promote conservation, but not necessarily with the livelihood outcomes envisaged. Wynberg and Van Niekerk (this volume) describe the indiscriminate application in South Africa and Lesotho of access and benefit sharing requirements under the Convention on Biological Diversity, with outcomes that are far from fair and equitable. Combined, these cases reaffirm the importance of assessing the impacts of international agreements and initiatives prior to their national implementation, and of tailoring their implementation in a manner that is intelligent and needs-driven.

Reflections and recommendations

The overwhelming message of this book, derived from a multitude of experiences, contrasting national contexts and diverse sectors, is that natural resource governance in sub-Saharan Africa is still driven from the centre by a complex mix of global and country-specific factors, is technocratic, and has largely failed to implement policy reforms on the ground. Despite commitments to a host of multilateral agreements, many of which incorporate good governance, human rights and sustainability principles, few countries have translated these principles into resource governance praxis. Furthermore, governance is constantly in a state of flux, with external and country-specific drivers influencing and being influenced by changing political and economic interests. Power dynamics, personal enrichment and political motivations play a key role in keeping local communities marginalized and disempowered. While some of the case studies point to various enabling factors that enhance governance practices (refer to Table 16.1), these are limited to factors such as support from NGOs, external financial support and an enabling policy environment. Many of the potential solutions to this situation exist only at a macro level, but important steps can be taken to begin to realize governance solutions that can deliver social justice and environmental sustainability.

Recognizing the multiscalar character of environmental governance

Although they may adopt the 'right' rhetoric, the processes of governance reform – broadening the network of actors, facilitating inclusive participation, building legitimate local-level institutions and developing appropriate rules – have not embraced a holistic, integrated, participatory, multicultural, evolutionary approach. Perhaps unsurprisingly, the multiscalar and multisectoral character of environmental governance, characterized by piecemeal, silo-like decisions taken by individual ministries mandated to govern a particular resource rather than overarching systems, has been challenging to implement. This is especially the case when competing interests are at stake.

There is an urgent need to put effort into developing approaches that can work for diverse actors with multiple worldviews, cultural contexts, social-legal systems, knowledge systems, needs and perceptions. Building upon existing local conditions and institutions, and fashioning governance from the bottom up, are underpinning principles for such approaches.

Embracing legal pluralism

Related to this is the need to embrace legal pluralism and work creatively within the framework of Africa's multiplicity of legal systems and governance actors. A diversity of systems of governance is part and parcel of the African reality and will become more so with the blurring of boundaries between the rural and urban. It is vital that governance systems evolve in such a way that they recognize and incorporate customary systems. This would involve assessing the extent to which customary rules are applied and complied with in practice (Techera 2008). The accommodation of different systems of law suggests a form of hybridization which allows customary law to continue to evolve and adapt rather than be subsumed by state law. The embrace of legal pluralism as a principle enables governance actors to identify legal frameworks that are informed by state and customary legal systems and are thus more likely to be supported.

Developing cultures of accountability

Accountability, at both individual and institutional levels, is a central pillar of governance systems for justice and sustainability. Mechanisms to ensure both upward and downward accountability need to be firmly entrenched in governance processes and institutions. Accountability can only be assured if democratically elected leaders are accountable to their constituencies and responsive to their needs and changing political, social, economic and ecological conditions.

Also required are the existence of multiple mechanisms including regular fair and transparent elections, agreed upon constitutions and modus operandi, full transparency in financial matters and the right to protest and voice objections (Ribot 2010).

Setting in place the social, human and knowledge infrastructure for implementation

As so many of the chapters in this book attest, the policies and norms to realize governance objectives of social justice and environmental sustainability are now largely in place in many African countries. This has been enabled, often through donor funding, by significant attention to policy formulation and an emphasis on inclusive participation in these processes. The largest lacunae now lie in the implementation of these policies.

Unlike policy formulation, which is considered the domain of all governance actors, implementation typically falls back into the hands of the government, which remains hampered by limited resources, poor leadership, constrained capacity and, in some cases, a lack of political will to draw in the range of actors required to realize policy objectives. It is crucial that more time be allocated in the process of policy formulation to the strategies, plans and institutional arrangements that may be required to give effect to policy. Intimate actor involvement is required to work out the how, who and when, as well as the measures and methods, of policy implementation. This in turn will create the continuity and accountability required to challenge application and interpretation. Hand in glove with such approaches should be the amalgamation of local, traditional and Western knowledge systems to enable new governance styles that are truly African in nature and appropriate to local needs.

It is vital that adaptive management, or 'learning by doing', underpins such processes. Whether at the policy, plan or project-implementation level, greater attention needs to be devoted to monitoring, learning and adapting policies and plans and revisiting management approaches and rules as understanding of the practical implementation challenges and realities becomes clearer. Ongoing monitoring and assessment of what is and is not working, and what is undermining policy intentions, is a logical step if good policies are to become good practice.

As Ojha et al (2013, p. 318) observe, the policy system is a learning system, and thus there is a need for participants to engage in 'action-reflection-review processes'. Viewing policy formulation and implementation as a learning process implies change as fresh understandings and insights evolve, new governance arrangements emerge and environmental changes occur. It also requires new forms of accountability as the range of governance actors involved in the policy system offer insights and ideas about how policy can be made more appropriate to context.

Developing the capacity to govern

The growing number of governance processes, actors, activities and legal and policy instruments relating to natural resource use and management, along with the multiplicity of institutions charged with their implementation, can be overwhelming, both for resource users and for those responsible for applying and interpreting law. Carefully designed capacity development initiatives, building on local skill sets and mentorship rather than externally imposed training programmes, need to be set in place to equip this diverse collection of governance actors to deal with the steering, regulating, monitoring and facilitation that accompanies policy implementation. Also needed are new, creative approaches to build an African leadership that can think about natural resource management in bold and visionary ways.

Strengthening and affirming a rights-based agenda

Lastly, an enormous resource lies within African citizens themselves. Awareness and confidence need to be developed and strengthened to ensure that local users are confident of their rights to natural resources, can effectively defend them, and have a battery of tools enabling them to seek help and support where needed. This should include partnerships with NGOs, researchers and development agencies that are well networked and supportive of the rights and needs of local communities and can collectively hold the state and private sector to account.

Note

1 Most countries in this volume have signed up or indicated commitment to a number of multilateral conservation and human rights agreements including the Convention on Biological Diversity (UNEP 1992a), the United Nations Framework Convention on Climate Change (UNCED 1992), the International Labour Organization Convention 169 (ILO 1989) and the United Nations Declaration on the Rights of Indigenous Peoples (2006).

Bibliography

Adger, W. N. and Jordan, A. (eds) (2009) *Governing Sustainability*, Cambridge University Press, Cambridge

African Progress Panel (2012) *Africa Progress Report 2012: Jobs, Justice and Equity: Seizing Opportunities in Times of Global Change*, African Progress Panel, Geneva, http://africaprogresspanel.org/en/publications/annual-reports/annual-report-2012/africa-progress-report-2012-documents/

Agrawal, A. and Gibson, C. C. (1999) 'Enchantment and disenchantment: The role of community in natural resource conservation', *World Development*, vol 2, no 4, pp. 629–49

Anstey, S. and Rihoy, L. (2009) 'Beacon and barometer: CBNRM and evolutions in local democracy in southern Africa', in B. B. Mukarumi, J. M. Manjengwa and S. Anstey (eds) *African Wildlife and Livelihoods: The Promise and Performance of Community Conservation*, James Currey, Oxford

Benda-Beckman, F. von and Benda-Beckman, K. von (2006). 'The dynamics of change and continuity in plural legal orders', *Journal of Legal Pluralism*, vol 8, pp. 1–44

Benjaminsen, T. A. and Bryceson, I. (2012) 'Conservation, green/blue grabbing and accumulation by dispossession in Tanzania', *Journal of Peasant Studies*, vol 39, no 2, pp. 335–55

Berkes, F., Colding, J. and Folke, C. (eds) (2003) *Navigating Social-Ecological Systems: Building Resilience for Complexity and Change*, Cambridge University Press, Cambridge

Berkes, F., Mathias, J., Kislalioglu, M. and Fast, H. (2001) 'The Canadian Arctic and the Oceans Act: The development of participatory environmental research and management', *Ocean & Coastal Management*, vol 44, pp. 451–69

Bernard, P. and Kumalo, S. (2004) 'Community-based natural resource management, traditional governance and spiritual ecology in southern Africa: The case of chiefs, diviners and spirit mediums', in C. Fabricius, E. Koch, H. Magome and S. Turner (eds) *Rights, Resources & Rural Development: Community-Based Natural Resource Management in South Africa*, Earthscan, London

Borrini-Feyerabend, G., Farvar, T., Nguinguiri, J. C., Ndangang, V. A. (2000) *Co-management of Natural Resources: Organising, Negotiating and Learning-by-Doing*, GTZ and IUCN Kasparek, Heidelberg, Germany

Bratton, M. and Rothchild, D. (1992) 'The institutional bases of governance in Africa', in M. Bratton and G. Hyden (eds) *Governance and Politics in Africa*, Lynne Rienner, Boulder, CO

Bromley, D. W. (2009) 'Abdicating responsibility: The deceits of fisheries policy', *Fisheries*, vol 34, pp. 280–90

Brown, K. (2009) 'Human development and environmental governance: A reality check', in N. Adger and A. Jordan (eds) *Governing Sustainability*, Cambridge University Press, Cambridge, pp. 32–51

Cinner, J. E. and Aswani, S. (2007) 'Integrating customary management into marine conservation', *Biological Conservation*, vol 140, nos 3/4 pp. 201–16

Cleaver, F. (1999) 'Paradoxes of participation: Questioning participatory approaches to development', *Journal of International Development*, vol 11, no 4, pp. 597–612

Cotula, L., Vermeulen, S., Leonard, R. and Keeley, J. (2009) *Land Grab or Development Opportunity? Agricultural Investment and International Land Deals in Africa*, IIED, FAO and IFAD, London and Rome

Davis, A. and Ruddle, K. (2012) 'Massaging the misery: Recent approaches to fisheries governance and the betrayal of small-scale fisheries', *Human Organization*, vol 71, no 3, pp. 244–54

Eyben, R. (2003) *The Rise of Rights*, IDS Policy Briefing, no 17, May, Institute of Development Studies, Brighton, United Kingdom

Fairhead, J., Leach, M. and Scoones, I. (2012) 'Green grabbing: A new appropriation of nature?', *Journal of Peasant Studies*, vol 39, no 2, pp. 237–61

Folke, C. (2006) 'Resilience: The emergence of a perspective for social-ecological systems analyses', *Global Environnemental Change*, vol 16, no 3, pp.253–67

Garcia, S. M., Allison, E. H., Andrew, N. J., Béné, C., Bianchi, G., de Graaf, G. J., Kalikoski, D., Mahon, R. and Orensanz, J. M. (2008) *Towards Integrated Assessment and Advice in Small-Scale Fisheries: Principles and Processes*, Fisheries Technical Paper No 515, Food and Agriculture Organization of the United Nations, Rome

German, L. A., Colfer, C. J. P., Barrow, E., Küchli, C., Blaser, J. and Wardojo, W. (2010) 'Forest governance and decentralization in Africa: Linking local, regional and global dialogues', in L. A. German, A. Karsenty and A. M. Tiani (eds) *Governing Africa's Forests in a Globalized World*, Earthscan, London, pp. 1–28

Greiber, T., Janki, M., Orellana, M., Savaresi, A. and Shelton, D. (2009) *Conservation with Justice: A Rights-Based Approach*, IUCN Environmental Law and Policy Paper No. 71. International Union for Conservation of Nature, Gland, Switzerland

Holmes, T. and Scoones, I. (2000) *Participatory Environmental Policy Processes: Experiences from North and South*, IDS Working Paper 113, Institute of Development Studies

ILO (International Labour Organisation) (1989) *Indigenous and Tribal Peoples Convention* C169, ILO

Isaacs, M., 2006. 'Small-scale fisheries reform: Expectations, hopes and dreams of "a better life for all"', *Marine Policy*, vol 30, no 1, pp. 51–9

Jentoft, S., Minde, H. and Nilsen, R. (eds) (2003) *Indigenous Peoples: Resource Management and Global Rights*, Eburon, Delft, Netherlands

Jones, B. (2010) 'The evolution of Namibia's communal conservancies', in F. Nelson (ed.) *Community Rights, Conservation and Contested Land: The Politics of Natural Resource Governance in Africa*, Earthscan, London

Karsenty, A. (2010) 'The new economic "great game" in Africa and the future of governance reforms in the forest sector', in L. A. German, A. Karsenty and A. M. Tiani (eds) *Governing Africa's Forests in a Globalized World*, Earthscan, London

Keeley, J. and Scoones, I. (2003) *Understanding Environmental Policy Processes: Cases from Africa*, Earthscan, London

Manor, J. (2004) 'Democratisation with inclusion: Political reforms and people's empowerment at the grassroots', *Journal of Human Development*. vol 5, no 1, pp. 5–29

Meinzen-Dick, R. and Pradhan, R. (2002) *Legal Pluralism and Dynamic Property Rights*, CAPRi Working Paper no 22, CGIAR Systemwide Program on Collective Action and Property Rights, Washington DC

Murombedzi, J. (2001) 'Committees, rights, costs and benefits: Natural resource stewardship and community benefits in Zimbabwe's CAMPFIRE programme', in D. Hulme and M. W. Murphree (eds) *African Wildlife and Livelihoods: The Promise and Performance of Community Conservation*, James Currey, Oxford

Murombedzi, J. C. (2010) 'Agrarian social change and post-colonial natural resource management interventions in southern Africa's "communal tenure" regimes', in F. Nelson (ed) *Community Rights, Conservation and Contested Land: The Politics of Natural Resource Governance in Africa*, Earthscan, London

Murphree, M. W. (2005) 'Congruent objectives, competing interests, and strategic compromise: Concept and process in the evolution of Zimbabwe's CAMPFIRE, 1984–1996', in J. P. Brosius, A. L. Tsing and C. Zerner (eds) *Communities and Conservation: Histories and Politics of Community-Based Natural Resource Management*, AltaMira Press, Walnut Creek, CA

Naughton-Treves, L. and Day, C. (2012) *Lessons about Land Tenure, Forest Governance and REDD+: Case Studies from Africa, Asia and Latin America*, University of Wisconsin–Madison Land Tenure Center, Madison, WI

Nelson, F. (ed.) (2010) *Community Rights, Conservation and Contested Land: The Politics of Natural Resource Governance in Africa*, Earthscan, London

Newell, P. (2011) 'The elephant in the room: Capitalism and global environmental change', *Global Environmental Change*, vol 21, no 1, pp. 4–6

New Zealand Law Commission (2006) *Converging Currents: Custom and Human Rights in the Pacific*, Study Paper 17, New Zealand Law Commission, Wellington

Ojha, H. R., Hall, A. and Sulaiman, R. (eds) (2013) *Adaptive Collaborative Approaches in Natural Resource Governance: Rethinking Participation, Learning and Innovation*, Routledge, London and New York

Ostrom, E. (1990) *Governing the Commons: The Evolution of Institutions for Collective Action*, Cambridge University Press, Cambridge

Petts, J. (2001) 'Evaluating the effectiveness of deliberative processes: Waste management case studies', *Journal of Environmental Planning and Management*, vol 44, no 2, pp. 207–26

Pinkerton, E. (2009) 'Partnerships in management', in K. L. Cochrane and S. M. Garcia (eds) *A Fishery Manager's Guidebook: Management Measures and Their Application, Second Edition*, John Wiley & Sons, Oxford

Pomeroy, R. and Douvere, F. (2008) 'The engagement of stakeholders in the marine spatial planning process', *Marine Policy*, vol 32, no 5, pp. 816–22

Ribot, J. C. (2004) *Waiting for Democracy: The Politics of Choice in Natural Resource Decentralization*, World Resources Institute, Washington, DC

Ribot, J. C. (2010) 'Forestry and democratic decentralization in sub-Saharan Africa: A rough review', in L. A. German, A. Karsenty and A. M. Tiani (eds) *Governing Africa's Forests in a Globalized World*, Earthscan, London, pp. 29–55

Roxburgh, C., Dörr, N., Leke, A., Tazi-Riffi, A., van Wamelen, A., Lund, S., Chironga, M., Alatovik, T., Atkins, C., Terfous, N. and Zeino-Mahmalat, T. (2010) *Lions on the Move: The Progress and Potential of African Economies*, McKinsey Global Institute, http://mckinsey.com/insights/africa/lions_on_the_move

Sowman, M. (2011) 'New perspectives in small-scale fisheries management: challenges and prospects for implementation in South Africa', *African Journal of Marine Science*, vol 33, pp. 297–311

Symes, D. (2006) 'Fisheries governance: A coming of age for fisheries social science?', *Fisheries Research*, vol 81, pp. 113–17

Tacconi, L, Siagian, Y., and Syam, R. (2006) *On the Theory of Decentralization, Forests and Livelihoods*, Environmental Management and Development Occasional Papers 9, Australian National University: Asia Pacific School of Economics and Government, Canberra, Australia

Techera, E. (2008) *Supporting the Role of Customary Law in Community-Based Conservation*, WP 2008, 26, Macquarie University School of Law, Sydney, Australia

United Nations (2006) *Declaration on the Rights of Indigenous Peoples*, United Nations General Assembly, sixty–first session, 2006

UNECE (United Nations Economic Commission for Europe) (1998) Convention on Access to Information, Public Participation in Decision-making and Access to Justice in Environmental Matters. Known as the *Aarhus Convention*, Aarhus, Denmark

UNEP (1992a) *Convention on Biological Diversity: Text and Annexes*, United Nations Environment Programme, Chatelaine, Switzerland

UNEP (1992b) *Rio Declaration on Environment and Development*, Rio de Janeiro, June 1992

UNFCCC (United Nations Framework Convention for Climate Change) United Nations Conference on Environment and Development, Rio de Janeiro, June 1992.

Van Sittert, L. (2002) '"Those who cannot remember the past are condemned to repeat it": Comparing fisheries reforms in South Africa', *Marine Policy*, vol 26, no 4, pp. 295–305

World Bank (2002) *Governance and Development*, World Bank, Washington, DC

WSSD (2002) Plan of Implementation of the World Summit on Sustainable Development, A/CONF.199/20, Chapter 1, Resolution 2, Johannesburg, September 2002

Index